Boston Studies in the Philosophy and History of Science

Volume 319

The series *Boston Studies in the Philosophy and History of Science* was conceived in the broadest framework of interdisciplinary and international concerns. Natural scientists, mathematicians, social scientists and philosophers have contributed to the series, as have historians and sociologists of science, linguists, psychologists, physicians, and literary critics.

The series has been able to include works by authors from many other countries around the world.

The editors believe that the history and philosophy of science should itself be scientific, self-consciously critical, humane as well as rational, sceptical and undogmatic while also receptive to discussion of first principles. One of the aims of Boston Studies, therefore, is to develop collaboration among scientists, historians and philosophers.

Boston Studies in the Philosophy and History of Science looks into and reflects on interactions between epistemological and historical dimensions in an effort to understand the scientific enterprise from every viewpoint.

More information about this series at http://www.springer.com/series/5710

Tilman Sauer · Raphael Scholl
Editors

The Philosophy of Historical Case Studies

Springer

Editors
Tilman Sauer
Institute of Mathematics
Johannes Gutenberg University Mainz
Mainz
Germany

Raphael Scholl
Department of History and Philosophy of Science
University of Cambridge
Cambridge
UK

ISSN 0068-0346 ISSN 2214-7942 (electronic)
Boston Studies in the Philosophy and History of Science
ISBN 978-3-319-30227-0 ISBN 978-3-319-30229-4 (eBook)
DOI 10.1007/978-3-319-30229-4

Library of Congress Control Number: 2016934433

Printed on acid-free paper

This Springer imprint is published by Springer Nature
The registered company is Springer International Publishing AG Switzerland

Contents

Contributors

Theodore Arabatzis Department of History and Philosophy of Science, University of Athens, Athens, Greece

Michael Bycroft Department of History, University of Warwick, Coventry, UK

Harry Collins Distinguished Research Professor of Sociology, Cardiff University, Cardiff, UK

Simone De Angelis Zentrum für Wissenschaftsgeschichte, Universität Graz, Graz, Austria

Allan Franklin Department of Physics, University of Colorado, Boulder, USA

Kostas Gavroglu Department of History and Philosophy of Science, University of Athens, Athens, Greece

Giora Hon Department of Philosophy, University of Haifa, Haifa, Israel

Katherina Kinzel Institut für Philosophie, Universität Wien, Vienna, Austria

Tarja Knuuttila Department of Philosophy, University of South Carolina, Columbia, USA

Andrea Loettgers Center for Space and Habitability, University of Bern, Bern, Switzerland; Department of Philosophy, University of Geneva, Geneva, Switzerland

Wolfgang Pietsch Munich Center for Technology in Society, Technical University Munich, Munich, Germany

Tim Räz FB Philosophie, University of Konstanz, Konstanz, Germany

Tilman Sauer Institute of Mathematics, Johannes Gutenberg University Mainz, Mainz, Germany

Raphael Scholl Department of History and Philosophy of Science, University of Cambridge, Cambridge, UK

Dana Tulodziecki Department of Philosophy, Purdue University, West Lafayette, USA

Adrian Wüthrich Institut für Philosophie, Literatur-, Wissenschafts- und Technikgeschichte, Technische Universität Berlin, Berlin, Germany

Chapter 1
Introduction

Tilman Sauer and Raphael Scholl

1.1 The Philosophy of Historical Case Studies

In her novel *Five Little Pigs*, Agatha Christie presents five different versions of the same murder. It falls to Hercule Poirot to sort out the conflicting accounts—to use "se little grey cells" in order to infer from the available facts what really happened. The famous Belgian detective thus finds himself in a similar predicament as historians and philosophers of science when they need to assess divergent reconstructions of the same historical episode. Whether explicitly or not, such reconstructions are always informed by philosophical positions about the character of science: Many episodes have been told and retold in different and often incompatible versions, none of which are manifestly correct or incorrect.

Although underdetermination is entertaining in mystery stories, it is a vexing challenge for history and philosophy of science. To illustrate, take one of our motivating instances: Semmelweis's work on the cause of childbed fever in the middle of the 19th century, long a textbook favorite not only in philosophy of science, but also in clinical research. It matters for our ongoing debates about scientific methodology whether Semmelweis proceeded by the hypothetico-deductive method, by inference to the best explanation, by experimental causal inference, or by flawed reasoning. All of these accounts have been defended in the literature, and it is surprisingly difficult to determine which of them offers the best balance between descriptive adequacy and philosophical insight.

T. Sauer (✉)
Institute of Mathematics, Johannes Gutenberg University Mainz, Mainz, Germany
e-mail: tsauer@uni-mainz.de

R. Scholl
Department of History and Philosophy of Science, University of Cambridge, Cambridge, UK
e-mail: raphael.scholl@gmail.com

T. Sauer and R. Scholl (eds.), *The Philosophy of Historical Case Studies*,
Boston Studies in the Philosophy and History of Science 319,
DOI 10.1007/978-3-319-30229-4_1

Similar problems present themselves in the reconstruction of many episodes in the history of science. Take as a second instance the well-known disagreement about the early history of gravity wave research between Allan Franklin and Harry Collins (discussed in this volume in Chaps. 6 and 7). Here the divergence is even more pronounced, and its implications reach even further. Franklin's account of an emerging rational assessment contrasts sharply with Collins's view of the social construction of consensus. Yet both authors are conceptually sophisticated and engage deeply with the relevant historical sources. Must we accept a stalemate, or can we develop philosophical tools and deepen our historical understanding until one of the two accounts emerges as clearly and justifiably more accurate?

The disagreements between Franklin and Collins about gravity wave research are typical for studies at the intersection of history and philosophy of science. The goal in such projects is not only to use philosophical concepts in order to make an historical episode intelligible, but also to use the historical episode in order to improve our understanding of these very same concepts. Thus, concepts from the theory of experiment, such as calibration and measurement, are crucial to how the history of gravity wave research is understood. More broadly, the interpretation of the case at once hinges on and speaks to the nature of scientific rationality, especially the role of scientific epistemology in consensus formation. It is unsurprising that authors with different views of these issues will initially interpret the historical sources differently.

However, we believe that the underdetermination problem of integrated history and philosophy of science is far from intractable. It is an invitation to think about methodology: to reflect carefully about how to relate philosophical concepts and historical cases to each other. When different philosophical concepts lead to different narratives of the same historical episode, can a close study of the empirical facts of history decide between the competing philosophical viewpoints? Is it possible to develop a new or refined philosophical account of experiment—or calibration, or measurement, or rationality—on the basis of the historical sources? Conversely, is philosophical analysis sometimes capable of adjudicating between competing historical narratives? In other words, can it be legitimate to prefer one historical reconstruction over another on philosophical grounds?

Renewed awareness of such methodological questions is relevant to many current areas of research. Case studies from contemporary and past science play a prominent role not only among historians and historically inclined philosophers of science, but also among the many current philosophers of science who engage in detail with scientific practice. Cases are routinely used to explore, illustrate, question, or test philosophical and historical points of view—they can even become the linchpins of controversies. It is thus worthwhile to engage in explicit discussion about how we can put cases to appropriate use in the study of science.

Philosophical presupposition about the essence and character of science are inextricably woven into each and every historical narrative. In some instances, these underlying assumptions are subtle. There are works where they exert their power quietly in stabilizing a traditional genre of historiography. Biographical studies are an example, or editorial projects. The very project of writing a scientist's biography, or of editing a corpus of source texts, is based on assumptions of the dynamics of science and of the quality of scientific work, even if the prefaces and introductions will not make their assumptions explicit in all cases. The different ways that biographies are written or editorial projects are organized continues to be determined by philosophical assumptions.

But there are also many instances where philosophical concepts stand out prominently and are exposed to critical evaluation, or where a historical narrative is questioned on conceptual grounds. Consider historiographical and philosophical categories with broad epistemological import that are directly applicable to the analysis of cases. Among those we find meta-level concepts like discovery, observation and observability, the reality of theoretical concepts, representation and modeling, and scientific methodology. When these meta-categories are in play, any discussion soon requires a jointly historical and philosophical analysis that is sensitive to the ways in which history and philosophy of science can and cannot be related to each other, and especially to the inferences that case studies do and do not permit. In short, a philosophy of case studies is needed.

In November 2013, we brought together a group of researchers to debate the relationship between philosophy and history of science at the University of Bern, at a workshop titled “The philosophy of historical case studies”. The present volume derives from that workshop and makes the points of view, arguments, and cases developed by the contributors available to a wider audience. In addition, it presents a number of further papers that were invited after the workshop.

All the essays in this volume reflect explicitly on the relation between philosophical concepts and the way we write our historical accounts. Most of them proceed from one or more examples of actual underdetermination, that is, from an instance where different historical narratives were in fact determined by different philosophical projects. The essays all contribute in some way to one of two broad goals: to improve our understanding of the methodological challenges of case studies, and to develop a framework for meeting these challenges.

We are far from anything like a canonical understanding of the underdetermination problem of integrated history and philosophy of science, not to mention a recipe-like methodology to meet its challenges. Nevertheless, we believe that the collection of essays in this volume illuminates in a particularly transparent way key aspects of the interplay between history of science and philosophy of science.

1.2 Overview

The present volume is divided into three parts. The first part is concerned with theory: Its contributions try to describe and advance the current state of the art in relating history and philosophy of science to each other. The second part revisits controversies: Its contributions take up cases where different philosophical approaches have produced conflicting histories, or where historical studies have failed to settle philosophical disagreements. Finally, the third part strives for application: It presents contributions that use case studies both to investigate historical questions and to expand philosophical concepts.

1.2.1 The Relations Between History of Science and Philosophy of Science

Michael Bycroft's contribution focuses on the historiographical and interpretative maxim of the symmetry principle. As he points out, the symmetry principle is central to the methodology of much of today's science studies. But clear statements of its meaning are difficult to find. The standard formulation according to which true and false beliefs should be explained in the same way is deceptively simple and ambiguous. Bycroft proposes and defends a version of the symmetry principle according to which historical investigation should not assume from the beginning that true beliefs are best explained rationally and false beliefs are best explained irrationally.

Giora Hon offers a distinction relevant for the historical reconstruction of concept formation. He proposes to distinguish between two kinds of knowledge of the scientific community about a certain problem at a certain time: shared "baselines" on the one hand, and personal "snapshots" on the other hand. The baseline knowledge is the common ground for every attempt at a given time to interpret experimental data or problem settings. But each author takes a different view of the shared knowledge. Individual "snapshots" are limited and selective and, more importantly, every author emphasizes different aspects of a problem over others. His paradigm example to illustrate the distinction and to demonstrate its significance is the famous opening sentence of Einstein's 1905 paper on the "Electrodynamics of Moving Bodies." Here Einstein confronts a general feature of Maxwell's electrodynamics with a peculiar feature that he takes as motivation and point of departure for his analysis of the foundations of kinematics. Hon exemplifies his distinction by discussing three different authors, Einstein, Lorentz, and Poincaré, with respect to the relevance of Kaufmann's cathode ray experiments of the velocity dependence of the electron mass for the evaluation of their different theories of electron dynamics. He therefore goes beyond suggesting a historiographical category. He contends that the actors themselves perceived their differences in terms of the "baseline" knowledge and their individual "snapshots" of it.

Wolfgang Pietsch reminds us that the issue of case study methodology is by no means restricted to the history and philosophy of science. Rather it is a much discussed topic in the social and medical sciences, and Pietsch draws on some of that literature to introduce distinctions relevant for history and philosophy of science. He distinguishes between a predictive and a conceptual mode of reasoning with case studies. The predictive mode, which is prominent in the medical sciences (Pietsch's example is AIDS), is less relevant for the history of science. Here, the conceptual mode of reasoning is more often found. It is related to the problem of analogical reasoning, and the generic problem of both is how to justify case-based generalizations.

Finally, Raphael Scholl and Tim Räz address both the foundational issue of why a combination of history and philosophy may result in any non-trivial insights at all, as well as the more difficult question of an adequate methodology for an integrated history and philosophy of science. They propose a typology of case studies for the purposes of integrated history and philosophy of science. Their classification, which is not intended as exhaustive, includes hard cases, paradigmatic cases, big cases, and randomized cases. Scholl and Räz further discuss the confrontation of philosophical concepts and historical cases, illustrating by example how to handle agreements and disagreements between historical cases and philosophical concepts, and how to adjust philosophical categories in response.

1.2.2 Controversies Reconsidered

The very existence of the essay by Allan Franklin and Harry Collins is a statement. Both authors have engaged in extensive—and influential—historical and sociological analysis of scientific enquiry. They have done so from different viewpoints and philosophical assessments of the essence and character of science, and they happen to have looked at the same historical episode: the first phase of gravitational wave research with a bar detector by Joseph Weber in the sixties and seventies. Not surprisingly, their assessments of why Weber's research program came to be rejected by the community differ strongly. According to Collins, sociological explanations, the analysis of interaction between the scientists and their mutual perceptions play a major role. Franklin, on the other hand, argued that Weber's results were rejected as a result of rational discourse along well justified methodological principles. Famously, their different assessments led to some acidic mutual polemic. Both authors have continued to defend their claims against their strongest critics—themselves—and have found the resistance they each met to be sincere and justified. As a result they have come to accept some of each other's criticisms. In the contribution to this volume, they lay out, for the first time, an agreement about their different points of view. And not only do they agree about what they do not agree upon. They reach a new agreement by highlighting insights that only emerged in the process of their extended debate.

Katharina Kinzel also takes as an example the case study of the early gravity wave experiments. She uses this case study, along with different accounts of the demise of the phlogiston theory, as examples to reflect on the possibility and restrictions of pluralism in historiography. She looks at the various evaluation criteria for differing historical accounts and finds that they are either basic and generic, or complex and specific to the case at hand. Her account is informed by concepts of literary theory, and she proposes to look at different historical accounts as structured by different narrative templates. Applying evaluation criteria based on selectivity of sources, theory-ladenness and narrativity, she evaluates on the one hand the different accounts of the early gravity wave research by Franklin and Collins, and on the other hand the accounts by Musgrave and Chang of the chemical revolution.

Tarja Knuuttila and Andrea Loettgers compare three different accounts of the emergence and early history of the Lotka-Volterra model for population dynamics. All three accounts focus on modeling and on Vito Volterra's work, and all three case studies deliver their points by contrasting Volterra's work with that of other scientists. Michael Weisberg discussed Volterra's work as an example of a special kind of theorizing, which he calls modeling as opposed to abstract direct representation. Weisberg contrasts Volterra's work with that of Darwin on coral reef formation to illustrate his point. Scholl and Räz also compare Volterra's work and Darwin's, but they see both as modelers and distinguish them from scientists using more "direct" approaches such as methods of causal inference. According to them, a key difference between the two modelers is that Darwin proceeded much farther on the path from a 'how possibly' to a 'how actually' model than Volterra. Knuuttila and Loettgers challenge both accounts. They point out that the difference between Weisberg and Scholl and Räz can in part be explained by the fact that the commentators focus on Volterra's works from different periods. In their own work, Knuuttila and Loettgers take an even larger view of the development of Volterra's thinking, and they contrast it with the work of Alfred Lotka.

Tim Räz reflects on a case of disagreement that is both highly specialized and, at the same time, raises methodological questions of broad significance. His topic is a disagreement among five Einstein scholars who put a great deal of joint research effort into the analysis and reconstruction of Einstein's so-called Zurich notebook. It was written between summer 1912 and spring 1913 and documents Einstein's and his friends Marcel Grossmann's search for a generally covariant field equation of gravitation. The notes document a learning curve from the very first acquaintance with elements of tensor calculus to rather sophisticated calculations of properties of tentative field equations. Along the way, Einstein and Grossmann famously wrote down the correct equations already—if only in linear approximation—only to discard them again. In the line-by-line reconstruction of Einstein's notes the five scholars agree on almost all details but nevertheless differ significantly in their assessment of what it was that induced Einstein to discard the right field equations. The disagreement crystallizes in a distinction between what they call coordinate conditions and coordinate restrictions, the former concept indicating a modern understanding, the latter concept identified in their reconstruction of the notes. The disagreement concerns the question of whether Einstein at the time of doing the calculations documented in the notebook was already aware of the concept of coordinate conditions. After laying out the problem, Räz analyzes the different positions and probes possible ways of furthering the debate and resolving the disagreement.

1.2.3 Integration in Practice

Theodore Arabatzis and Kostas Gavroglu analyze the relationship between history and philosophy of science by questioning the concept of discovery as a simple historiographic category. In a naive understanding, discoveries are localizable in space and time and one can also identify the object of discovery and the subject, the discoverer. One can ask the four "w"-questions—the what, who, where, and when—and expect unambiguous answers for any genuine discovery. But as Arabatzis and Gavroglu show, in actual historical cases, we need to be prepared that neither of those question can be answered unambiguously. Discoveries, they maintain, are extended historical processes. Their example is the discovery of the inert gas argon by Lord Rayleigh and William Ramsay in the late 1890s. Closer historical analysis of the discovery reveals that the major difficulty in accepting the experimental data as indicating a new chemical element rather than as a discrepancy, was a necessary revision of the concept of chemical element itself. Prior to the discovery of Argon, the concept of "element" implied that substances that were identified as elements would be reacting with each other. But argon was chemically inert. The process of turning a discrepancy into a discovery involved the revision of the general concept of chemical element, in order to accommodate the chemical properties of argon as properties of a new element. Their case study demonstrates a philosophical point about the nature of scientific progress. Analyzing the discovery of argon with a skeptical stance toward the received meaning of the philosophical category of "discovery" reveals the inner workings of such a process and gives clues as to how the category should be used in a descriptively adequate way. The authors claim that the confrontational model of integrating history and philosophy of science should be replaced by a model where the historiographical categories should be judged by the historical narratives that they enable.

Simone de Angelis studies the interplay of *senso* (sense perception) and *discorso* (reasoning) in Galileo's observation of lunar mountains between 1609–1611. He extends previous discussions that either considered the finding of mountains as a straightforward observational fact or focused only on Galileo's geometrical, model-based reasoning. De Angelis argues that historical episodes should be understood as integrating a much greater range of historical and conceptual material. To get a full understanding of the episode of the lunar mountains, we must consider not only the different types of texts and forms of representation that Galileo produced, but also the critical context in which particular arguments and models were presented and received. Further, we need to take account of the various epistemic strategies that Galileo employed, including instruments, observations, theories, models, arguments and one experiment. Only if we integrate these many aspects of the episode can we understand the epistemic situation in which Galileo worked, and how he was able to conclude that the moon is mountainous.

Tilman Sauer considers the Stern-Gerlach experiment of 1922, focusing on the different perspectives that have been adopted with regard to the experiment and its outcome. In the experiment, individual silver atoms were sent through an inhomogeneous magnetic field and their deflection was observed. The historical actors considered this an *experimentum crucis*: while classical physics predicted a broadening of the beam of silver atoms, Bohr's quantum theory predicted a splitting of the beam into two components. The experiment spoke for Bohr's theory by demonstrating discrete deflection of silver atoms. However, the Stern-Gerlach experiment is now seen in a different light. First, it is seen as a confirmation of angular momentum projection quantization of the silver atom's electron spin. Second, it is seen as a demonstration of the dynamical collapse of the wave function in a quantum measurement, an interpretation which Sauer traces through Einstein, Ehrenfest and Bohm. Sauer argues that an adequate understanding of the Stern-Gerlach experiment requires us not only to take account of the various perspectives on the experiment that have been adopted, but also on how one perspective gave way to the other over time.

Dana Tulodziecki examines the transition from zymotic views of disease to germ views in the 19th century. In the debate about scientific realism, the case is seen as an instance supporting the pessimistic meta-induction: even though the zymotic theory was predictively successful, it was eventually abandoned in favor of the more successful germ theory. Faced with this data, anti-realists would argue that the radical discontinuity shows success to be no indication of truth. Realists would counter that the discontinuity is only apparent: those elements of the zymotic view which were responsible for the theory's success were, in fact, retained in the germ theory. Thus, predictive success remains an argument for truth in selective realism. Tulodziecki examines the historical sources closely and concludes that neither the anti-realist emphasis on radical discontinuity nor the realist emphasis on continuity of key elements allows an adequate understanding of the transition from zymes to germs. There was never a discrete choice between two theories, each of which had arguments in favor and against. Instead, there was a gradual evolution from one theory to the other, with elements being replaced step by step. The historical sources thus show the debate about scientific realism to be based on false assumptions. A reframed question emerges from this discussion: How can we adequately describe and explain the actual, gradual transition from the zymotic to the germ view of disease, and how does this actual transition relate to the question of scientific realism?

Adrian Wüthrich takes issue with the concept of unobservability and the alleged role that the maxim of eliminating unobservables played for Heisenberg's *Umdeutung* of kinematic and mechanical relations in the foundation of matrix mechanics. Wüthrich doubts that observability is a sharp concept and is sceptical as to whether a clear distinction between observable and unobservable can be upheld. In the abstract to his seminal 1925 paper, Heisenberg gives prominence to the maxim of eliminating from the theory unobservable quantities like electron orbits. The actual relevance of this methodological maxim has been questioned, and Wüthrich agrees with Mara Beller's criticism of Heisenberg's claim. In his paper, however, he takes Beller's critical attitude toward the rhetoric strategy of Heisenberg as a challenge to interpret Heisenberg's self-proclaimed method in a way that gives it a positive turn. Rather than assuming that no explicit methodology was at play, he argues that the actual methodological strategy that the actors were using was to ask for minimal causal explanations in the sense that a theory should only posit such entities as are required for an adequate explanation.

Acknowledgments The editors wish to thank the University of Bern and its Intermediate Staff Association (Mittelbauvereinigung) for generous funding of the workshop, Claus Beisbart and the Institute for Philosophy at the University of Bern for institutional support, the contributors for their enthusiasm, their essays, and their patience, and the staff at Springer for competent and productive cooperation. Raphael Scholl was supported by a grant from the Swiss National Science Foundation, (grant number P300P1_154590).

Part I
The Relations Between History of Science and Philosophy of Science

Chapter 2
How to Save the Symmetry Principle

Michael Bycroft

Abstract The symmetry principle is a central tenet of science studies, but clear statements of the principle are hard to find. A standard formulation is that *true and false beliefs should be explained in the same way*. This claim is multiply and harmfully ambiguous. The aim of this paper is to identify the main ambiguities and defend a more precise version of the symmetry principle. I argue that the principle should refer to types of cause not causes *in general*, that the relevant types are rational and irrational causes not social and non-social ones, that true and false beliefs should be explained impartially not identically, and that impartiality does not imply a ban on truth as an explanation of belief. The symmetry principle that emerges from these choices is that *historians should not assume in advance of historical inquiry that true beliefs are best explained rationally and that false beliefs are best explained irrationally*. I argue that this principle does what all symmetry principles should do: it is conducive to good historical writing, protects us from a genuine threat, makes room for the sociology of true beliefs, does not cast doubt on legitimate projects such as internal history of science, and does not commit us to controversial philosophical positions such as skepticism about present-day scientific theories.

2.1 Introduction

The symmetry principle is a central tenet of science studies—perhaps *the* central tenet of science studies—but clear statements of the principle are thin on the ground. According to a standard formulation, the principle is that *true and false beliefs should be explained in the same way*. This statement is multiply and harmfully ambiguous. Does it mean that all beliefs should be explained causally rather than acausally? Or does it mean that they should be explained using the same types of cause? If the latter, what kinds of cause do we have in mind? And what does "in the same way" mean? Should we really explain all beliefs in the same way, or should we

M. Bycroft (✉)
Department of History, University of Warwick, Coventry CV4 7AL, UK
e-mail: m.bycroft@warwick.ac.uk

T. Sauer and R. Scholl (eds.), *The Philosophy of Historical Case Studies*, Boston Studies in the Philosophy and History of Science 319,
DOI 10.1007/978-3-319-30229-4_2

simply keep an open mind about which explanations hold in any given case? Also, does the symmetry principle imply that some kinds of explanation are illegitimate? Or does it simply ask us to distribute our explanations evenly across true and false beliefs? Finally, how do we reconcile our equal treatment of past beliefs with our conviction—which most of us have—that all beliefs are *not* equal? Is it enough to treat the principle as a heuristic with no epistemological consequences, or is a more substantial response required? If the latter, what is the best response?

The aim of this paper is to distinguish between the various answers to these questions and to defend a particular answer to each. The result will be what I hope is a clearer version of the symmetry principle. To anticipate, the principle is the following: *historians should not assume in advance of empirical inquiry that true beliefs are best explained rationally and that false beliefs are best explained irrationally*. In short, historians should not use truth as a guide to rationality. I shall call this the Symmetry Principle, or the Principle for short (note the capital letters). Saving the symmetry principle means rescuing the Symmetry Principle from the many inferior maxims that go by that name.

I shall argue that the Symmetry Principle is more successful than other versions in meeting the following requirements. Firstly, it is sound. Historians who follow the Symmetry Principle will, all else being equal, give more accurate accounts of past and present science than those who routinely violate the Principle. Secondly, it is necessary in the sense that it protects us against an error that we are otherwise likely to commit. Thirdly, it performs the function for which the phrase "symmetry principle" was coined, namely to make room for sociological explanations of established scientific beliefs, as opposed to sociological explanations of scientific institutions or of discredited beliefs. Fourthly, the Symmetry Principle performs this function without prejudice to other goals that historians and sociologists of science can legitimately pursue. In particular, the Symmetry Principle says nothing against the practice of internal history of science. Finally, the Symmetry Principle does not require us to take sides in debates that are live ones in mainstream philosophy of science. I shall say more about these requirements when I invoke them in the course of my argument.

Given the number of articles and chapters that have been written on the symmetry principle, readers may wonder why another one is necessary. The short answer is that most of those articles and chapters have been written by sociologists, philosophers and scientists rather than by historians. As a result, the symmetry principle is usually discussed as part of larger debates about the promise of one or other sociological programme or about the viability of scientific realism. The principle is less often discussed as a tool for historical research, with the result that the second, fourth and fifth criteria in the previous paragraph are rarely taken into account. When historians invoke the symmetry principle, we tend to take it for granted, referring the reader to sociologists and philosophers for a more detailed defence and definition of the principle (e.g. Golinski 2005, p. x). Admittedly, there are overlaps between the historian's interest in the symmetry principle and that of the philosopher or sociologist. My debts to existing literature will be especially apparent in sections two and three below. However even in those sections I hope to give a historiographical twist to old debates.

2.2 Human Action Versus Types of Human Action

Does the symmetry principle state that all beliefs should be explained, at least partly, as the consequences of human action? Or does it state that all beliefs should be explained using the same range of human activities? Both versions can be found in the first detailed exposition of the symmetry principle, Barry Barnes' *Scientific Knowledge and Sociological Theory*. On the one hand, Barnes says that his target is the practice of "treating truth as unproblematic and falsehood as needing causal explanation" (Barnes 1974, p. 3). To treat true beliefs as "unproblematic" is to suppose that they "derive directly from awareness of reality" or that they "are the consequence of direct apprehension rather than effort and imagination" (p. 2). These statements suggest that Barnes is out to discredit sociologists who recognise no causal explanations of true beliefs, or who recognise only a trivial kind of causal explanation whereby states of affairs completely explain why people believe those states of affairs. Barnes is attacking the idea, for example, that the fact that the moon is mountainous is a sufficient explanation of Galileo's belief that the moon is mountainous.[1]

On the other hand, there are passages in which Barnes seems to say that true beliefs are routinely explained in causal terms, and moreover that these causes include human activities. Barnes devotes several pages to a survey of philosophers' accounts of "how beliefs actually can arise" through such causal processes as "sensory inputs, memory, induction and deduction" (p. 7). Barnes contrasts these causes with the ones usually invoked by sociologists to explain false beliefs, such as "inferior or impaired mentality, stupidity, prejudice, bigotry, hypocrisy, ideology, conditioning and brain-washing" (p. 2). On this showing, Barnes' complaint is not that sociologists have ignored the human activities that give rise to true beliefs. Instead it is that sociologists have explained true beliefs in terms of the former cluster of activities (sensing, deducting, and so on) rather than the latter cluster of activities (being stupid, prejudiced, and so on).

This ambiguity has not gone away in subsequent expositions of the principle. The peak of clarity came in David Bloor's 1976 account of the Strong Programme in the sociology of knowledge, where he distinguished between the principle that true and false beliefs both "require explanation" and the principle that true and false beliefs require explanation in terms of "the same types of cause." Bloor called the former the principle of "impartiality" and the latter the principle of "symmetry" (p. 7). This distinction did not last long, however. In (1981) Bloor referred to studies "in which both true and false beliefs are treated 'symmetrically,' i.e. as equally in need of explanation" (p. 392; cf. Barnes and Bloor 1982, p. 23). Harry Collins is a similar case.

[1] Cf. Barnes (1972), esp. pp. 376, 378.

In several places he advises sociologists to assume that "the natural world in no way constrains what is believed to be" (Collins 1981a, p. 3, 1981b, p. 218, 1982, p. 140). This suggests that Collins' project is to introduce human activities into our explanations of the beliefs of scientists. In other places, however, Collins has associated the symmetry principle with the project of "showing the interpretative flexibility of experimental data." Here the targets of Collins' relativism do not appear to be historians who ignore human activities altogether, but rather those who concentrate on a particular kind of activity, namely carrying out experiments and inferring theories from the results of those experiments. According to Collins, these activities are not the "decisive" ones in the emergence of scientific consensus (Collins 1981a, pp. 3–4, 7, cf. 1987, p. 825). Even the critics of the symmetry principle have sometimes been guilty of equivocating between explanations that appeal to truth and those that appeal to human activities. For example, Jean Bricmont and Alan Sokal, in a recent paper attacking the symmetry principle, slide between two versions of the view they are attacking. Initially it is the view that the *truth* of a belief cannot explain the belief; later it is the view that the *evidence* in favour of a belief cannot explain the belief.[2]

What do historians make of all of this? Jan Golinski's *Making Natural Knowledge* is a good place to look for an answer, since Golinski is sympathetic to the Strong Programme but identifies himself as a historian rather than a sociologist (Golinski 2005, pp. x, xix–xx, 5). Golinski's overall historical approach, which he calls "constructivism," was "inaugurated by a determination to explain the formation of natural knowledge without engaging in assessment of its truth or validity." This attitude of epistemic neutrality is just what he calls the "symmetry postulate" (p. 7). His phrasing of that postulate does not reveal whether he is urging the use of human activities *tout court*, or rather a particular kind of human activity, to explain true beliefs. However his definition of constructivism suggests that he has the former in mind. The constructivist "regards science as a human product, made with locally situated cultural and material resources, rather than as simply the revelation of a pre-given order of nature" (pp. xvii, 6).

To save the symmetry principle we need to reinstate Bloor's 1976 distinction between explaining all beliefs with (human) causes and explaining them all with the same types of (human) cause. As I shall put it, we need to distinguish between the "causal" and "multicausal" readings of the symmetry principle. One reason for this is to do justice to internal history of science. Traditionally, internal history of science has concerned itself with what Barnes called "sense perception, memory, deduction and induction." One consequence of the equivocation that I have been describing is that internal historians of science are lumped together with those who believe that theories "derive directly from awareness of reality." The danger of this conflation is that the sins of the latter will be unfairly attributed to the former. Barnes, Bloor and Collins never explicitly make this attribution. However a reader of their works could be forgiven for thinking that internal historians of science are guilty of some kind of

[2] Compare Bricmont and Sokal (2001a, p. 40, 2001b, p. 245). The equivocation is partly resolved at Bricmont and Sokal (2001b, p. 246).

explanatory subterfuge, and that the only way to give genuinely *causal* accounts of past science is to become a social historian of science.

Distinguishing the causal and multicausal readings of the symmetry principle has the added advantage of enabling us to reject the former. This is necessary because the causal reading does not protect us against a genuine threat. Few historians of science, past or present, have tried to explain past theories without reference to human activities of one kind or another.[3] This generalisation may seem rash, but it becomes plausible as soon as we see what it amounts to. An example may help to illustrate the point. Consider William Whewell, the nineteenth-century polymath whose *History of the Inductive Sciences* is one of Golinski's examples of a pre-constructivist work. Consider, in particular, a randomly chosen passage in which Whewell explains Humphrey Davy's theory that chemical and electrical attractions have the same cause (Whewell 1837, vol. 3, pp. 154–162). By my count, Whewell refers to 18 separate human actions in the course of his 9-page explanation. These include such things as: Davy's acquisition of a battery of great power in 1801; Davy's conjecture that in all cases of chemical decomposition, the elements are related to each other as electrically positive and negative; William Wollaston's demonstration that the Voltaic pile is always accompanied by oxidation or other chemical changes, and his conclusion that the pile cannot be explained solely in terms of contact between different metals; and Davy's equivocations about exactly what he meant by his electro-chemical theory. Acquiring an object, making a conjecture, drawing an inference, equivocating—surely these are human activities in the same sense that pursuing a class interest or upbraiding a colleague are human activities. Histories of science have always referred to such activities. Indeed, it is hard to imagine how one could write history of science without such references.

Why then have twentieth-century authors so often claimed the contrary? One plausible answer is that the authors in question have confused the claim that scientific theories have no human causes with other, superficially similar claims. For example, Golinski points out, rightly, that Whewell believed that the natural sciences make steady progress over time, and that they do so using a single method that is common to them all (2005, pp. 3–5). These beliefs may be false, but they do not imply that Whewell believed scientific theories arise independently of human action. On the contrary: Whewell recognised at least one activity that scientists perform and that is causally responsible for their beliefs, namely the act of implementing their method. There is another confusion lurking in Golinski's claim that eighteenth- and nineteenth-century historians of science saw the mind as a "mirror of nature." Golinski names Priestley and Whewell as holders of this view. No doubt these men believed that truth consists in a correspondence between mind and nature, and that truth is something that scientists regularly attain. But both of these beliefs are compatible with the view that scientists need to *do* things—including complex, difficult and time-consuming things—in order to acquire true beliefs.

Another source of confusion is that philosophical disagreements do not always have serious historiographical consequences. I have in mind the disagreement

[3] Laudan (1981b, p. 178) makes the same point about philosophers of science.

between those who recognise a class of nonmaterial facts, namely the facts about which inferences are objectively correct, and those who think that the only facts about inferences are the psychological ones about people endorsing this or that inference. John Worrall has defended the former view, which Bloor firmly opposes (Worrall 1990, pp. 313–318; Bloor [1976] 1991, pp. 178–79). According to Worrall, nonmaterial facts not only exist but can be legitimately used by historians to explain some of the inferences that we observe in the historical record. As both Bloor and Worrall recognise, their disagreement is real and fundamental. But what difference does it make to the way they do history? A glance at Worrall's historical papers suggests that it makes little difference, at least not with regards to his willingness to explain the outcomes of scientific debates in terms of the spatio-temporal activities of the scientists involved. His papers are awash with scientists whose hands manipulate objects and whose brains organise data and draw inferences (e.g. Worrall 1976, 1990).

No doubt Bloor's account of the same episodes, if he were to write one, would be different from Worrall's. But the difference between the two accounts would probably not lie in the amount of human activity they describe. More plausibly, it would lie in the *kind* of human activities they describe and that they consider causally significant. Worrall would focus on "sensory inputs, memory, induction and deduction," to borrow Barnes' list, whereas Bloor would focus on social interests and conventions. For want of better terms, Worrall would focus on "rational" causes and Bloor on "social" causes. A symmetry principle based on a distinction such as this one—a distinction between two different types of cause—is more promising than a principle urging causal explanations of all beliefs. The latter principle is sound but unnecessary.

2.3 Social Versus Rational

But what types of causes should we focus on here? Is the distinction between social and rational causes the right one for the job? The fact that many authors fail to distinguish between the causal and the multicausal readings of the symmetry principle means that it is not easy to know how they answer this question. Nevertheless, the standard answer seems to be that the social/rational distinction is dispensable, if not illusory. As many people have pointed out, social causes and rational ones are not mutually exclusive. Social causes usually involve cognition of some kind—after all, a scientist has to identify his interests in order to act upon them, and this identification requires both reason and experience. Conversely, reason is a social phenomenon in the obvious sense that it is usually carried out by groups of individuals who interact with one another. Moreover, the way in which these groups are organised—in small teams rather than large ones, for example—can effect the methods they pursue and the theories they adopt.[4]

[4]These points are sometimes framed as a debate about the validity of the distinction between "internal" and "external" factors, e.g. Barnes (1974, Chap. 5), Shapin (1992).

These overlaps leave us with two choices.[5] Firstly, we could revert to the distinction between causes that are social and those that are not. These categories are, by definition, mutually exclusive; and we can safely assume that the latter category is not empty, since it is surely not the case that social causes are the only kind of cause at work in past science. Secondly, we could revert to the distinction between rational and non-rational causes. When they have expressed an opinion on the matter, sociologists have typically chosen the first option. That is, they usually frame the symmetry principle as the view that all beliefs should be explained in terms of "social causes," "socialisation," the "social dimension" of science, or the "socially negotiated character" of science.[6] In order to save the symmetry principle, I suggest, we need to reject the first option and adopt the second.

The reason for this is that only the rational/irrational distinction gives us a symmetry principle that protects us from a genuine threat. Critics of the sociology of science have rarely maintained that social factors, *as social factors*, cannot help to explain the formation of a true belief. Insofar as they have denied a role for social factors, they have done so not because they perceived those factors to be social but because they perceived them to be irrational. Admittedly, this is a claim about the background motives of the critics in question, and since those motives are often tacit they are not easy to analyse.[7] However we can do worse than consider the case of Larry Laudan, one of the staunchest and most persistent critics of the Strong Programme. Laudan once argued that a historian should only consider social factors as an explanation for a scientist's belief if she has been unable to find a rational explanation for the belief (Laudan 1978, pp. 201–10). On this showing, Laudan's view seems clear-cut: "sociology is only for deviants," as Newton-Smith put it (1981, p. 238).

If we read carefully, however, we find that Laudan has plenty of time for the sociology of rational beliefs:

> The flourishing of rational patterns of choice and belief depends inevitably upon the pre-existence of certain social structures and social norms. (To take an extreme example, rational theory choice would be impossible in a society whose institutions effectively suppressed the open discussion of alternative theories.) ... *we need further exploration into the kinds of social structures which make it possible for science to function rationally* (when it does so) (1978, pp. 209, 222, original emphasis).

Clearly Laudan is not opposed to social explanations per se. Instead he is opposed to a particular kind of social explanation, namely those that compromise the rationality of the beliefs that are so explained. Since Laudan wrote, at least four philosophers of science have echoed his call for more studies of the social dimension of rationality (Papineau 1988; Worrall 1990, p. 314; Bird 2000, p. 275; Lewens 2005, pp. 567–68).

Unlike the social/nonsocial distinction, the rational/irrational distinction gives real bite to the symmetry principle. If we plug the latter distinction into the standard

---

[5] Some would add a third option, which is to formulate the symmetry principle without reference to the "social", the "rational", or related concepts. Latour (1993, pp. 91–97) seems to take this option.

[6] E.g. Barnes (1974, p. 6), Bloor ([1976] 1991, p. 6), Collins (1981a, p. 4), Golinski (2005, p. xx).

[7] Of course, this caution also applies to the rival claim that the social rather than the rational has been the main bone of contention.

formula, we end up with a principle that true beliefs can be explained using irrational causes and false beliefs using rational ones, even when the true and false beliefs in question are rival beliefs. This principle has teeth because it cuts the link between truth and rationality that we all rely on when assessing beliefs. How do we decide whether climate change is man-made, whether there is life on distant planets, or whether it will rain tomorrow? We consider the evidence, weigh the arguments for and against, evaluate our sources, search for new sources, and perhaps assess the social structure of any relevant expert communities—in short, we exercise our rationality. We do all this because we think that the more diligently we do it, the more likely we are to make a correct assessment of the belief. In other words, we assume that rationality is a good guide to truth—no doubt a fallible guide, but the best guide we have, and better than no guide at all. Now, if rationality is a good guide to truth, then the reverse must be true: truth must be a good guide to rationality. Hence, when we study past science it is natural for us to assume that true beliefs have rational origins and that false beliefs have irrational ones—or at least that the overall balance of rationality lies with true beliefs. This tendency is so natural that it is worth having a principle to guard against it.

Several objections might be raised against this version of the symmetry principle. One is that it strays too far from the original purpose of the principle, which was to make room for sociological explanations of true beliefs. Admittedly, the social/nonsocial distinction does a better job of serving the purposes of Barnes and Bloor than the rational/irrational distinction. Only the former distinction gives us a principle that explicitly states that true beliefs can be explained sociologically. The term "social" does not appear in the principle I am advocating. Nevertheless, my principle certainly makes room for sociological explanations. Moreover, it makes room for precisely those sociological explanations that trouble philosophers, namely those that are irrational.

The term "rationality" may be a stumbling block for some readers. Have not historians and sociologists shown just how problematic this term is? In particular, have they not shown that rationality is context-dependent, in the sense that different problems or subject-matters call for different methods; that it is non-consensual, in the sense that different people endorse different methods when presented with the same problem or subject-matter; and perhaps even that it is relative, in the sense that no person's notion of rationality is objectively better than anyone else's? Let us suppose, for the sake of argument, that all of these claims about rationality are true. Even then, they do not cause problems for the Symmetry Principle. All that is required is that each historian, given a context and a set of belief-forming processes, is able to sort the processes into those that, in her judgement, are conducive to true beliefs in that context and those that are conducive to false beliefs in that context. It is not necessary that the historian would make the same judgement given a different context, or that her judgements are the same as any other historian's, or that they can be objectively ranked alongside those of other historians.

But isn't rationality—even rationality in the meagre sense I have just outlined—a normative notion? And is there any place for normative notions in the descriptive discipline of history? The answer to both questions is "yes." Rationality is a normative notion even if it is context-dependent and non-consensual; and even if it is relative, and even if the historian believes it to be relative, it may be sufficiently normative to raise her hackles as a historian. But there is a place for this normative notion in history, because it occupies a small and self-deprecating place when enshrined in the Symmetry Principle. That Principle does not require us to make any normative statements in our books and articles, or even in the course of our research. On the contrary: it enjoins us *not* to make normative judgements when we go about explaining past beliefs. The only reason the Symmetry Principle refers to a normative notion is to denigrate that notion as a guide to historical research. The notion of "rationality" in my principle is as innocuous as the notion of "rationality" in earlier versions of the principle, such as the one stated by Barry Barnes and David Bloor in 1982.[8]

Scientific realists might worry that my principle has *too much* bite. If truth is a poor guide to the rationality of past beliefs, as the Symmetry Principle maintains, why should rationality be a good guide to truth in the present? And if rationality is indeed a poor guide to truth in the present, then there is no reason to think that our best current scientific theories are anywhere near the truth. This conclusion is absurd, the realist might argue, so my principle must be abandoned. I agree that the conclusion is absurd, but not that it follows from the Symmetry Principle. My defense depends in part on the resolution of a third ambiguity that muddies much of the literature on the symmetry principle.

2.4 Restrictive Versus Permissive

The instruction "explain all beliefs in the same way" can be followed in two quite different ways. The historian can assume that all beliefs really can be explained in the same way. Or she can suspend judgement about how they can be explained until she has done enough historical research to make this judgement. In short, the historian can treat beliefs identically or impartially.[9] The difference between the two approaches—and it is a big difference—is that the former rules out the possibility that different beliefs are susceptible to different sorts of explanation, whereas the latter leaves this possibility open. For this reason I shall call the former the "restrictive" approach and the second the "permissive" approach.

Which of these approaches do we find in canonical statements of the symmetry principle? If we turn to the early manifestos of Barnes and Bloor, we find restrictive versions of the principle. For example, Bloor writes that a reformed sociology of knowledge "would be symmetrical in its style of explanation. The same types of cause would explain, say, true and false beliefs" (Bloor [1976] 1991, p. 7). There is

[8]"Regardless of whether the sociologist evaluates a belief as true or rational, or as false and irrational, he must search for the causes of its credibility" (Barnes and Bloor 1982, p. 23).

[9]Here "impartially" means simply "without bias," and is unrelated to Bloor's principle of "impartiality" (on the latter see above, Sect. 2.2).

no room here for the possibility that true and false beliefs sometimes have different causes. The sociologist is assured that all beliefs have the same types of cause and is advised to seek them out. Bloor is just as strident in the 1982 paper he co-authored with Barnes (p. 23). Perhaps it would be unfair to rely too heavily on these slogans, however. To get a more nuanced view we might consider how Barnes and Bloor clarified their principle and how they and their followers have applied it to historical cases. Unfortunately, these two considerations point in opposite directions.

On the one hand, Barnes and Bloor both soften their initial statement of the principle. Barnes does so in a chapter on the role of "external" and "internal" factors in the history of science. Barnes glosses the former as "socio-economic" and the latter as "intellectual" or "technical." Barnes is refreshingly permissive about internalist historians, saying that he does not "take any a priori objection to their rejection of the significance of external or non-intellectual factors … We may proceed with an open mind to examine the [empirical] case against the externalists" (Barnes 1974, pp. 104–5). Bloor makes a similar concession. It is "surely correct," he writes, "that only some, and not all, episodes in the history of science are found to be crucially dependent on particular, social interests." The social component of knowledge is "always present" in science, but it is not necessarily "the trigger of any and every change" (Bloor [1976] 1991, pp. 166–67; cf. Ben-David 1981).

These statements are clear enough, but they are belied by the way that the symmetry principle is usually used to praise or blame a piece of historical work. When a study is praised as "symmetric," this is usually because it explains true and false beliefs in the same way. For example, Bloor praises J.B. Morrell for his "conspicuously symmetrical" account of two nineteenth-century chemical research schools led by Justus von Liebig and Thomas Thomson. Morrell sets out to explain why Liebig's school achieved international fame while Thomson's fell into obscurity. By "symmetrical" Bloor means that Morrell explains the plight of *both* schools in terms of the *same* set of factors—their interaction with the physical world in their laboratories, the personalities of Thomson and Liebig, their financial arrangements, and so on (Bloor [1976] 1991, pp. 34–36). Similarly, Barnes and Shapin congratulate Brian Wynn on his refusal to find "asymmetry" in the work of late-Victorian physicists at the University of Cambridge. By this they mean that Wynn considered both social and intellectual factors in his study, and that he found "no empirical basis for giving the one priority over the other" (Barnes and Shapin 1979, p. 95). Praise such as this gives the impression—intended or otherwise—that historians violate the symmetry principle whenever they give unequal weight to social and intellectual factors in their explanations of a belief.

Criticism sometimes conveys the same message as praise. For example, Shapin considers "profoundly asymmetrical" a paper by Charles C. Gillispie on Denis Diderot and other eighteenth-century thinkers who drew moral lessons from nature (Shapin 1980, p. 122). Some of Diderot's contemporaries, such as Voltaire, thought that nature bears no such lessons. It is clear from the paper that Gillispie sides with Voltaire on this matter. Shapin's complaint is that Gillispie explains Diderot's view as the product of a political ideology; when Gillispie explains Voltaire's view, by contrast, he appeals to the fact that Voltaire read and understood Newton's scientific works. Only those who adopt the restrictive view of the symmetry principle will find this a reasonable complaint. Those who adopt the permissive view will be open to the possibility that Voltaire was right for a good reason (consulting the opinion of an expert) whereas Diderot was wrong for a bad reason (adjusting his metaphysics to fit his politics). On the permissive view, Gillispie's account is asymmetric but need not be viciously so. On that view, what matters is the symmetry of the reasoning that led to his explanation, not the symmetry of the explanation itself.

It is hard to imagine how anyone could go about defending the restrictive view of the Symmetry Principle. Such a defence would require an a priori demonstration that, in every past scientific debate, the reasons on each side of the debate have been equally good. Perhaps an argument can be detected in the oft-repeated claim that science is "constitutively social" and that it is a "form of culture like any other." These phrases remind us that social phenomena are not optional additions to scientific life but indispensable components of it. It does not follow, however, that social and non-social factors are evenly distributed across true and false beliefs; and even if this did follow, it would not imply that rational and irrational factors are so distributed.

For the rest, the permissive reading sits well with two premises that most sociologists of science share with most historians of science. One is that empirical research is a more reliable source of data about past and present science than a priori speculation, at least if our aim is to describe rather than to evaluate the beliefs of scientists. The other premise is that historians cannot safely assume that the past resembles the present, or that a given period in the past resembles any other period in the past. These premises are hardly compatible with the restrictive reading of the symmetry principle, which rules out some phenomena a priori and treats a certain kind of symmetry as a historical constant.[10]

2.5 Equivalence Versus Exclusion

On the face of it, symmetry principles do not prohibit any explanations, whether social, non-social, rational, irrational, or whatever. They simply prohibit some ways of distributing these explanations across true and false beliefs. Nevertheless, prohibitions of the former kind are a recurring theme in literature on the symmetry principle. Indeed, symmetry principles have always made a double recommendation: all beliefs should be treated in the same way (equivalence), and certain treatments should not be applied to any beliefs (exclusion). The aim of this section is to untangle these two recommendations and to dissociate exclusions from the symmetry principle.

The exclusions in question are of two broad kinds. Some downplay the causal significance of certain human activities; others impose a total ban on certain explanatory resources that are not human activities, such as laws of nature and absolute standards of rationality. Harry Collins' writings illustrate both kinds of exclusion. As noted above, Collins downplays the role of experimentation, and especially the practice

[10]Cf. Laudan (1981b, p. 191).

of replicating experiments, in the resolution of scientific disputes. Collins seems to think that replication makes *some* difference to the beliefs of scientists, but he insists that it is not "decisive" and that it explains less than do social circumstances such as the relations of trust between scientists. Collins engages in a different kind of exclusion when he rejects "truth, rationality, progress and success" as valid explanations for any scientific beliefs, even when they are supplemented with other sorts of explanation. Similarly, Bloor would like to excise all "teleological" explanations from science studies, and Barnes and Shapin rail against "lazy references to reality, to nature, or logic, or necessity" (Bloor [1976] 1991, pp. 11–12; Barnes and Shapin 1979, p. 10, cf. p. 187).

Some of these exclusions are, to put it mildly, controversial. Consider Collins' argument for the indecisiveness of replication. This argument is either purely empirical or partly philosophical. If the former, we should be wary of assuming that it applies outside the cases that Collins has so far considered, which amount to a tiny fraction of twentieth-century science. The argument may apply outside those cases, but this is not something that historians can assume in advance of inquiry, much less enshrine in a methodological principle. If Collins' argument is partly philosophical, it rests on a dubious interpretation of the Quine-Duhem thesis, according to which theories can always be legitimately rescued from counter-examples by suitable adjustments to other parts of the theorist's belief system.[11]

Truth is another controversial exclusion. It is not at all evident that facts about nature can play no legitimate role in explanations of scientists' beliefs. The standard arguments to the contrary are that nature is the same for all actors, so it cannot explain differences between actor's beliefs; and that truth cannot explain the outcome of scientific disputes because truth is precisely what is under dispute. Neither of these arguments is persuasive. As Nick Tosh has pointed out, similarities in conditions *can* explain differences in outcomes. To use one of Tosh's examples, the existence of the atmosphere helps to explain why a lead weight falls faster than a feather. As for the second argument, it conflates the *justification* that an actor offers for a belief with the historian's *explanation* of the actor's belief. Obviously, Galileo did not include the claim "the moon is mountainous" in his arguments for the thesis that the moon is mountainous. That would have been circular. However it does not follow that the mountains on the moon played no causal role in bringing about Galileo's belief that those mountains exist (Tosh 2006, pp. 684–92, 2007, pp. 187–91; private communication).

It is tempting to conclude from these debates that we should give up the symmetry principle as a bad job.[12] A better response is to recognise that the symmetry principle, properly understood, does not commit us to any controversial exclusions. We can fairly distribute our explanations without thereby denigrating entire classes of explanation. Why would anyone think otherwise? Collins gives an argument that is

[11]On the history of dubious interpretations of the Quine-Duhem thesis, see Zammito (2004, pp. 17–25, 148, 150, 159, 163, 173, 180).

[12]If I understand them right, this is the conclusion of Tosh (2006, 2007), Bricmont and Sokal (2001a, b).

plausible but flawed. For him, the symmetry principle states that all beliefs should be explained in the same way regardless of whether they are TRASP (Collins' acronym for "true, rational, 'successful, or progressive'). Collins posits that only beliefs that are TRASP can be explained in terms of the fact that they are TRASP. From this posit, it follows that historians are bound to violate the symmetry principle unless they omit TRASP-ness altogether from their explanations (Collins 1981b, p. 217).

Collins' posit may be partially conceded—but only partially—in the case of truth. The mountains on the moon can explain Galileo's belief that there are mountains on the moon. But the perfectly smooth surface of the moon cannot explain the belief, held by Galileo's contemporary Ludovico delle Colombe, that the moon has a perfectly smooth surface, because the moon does *not* have such a surface.[13] More generally: the truth of a true belief can explain why people hold the belief; but the truth of a *false* belief cannot explain why people hold that belief, for the simple reason that there is no such thing as the truth of a false belief.

There are two reasons why this concession to Collins is only partial. Firstly, it does not apply to the truth of all beliefs but only to the truth of whatever belief we are trying to explain. The smoothness of the moon cannot explain the belief that the moon is smooth; but *other* facts about nature, ones recognised by modern science, can plausibly explain this belief. For example, Colombe's error about the moon's surface was partly due to his conviction that the moon is more perfect than terrestrial bodies, a belief due in part to the fact that the moon revolves around the earth roughly once every twenty-seven days. The second caveat is that the truths that can explain false beliefs include the truths of *rival* beliefs. Consider another of Colombe's false beliefs, namely that the blotches on the moon's surface, as seen through Galileo's telescope, are due to the uneven density of the moon's interior. The observed blotches were due precisely to the phenomenon—the mountains on the moon—that Colombe set out to refute with his density theory. In general, there are no truths that can explain a given true belief but cannot, at least in principle, explain a rival false belief. It follows that, although appeals to truth lead to asymmetry, this is an asymmetry of a particularly mild kind. The difference between true and false beliefs does not lie in the number or kind of truths that can explain them, since the same set of truths are candidates as causes of both true and false beliefs. The difference lies rather in the *relation* between the beliefs explained and the truths that explain them. True beliefs can enjoy a relationship with the truths that explain them that is unavailable to false beliefs. The relationship is that of correspondence: true beliefs can be explained by *their own* truth, whereas false beliefs cannot be explained by their own truth.

In the case of rationality there is no asymmetry, not even of the relational kind that applies to truth. Asymmetry arises for truth because truths are tenseless: if "the moon was mountainous in the seventeenth century" is a true proposition in 2014, that means that it was also a true proposition in 114 and 1614, and that it will still be a true proposition in 10014. The same cannot be said about rationality. In 2014 it is rational to believe the statement "the moon was mountainous in the seventeenth century."

[13]On Colombe's view on the moon, see e.g. Heilbron (2010, pp. 172–73).

But it is conceivable, and is arguably the case, that this belief was *not* rational in 114 or even in 1514. Beliefs that are rational today may have been irrational in the past, for the simple reason that the available evidence has changed over time. As a result, explanations of a belief that appeal to the rationality of the belief are not doomed to asymmetry. Such explanations can be applied *both* to beliefs that are rational today *and* to beliefs that are irrational today. The symmetry principle gives us no grounds for writing off rationality as an explanatory resource.

This conclusion assumes the time-dependence of rationality, but only a very mild kind of time-dependence. No paradigm-shifts are required, no dramatic changes in the meanings of the terms of theories or the criteria used to assess those theories. All that is required is that reasonable people can change their minds. An everyday example will suffice to illustrate the kind of process I have in mind. On Saturday evening I consult a competent weather report, one that has served me well in the past, and I learn that there will be no rain in my region on Sunday morning. When I wake up on Sunday morning, I hear a constant drumming on the roof and observe through a wet window that there are large puddles on my driveway. On Saturday evening I believed, rationally, that it would not rain on Sunday morning. On Sunday morning I believed, rationally, that my earlier belief had been mistaken. My metaphysics and epistemology did not change overnight. All that changed was the state of the evidence, and my beliefs about Sunday's rain changed accordingly. This is probably not the only sense in which rationality has changed in the course of past science. But we do not need to agree on any of the more controversial kinds of time-dependent rationality in order to agree that explanations that appeal to rationality are not inherently asymmetric.

Let us consider truth again. Given that some appeals to truth are inherently asymmetric, albeit mildly so, should we ban the asymmetric ones from historical explanations? No. Firstly, there is a perfectly good symmetry principle that makes no reference to truth-based explanations. According to the Symmetry Principle, the truth-value of a belief is a poor guide to its rationality. This principle requires that we explain some false beliefs in terms of their rationality. But it does not require that we explain some false beliefs in terms of their *truth*, and that in the absence of such cases we explain *no* beliefs in terms of their truth. Secondly, and most importantly, we gain nothing of value from the latter requirement that we do not already get from the former. What, after all, is the point of the latter requirement, in the eyes of Barnes and Bloor and Collins and Shapin? Their aim is partly to prevent historians from explaining true beliefs solely in terms of their truth. But very few historians of science have done this; not even Whewell did this, as I argued in the first section of this paper. Their other aim is to promote epistemic openmindedness. This means checking our tendency to explain true beliefs in terms of reason and sense experience and false beliefs in terms of sloppiness and self-interest. But the Symmetry Principle already achieves this goal. There is no need to add an extra clause banning appeals to truth in historical explanation.

2.6 The Symmetry Principle and Scientific Realism

The observation that the rationality of a belief can change over time helps to free the symmetry principle from controversial exclusions. However the same observation appears to lead to the equally controversial conclusion that we have no reason to believe the claims of present-day scientists. To recapitulate the argument at the end of Sect. 2.3: as per the Symmetry Principle, truth-value is a poor guide to the rationality of past beliefs; therefore rationality is a poor guide to the truth-value of past beliefs; therefore rationality is a poor guide to the truth-value of *present* beliefs; therefore there are *no* good guides to the truth-value of present beliefs. This conclusion will be a *reductio ad absurdum* of the Symmetry Principle for anyone—and this is surely just about everyone—who thinks that there are good reasons to believe many of the things that present-day scientists tell us about nature. Most of us believe that the moon *is* mountainous, and moreover that this belief is justified. The Symmetry Principle is in trouble if it implies otherwise.

This worry should not be confused with two other worries that philosophers have entertained about the symmetry principle. David Papineau once wondered whether the sociology of science "discredits science." He concluded that it does not, but his reason for thinking that it *might*was that the sociology of science—and in particular the symmetry principle—seems to tell us that scientists are not rational and hence that their conclusions cannot be trusted (Papineau 1988). By contrast, my worry is that the Symmetry Principle tells us that scientists cannot be trusted even when they *are* rational. This worry should also be distinguished from Larry Laudan's thesis that the predictive and explanatory success of present-day theories is a poor guide to their truth. Laudan built his case on the historical observation that many theories that were successful in the past are now considered false, and indeed have been superceded by theories that were, at some periods in the past, less successful than they. Laudan argued that today's successful theories will suffer the same fate as their forebears (Laudan 1981a, cf. 1981b, p. 186).

This "pessimistic induction," as it is now known, is similar to the argument I outlined two paragraphs ago. The difference is partly one of scope. Laudan intended his argument to apply only to scientists' beliefs about unobservable entities, such as electrons and ethers; and only to one criterion for assessing those beliefs, namely their predictive and explanatory success. My argument is wider. Indeed it is as wide as the Symmetry Principle, which applies to all the beliefs of scientists and to all rational criteria that they use to assess those beliefs. The two arguments also differ in their conclusions. Laudan concluded that we have no reason to believe today's theories about unobservable entities. My argument assumes that such doubt is implausible if broadened to include all the beliefs of today's scientists, and concludes that the premise—the Symmetry Principle—must be at fault. I shall call this argument the "pessimistic reduction," a name that signifies its similarity to the pessimistic induction and its aim of reducing the Symmetry Principle to absurdity. The aim of the rest of this section is to defend the Symmetry Principle against the pessimistic reduction.

One option is to take a leaf from Collins' book and insist that the Symmetry Principle is merely a heuristic device and hence that it says nothing about what the world contains or about how much we know about its contents. In other words, it is methodologically sound but epistemically and metaphysically innocent. It asks not that historians become anti-realists, but merely that they suspend their realism when examining past science. This is now the orthodox interpretation of the symmetry principle, endorsed by Shapin, Bloor and Golinski as well as Collins.[14] Unfortunately, methodological advice cannot be so easily separated from substantive claims about the past, and such claims can have epistemic consequences.[15] The way to disarm the pessimistic reduction is not to deny that the Symmetry Principle has any consequences for epistemology but to formulate the Principle in such way as to limit the damage that those consequences do to our epistemic intuitions.

Plausibly, one way to soften the consequences of the Symmetry Principle is to place a constraint on the kinds of theories to which it applies. The most obvious option would be to restrict the Principle to claims about unobservable entities. This would license historians to use truth-value as a guide to their explanations of a great number of past beliefs: claims about particular objects and events, from the forms of plants to the size of earthquakes; empirical laws, such as the sine law of refraction and the value of the mechanical equivalent of heat; and, depending on how one defines "unobservable entity," claims about distant galaxies and geological events that occurred in the remote past. Much of the history of physics and chemistry would become immune to the Principle, and historians of anatomy and physiology would be almost entirely beyond its grasp.

The problem is that, despite these massive concessions from history to philosophy, the Symmetry Principle would still be controversial. After all, scientific realism is still a respectable thesis among philosophers of science. Many philosophers believe that there are grounds for confidence not just in particular observations and empirical laws but also in unobservable entities such as proteins, quarks and oxygen molecules. These philosophers would likely reject any version of the Symmetry Principle that casts doubt on scientist's claims about unobservable entities, even if it leaves the rest of science in tact. Granted, scientific realism may be false. However, historians should not adopt methodological principles that pre-empt this conclusion. In doing so they would alienate a large portion of their potential readership, forfeit the right to use their research as an independent test of scientific realism, and incur the obligation of revising all their publications should scientific realism turn out to be true. In short, limiting the Principle to unobservable entities is not going to please anyone. It is too great a constraint for the average historian, and too small a constraint for the average philosopher.

---

[14]This interpretation is so orthodox that some writers now say that is has always been orthodox (Shapin 1999, p. 4; Golinski 2005, p. 8). This claim is hard to reconcile with passages in the early writings of Barnes (1974, p. 154), Barnes and Bloor (1982, p. 27), and Collins and Cox (1976).

[15]In this I agree with Bricmont and Sokal (2001a, pp. 38–43), though my reasons are not identical to theirs.

A more promising approach is to observe that only the restrictive version of the Symmetry Principle is prey to the pessimistic reduction.[16] To see this, consider the historical theses that we commit ourselves to if we adopt these two versions of the Principle. The permissive version commits us to a view about the state of our knowledge about the past. It implies that we currently have no reason to think that the balance of rationality has always lain with true beliefs. If we did have such a reason, then the permissive Principle would be unnecessary, since the inference that Principle disallows would be a legitimate inference. In such a world, the permissive Principle would not, it must be said, lead historians into error. But it would lead them to truth more slowly than necessary, since it would force them to do time-consuming empirical research to answer a question—where did the balance of rationality lie in past debates?—that they could have answered from their armchairs. When we endorse the permissive Principle we commit ourselves to the belief that we are not in such a world.

What are the epistemological consequence of this commitment? Merely that infallible scientific realism is false, i.e. that we cannot be absolutely certain that scientists' current beliefs about nature are true. This is a perfectly respectable consequence that most scientific realists would happily accept. Most realists, and perhaps all of today's realists, are fallibilists. They concede that there is a possibility that any given present-day theory is false. They simply insist that the probability of this being the case is rather small. This is consistent with the belief that there are many historical debates in which the balance of rationality did not lie with the true theory.

It might be objected that fallible scientific realism is *not* consistent with the view the balance of rationality lay with the false theory in *most* historical debates. Indeed, fallible realism seems to require that rationality lay with the true theory in the *vast majority* of historical debates. Fortunately, this realist thesis does not render the permissive Principle unnecessary. Given any historical case, the realist historian can be fairly confident that rationality lies with the true theory. However he can be *much more* confident in this judgement if he examines the historical record. And if his empirical research shows that this was one of the rare cases in which rationality lay with the false theory, his data trumps his a priori expectation. The realist historian who examines a past debate is like a policeman in a quiet neighbourhood who examines a driver for excessive drinking. The policeman believes on good evidence that most drivers in the neighbourhood are sober, and hence that this particular driver is unlikely to be drunk. Nevertheless he checks the driver's breath, as thoroughly as he can, because he knows that, in individual cases, checking is better than guessing.

This response to the pessimistic reduction is not available to those who endorse the restrictive version of the Symmetry Principle. The realist requires that rationality lay with the true belief in the vast majority of historical debates. The restrictive Principle flatly contradicts this requirement. It states that in every historical debate worthy of

[16]There are other promising approaches. One is to distinguish between live scientific debates, where many relevant experts disagree on a question, and settled debates, in which there is a wide though not universal consensus on the question. Here I focus on the restrictive/pessimistic distinction in order to show the importance of that distinction.

the name, the side that turned out to be wrong had as many good reasons for their position as the side that turned out to be right. There is no question of adopting the restrictive Principle merely as a methodological tenet. If scientific realism is true, then historians who follow the restrictive Principle are in great danger of giving false accounts of past science. The permissive Principle runs no such risk.

2.7 Conclusion

One version of the symmetry principle is that there are a great variety of beliefs and that all of them are equal. The moral of this chapter is that there are a great variety of symmetry principles and that they are *not* all equal. To save the symmetry principle we need to distinguish it from the many dubious maxims with which it has become entangled over the course of its long career. To begin with, the principle should not urge the inclusion of human activities in explanations of the beliefs of scientists. This is not because human activities have no explanatory role, but because very few historians of science, past or present, have maintained such a perverse doctrine. Similarly, the principle should distinguish between rational and irrational beliefs rather than social and nonsocial ones, since very few historians or philosophers have denied that social factors can help to explain true beliefs. The principle should be permissive rather than restrictive, at least until someone demonstrates that all scientific battles without exception were fought with equally powerful arguments on both sides. In the meantime, the permissive view is in tune with the empiricism and historicism espoused by many sociologists and historians. The principle should instruct us to distribute our explanations in an impartial manner, but it should not exclude any types of explanation from historical practice. Some such exclusions may be justified, but it is not the principle that justifies them, and in particular the principle does not exclude appeals to truth in historical explanations. Finally, the principle should not force us to take sides on controversial philosophical questions such as the truth or otherwise of scientific realism. The principle that results from these choices is what I have been calling the Symmetry Principle. It states that historians should not assume in advance of empirical inquiry that true beliefs are best explained rationally and that false beliefs are best explained irrationally. Stated briefly—perhaps too briefly to avoid further confusion—historians should not use truth as a guide to rationality.

References

Barnes, B. 1972. Sociological explanation and natural science: A Kuhnian reappraisal. *European Journal of Sociology/Archives Européennes de Sociologie* 13(2): 373–391.

Barnes, B. 1974. *Scientific knowledge and sociological theory*. London: Routledge and Kegan Paul.

Barnes, B., and D. Bloor. 1982. Relativism, rationalism, and the sociology of knowledge. In *Rationality and relativism*, ed. M. Hollis, 21–47. Cambridge, MA: MIT Press.

Barnes, B., and S. Shapin (eds.). 1979. *Natural order: Historical studies of scientific culture*. Beverly Hills: Sage Publications.
Ben-David, J. 1981. Sociology of scientific knowledge. *The state of sociology: Problems and prospects*, 40–59. Beverly Hills: Sage.
Bird, A. 2000. *Thomas Kuhn*. Chesham: Acumen.
Bloor, D. 1981. Sociology of (scientific) knowledge. In *Dictionary of the history of science*, ed. J. Browne, W. Bynum, R. Porter, 391–393. London: MacMillan.
Bloor, D. [1976] 1991. *Knowledge and social imagery*. Chicago, IL: University of Chicago Press.
Bricmont, J., and A. Sokal. 2001a. Science and sociology of science: Beyond war and peace. In *The one culture?: A conversation about science*, ed. H. Collins, and J. Labinger, 27–47. Chicago, IL: University of Chicago Press.
Bricmont, J., and A. Sokal. 2001b. Reply to our critics. In *The one culture?: A conversation about science*, ed. H. Collins, and J. Labinger, 243–254. Chicago, IL: University of Chicago Press.
Collins, H. 1981a. Introduction: Stages in the empirical programme of relativism. *Social Studies of Science* 11(1): 3–10.
Collins, H. 1981b. What is TRASP? The radical programme as a methodological imperative. *Philosophy of the Social Sciences* 11(2): 215–224.
Collins, H. 1982. Special relativism: The natural attitude. *Social Studies of Science* 12(1): 139–143.
Collins, H. 1987. Pumps, rock and reality. *The Sociological Review* 35(4): 819–828.
Collins, H., and G. Cox. 1976. Recovering relativity: Did prophecy fail? *Social Studies of Science* 6(3/4): 423–444.
Golinski, J. 2005. *Making natural knowledge: Constructivism and the history of science*, 2nd ed. Chicago: University of Chicago Press.
Heilbron, J. 2010. *Galileo*. Oxford: Oxford University Press.
Latour, B. 1993. *We have never been modern*. Harvard, MA: Harvard University Press.
Laudan, L. 1978. *Progress and its problems: Towards a theory of scientific growth*. University of California Press.
Laudan, L. 1981a. A confutation of convergent realism. *Philosophy of the Social Sciences* 48(1): 19–49.
Laudan, L. 1981b. The pseudo-science of science? *Philosophy of the Social Sciences* 11: 173–198.
Lewens, T. 2005. Realism and the strong program. *The British Journal for the Philosophy of Science* 56(3): 559–577.
Newton-Smith, W. 1981. *The rationality of science*. Routledge and Kegan Paul.
Papineau, D. 1988. Does the sociology of science discredit science? In *Relativism and realism in science*, ed. R. Nola, 37–57. Dordrecht: Kluwer Academic Publications.
Shapin, S. 1980. The social uses of science. In *The ferment of knowledge: Studies in the historiography of eighteenth-century science*, ed. G.S. Rousseau, and R. Porter. Cambridge: Cambridge University Press.
Shapin, S. 1992. Discipline and bounding: The history and sociology of science as seen through the externalism-internalism debate. *History of Science* 30(90): 333–369.
Shapin, S. 1999. Rarely pure and never simple: Talking about truth. *Configurations* 7(1): 1–14.
Tosh, N. 2006. Science, truth and history, part I. Historiography, relativism and the sociology of scientific knowledge. *Studies in History and Philosophy of Science Part A* 37(4): 675–701.
Tosh, N. 2007. Science, truth and history, part II. Metaphysical bolt-holes for the sociology of scientific knowledge? *Studies in History and Philosophy of Science Part A* 38(1): 185–209.
Whewell, W. 1837. *History of the inductive sciences*. London: John W. Parker.
Worrall, J. 1976. Thomas Young and the 'refutation' of Newtonian optics. In *Method and appraisal in the physical sciences*, ed. C. Howson. Cambridge: Cambridge University Press.
Worrall, J. 1990. Rationality, sociology and the symmetry thesis. *International Studies in the Philosophy of Science* 4(3): 305–319.
Zammito, J. 2004. *A nice derangement of epistemes: Post-positivism in the study of science from Quine to Latour*. Chicago: University of Chicago Press.

Chapter 3
"Baseline" and "Snapshot": Philosophical Reflections on an Approach to Historical Case Studies

Giora Hon

Abstract Logically, generating knowledge requires a fixed set of presuppositions, anchored in a given conceptual framework. Scientists may or may not be aware of all the elements that are involved in the process of generating knowledge but, whether the elements are assumed explicitly or implicitly, they have to be fixed for the production of knowledge to be coherent. I distinguish between two sets of elements of knowledge, which I call a "baseline" and a "snapshot." The baseline represents the sum of what is, in principle, available to the community of practitioners in the field. In contrast, a snapshot is personal, that is, it is the result of applying some rules of selection to the baseline. A snapshot includes, in addition to the selected elements, idiosyncratic assessments of the elements; such assessments may not be found in the standard literature. I analyze two case studies, theoretical and experimental, in which the practitioners themselves presupposed the distinction here proposed. I show that the distinction is an effective tool in the presentation of case studies with the goal of throwing light on how scientific knowledge is modified and changed. What is illuminating in the cases at hand is the fact that the scientists themselves exhibited in their works the dynamics of "baseline" and "snapshot," in parallel to the practice of the historians and the philosophers of science.

3.1 Introduction

Scientific knowledge is in constant flux: sometimes the change is fundamental, sometimes it is incremental; despite the important differences between these two kinds of changes, we find consistent features in the way the new knowledge relates to the antecedent state of a discipline. Logically, generating knowledge requires a fixed set of presuppositions, anchored in a given conceptual framework. The practitioner may or may not be aware of all the elements that are involved in the process of generating knowledge but, whether the elements are assumed explicitly or implicitly, they have

G. Hon (✉)
Department of Philosophy, University of Haifa, Haifa, Israel
e-mail: hon@research.haifa.ac.il

T. Sauer and R. Scholl (eds.), *The Philosophy of Historical Case Studies*,
Boston Studies in the Philosophy and History of Science 319,
DOI 10.1007/978-3-319-30229-4_3

to be fixed for the production of knowledge to be coherent. In other words, the scientist determines the relevant background and keeps it fixed throughout the episode during which he or she seeks to contribute to some aspect of scientific knowledge.

There is thus a variety of background knowledge and, generally, we distinguish between two sets of elements of knowledge, which we call a "baseline" and a "snapshot" (see Hon and Goldstein 2009). A baseline captures scientific knowledge at a certain time and it is relatively stable for some given duration. The baseline represents the sum of what is, in principle, available to the community of practitioners in the field. Hence, this kind of background knowledge has no nuances and exhibits no preferences, for it is just an inventory of elements. In contrast, a snapshot is personal, that is, it is the result of applying some rules of selection to the baseline, separating the wheat from the chaff as seen in the context of a specific conceptual framework and metaphysical outlook. A snapshot is directly related to a baseline but it is not simply a subset since it includes, in addition to the selected elements, individual assessments of the elements; such assessments may not be found in the standard works of the relevant field in the public domain, for they reflect the idiosyncratic view of a practitioner. Evaluations, which are personal to a large extent, create a tension, or a problem, which the scientist then seeks to address. In sum, the baseline is public and more or less explicit: what all practitioners are expected to know in a given domain. By contrast, the snapshot is unique to the individual scientist and often it is not fully articulated by the practitioner; rather, it is frequently the case that the historian (or philosopher) identifies implicit elements of the snapshot that were taken for granted by the scientist.

To write a history of scientific change, and more specifically, conceptual change, the historian of science must establish a set of references. Change is relative, for it can be determined only through comparison. So what methodology facilitates such a comparison? In other words, how does the historian obtain the baseline of the domain under discussion at a particular period, and the snapshot for a specific practitioner?

The availability of dictionaries is useful, but key classical texts, encyclopedias, and review articles in learned journals are the best sources of evidence for the historian who wishes to determine the relevant baseline. For example, Diderot's *Encyclopédie* is a classical case in point (Diderot et al. 1751–1765). In fact, scientists themselves discovered the importance of establishing a baseline, and on several occasions in the modern era a group of scientists committed themselves to a comprehensive recording of the edifice of knowledge of the discipline under scrutiny. A case in point is the *Encyklopädie der mathematischen Wissenschaften mit Einschluss ihrer Anwendungen* of 1898–1904, which Klein initiated (see Dyck 1904).

By its very nature, as personal and idiosyncratic, a snapshot is much more difficult to determine than the relevant baseline. Here the historian has to exercise a great measure of ingenuity and to engage sympathetically with the author in question, seeking in the writings of the scientist clues for new elements and assessments. Some historians turn to private records (e.g., Franklin's (1981) use of Millikan's laboratory notebooks), others seek published commentaries which throw light on the novel contribution, still others search for autobiographical notes and reminiscences (including those by colleagues), or even circumstantial evidence that may reveal the scientist's sources and assessments at the initial stages of research. But one should not underestimate the vast amount of information that can be gleaned from the very publications of the practitioner under consideration.

In what follows, I analyze two case studies, theoretical and experimental, in which the practitioners themselves, the very actors, presupposed the distinction here proposed. To anticipate my conclusion, the distinction between baseline and snapshot is an effective tool in the presentation of case studies with the goal of throwing light on how scientific knowledge is modified and changed.

3.2 A Case in Point: Theory

Here is a well-known opening sentence of a published paper in an established journal. In effect the author begins his paper by drawing the baseline and sketching his snapshot. The distinction is explicit, albeit presented in a most compact way:

> It is known that Maxwell's electrodynamics—as usually understood at the present time—when applied to moving bodies, leads to asymmetries which do not appear to be inherent in the phenomena.[1]

This is how Einstein opens his celebrated paper of 1905, "On the Electrodynamics of Moving Bodies."

Einstein commences his paper by calling attention to the fact that Maxwell's electrodynamics—as usually understood at the present time—leads, when applied to moving bodies, to asymmetries which do not correspond to anything in the phenomena. One may be tempted to think that Einstein directs his criticism, right at the outset of the paper, at the formalism of Maxwell's theory, for this formalism is responsible for these asymmetries. It would seem that one need to address the form of the equations, either modify or replace it, with the goal of making the equations correspond faithfully to the phenomena. But then one takes Einstein's expression "as usually understood at the present time" to mean that the fault lies in the way the theory had been understood, not the equations. After all, later in the paper Einstein introduces the Maxwell-Hertz equations lock, stock, and barrel, and does not question their validity in any respect. Indeed, the evidence in support of Maxwell's equations at the turn of the 20th century was overwhelming; they satisfactorily described a large number of phenomena. Thus, the formalism is taken to be correct without comment. The expression "Maxwell's theory" or, to be precise, "Maxwell's electrodynamics," refers then not merely to the equations but to the equations together with a host of assumptions, derivations, and interpretations.

If we recast Einstein's claim positively, it may yield a new perspective: even in cases where the electrodynamic (or electromagnetic) phenomena arise from the same relatively moving bodies, the theory leads to asymmetries, for it distinguishes

[1]"Daß die Elektrodynamik Maxwells—wie dieselbe gegenwärtig aufgefaßt zu werden pflegt—in ihrer Anwendung auf bewegte Körper zu Asymmetrien führt, welche den Phänomenen nicht anzuhaften scheinen, ist bekannt" (Einstein 1905, p. 891).

different cases based on the choice of which body is moving and which is at rest, whereas only the relative motion is relevant. Einstein takes for granted that the phenomena should be given precedence over the theory and proceeds to analyze the famous thought experiment of the magnet and the conductor.

There is much to be admired in the lean account that Einstein gives in his research papers. He expresses his argument tersely and one wishes to unpack it. So we return to the original sentence: "It is known that Maxwell's electrodynamics—as usually understood at the present time—when applied to moving bodies, leads to asymmetries which do not appear to be inherent in the phenomena." On the one hand, Einstein acknowledges a large body of literature which is "Maxwell's electrodynamics" that forms the "baseline," but the addition, the latter part of the sentence, this is Einstein's take on this literature; he is bothered by the asymmetries which are the consequences of this theory: this is the "snapshot." The asymmetry is an "artifact" of the theory; it is not "inherent" in the phenomena.

The issue for Einstein concerns the theory, not just the equations; thus, formal solutions were not in fact solutions, for they were merely manipulations of formulas. This is implied in Einstein's explicit reference in the opening sentence to Maxwell's theory and not to its equations. The problem is that the theory distinguishes between electricity and magnetism whereas, in fact, these phenomena are different manifestations of the same thing—the electromagnetic field. In a word, indistinguishability is the key concept, not interchangeability.

Einstein refers to Maxwell's electrodynamics, that is, he does not address the formalism, but the theory as it is commonly interpreted. Thus, "as usually understood" refers to the assumption of the ether and the further assumption that electricity and magnetism are distinct phenomena, to be sure interchangeable, but still distinct. For Einstein the issue is not the symmetry or the asymmetry of the equations—which was a central theme for Hertz, Heaviside, and some other notable physicists—but the removal of asymmetries that are "artifacts" of the theory and have no objective reality. In sum, in a stroke, in one opening sentence, Einstein—in a most dense way—marks the baseline of electromagnetism at the turn of the last century, as well as characterizes his snapshot of this common understanding of Maxwell's electromagnetism.

In a sense, Einstein is doing here the work of the historian and philosopher of science: he positions himself with respect to the history of the discipline and outlines his philosophical critique. What Einstein then develops with his methodological commitment to physical rather than formal argument, in relation to the baseline of Maxwell's electromagnetism, is considered revolutionary (see Hon and Goldstein 2005).

3.3 A Case in Point: Experiment

I now turn to a story of experiment, the Kaufmann experiment. The year is 1906. Walter Kaufmann, the renowned and well respected German experimental physicist, who a few years earlier had been very close to the discovery of the electron, published his definitive results of a set of experiments in which he tested several theories of electron based on different assumptions regarding the constitution of this newly discovered particle. Indeed, Kaufmann entitled his paper (1906), "On the Constitution of the Electron."

Between 1898 and 1906 Kaufmann executed several sets of experiments on the relation between the mass of a charged particle and its velocity. In his initial experiments he had subjected cathode rays to electric and magnetic fields, but the crucial innovation came in the second phase when he introduced β-rays into the experimental set up. Kaufmann measured the observed deflections of the rays as they were recorded on photographic plates.

Essentially, what is impressive in this story is not so much the experiment itself but the reactions it received. Here too the distinction between baseline and snapshot is applicable with illuminating consequences. The baseline consists of the experimental result, or rather results, as they were published in a series of papers from the turn of the last century till the definitive paper in 1906.

We stay with the relativity theory for this is what it was about. In fact, Kaufmann was the first physicist to cite Einstein's (1905) relativity paper. He stated the conclusion of his series of experiments in no uncertain terms:

> The ... results speak decisively against the correctness of Lorentz's theory and consequently also of that of Einstein's theory. If one were to consider these theories as thereby refuted, then the attempt to base the whole of physics, including electrodynamics and optics, upon the principle of relative motion would have to be regarded at present as also unsuccessful.[2]

It is noteworthy that the first response in print to Einstein's relativity paper was an experimental refutation, not some logical or conceptual critique of the theory.

Kaufmann began his paper with a theoretical discussion of various theories of electron, claiming that Einstein's formulae amounted to those of Lorentz; indeed, he regarded Lorentz's and Einstein's theories as two theories which assume the same electron constitution, which he called the Lorentz-Einstein theory of the electron. In concluding the theoretical introduction, Kaufmann anticipated the general result of the experiment:

> The results of the measurements are not compatible with the Lorentz-Einstein fundamental assumption. The Abraham and the Bucherer equations depict equally well the results of the observations. For the present a decision between these two theories by a measurement of the transverse mass of β-rays appears to be impossible.[3]

---

[2]"Die vorstehenden Ergebnisse sprechen entschieden gegen die Richtigkeit der Lorentzschen und somit auch der Einsteinschen Theorie; betractet man diese aber als widerlegt, so wäre damit auch der Versuch, die ganze Physik einschließlich der Elektrodynamik und der Optik auf das Prinzip der Relativbewegung zu gründen, einstweilen als mißglückt zu bezeichnen." (Kaufmann 1906, p. 534), quoted in (Hon 1995, pp. 170–171).

[3]"Die Messungsergebnisse sind mit der Lorentz-Einsteinschen Grundannahme nicht vereinbar. Die Abrahamsche und die Buchererschen Gleichungen stellen die Beobachtungsresultate gleich gut dar. Eine Entscheidung zwischen beiden durch Messung der transversalen Masse der β-Strahlen erscheint einstweilen als unmöglich." (Kaufmann 1906, p. 495), quoted in (Hon 1995, p. 190).

According to Kaufmann, the equations for the motion of the electron given by Lorentz and Einstein differ "very considerably from that of Abraham."[4] Kaufmann therefore expressed some surprise that, as he remarked,

> an application of the equations to my earlier measurements by Herr Lorentz led to the ... result that my observations could be represented by him with the same accuracy as by the Abraham equations for the rigid electron.[5]

As a consequence of Lorentz's re-interpretation, Kaufmann's experiment lost its conclusiveness. Kaufmann therefore redid the experiment in an attempt to force a decision primarily between the theory of Abraham and that of Lorentz. Given sufficient experimental accuracy, the results of the various theories could be separated. Indeed, Kaufmann pointed out that in the new set of experiments there was a 5–7 % difference between the velocities yielded by the theories of Abraham and Lorentz for each measured curve-point. Hence, Kaufmann claimed, "a way was provided for differentiating between the two theories."[6]

In view of his experimental results, Kaufmann argued that physics cannot be based on the principle of relative motion. He could thus uphold the assumption that "physical phenomena depend on the movement relative to a quite definite coordinate system which we call *absolute resting ether*."[7] The fact that it had not been possible to demonstrate experimentally the existence of this fixed coordinate system did not deter Kaufmann from concluding that "a decision may not be made as to the impossibility of such a proof."[8]

Kaufmann did not consider his experiment a direct test of Einstein's relativity theory; rather, he intended his experiment to be an *experimentum crucis* of what was one of the most contested issue in physics at the turn of the last century, namely, the theory of electron. At stake were three theories, that of Abraham, Bucherer, and Lorentz; Kaufmann's experiment should have functioned then as an arbiter. The purported refutation of Einstein's theory is considered in the conclusion of Kaufmann's paper a consequence of the incompatibility of Lorentz's theory of electron

---

[4]"... sich von den Abrahamschen Gleichungen sehr wesentlich unterscheiden." (Kaufmann 1906, p. 493), quoted in (Hon 1995, p. 190).

[5]"Eine Anwendung der Gleichungen auf meine bisherigen Messungen durch Hrn. Lorentz führte zu dem überraschenden Resultat, daß meine Beobachtungen durch sie mit derselben Genauigkeit darstellbar seien, wie durch die Abrahamschen Gleichungen für das starre Elektron." (Kaufmann 1906, p. 493), quoted in (Hon 1995, p. 190).

[6]"... war zugleich ein Weg gegeben, zwischen beiden Theorien zu entscheiden." (Kaufmann 1906, p. 493), quoted in (Hon 1995, p. 190).

[7]"Wir werden vielmehr einstweilen bei der Annahme verbleiben müssen, daß die physikalischen Erscheinungen von der Bewegung relativ zu einem ganz bestimmten Koordinatensystem abhängen, das wir als den *absolut ruhenden Äther* bezeichnen." (Kaufmann 1906, p. 535), quoted in (Hon 1995, p. 194, italics in the original).

[8]"Wenn es bis jetzt nicht gelungen ist, durch elektrodynamische oder optische Versuche einen derartigen Einfluß der Bewegung durch den Äther nachzuweisen, so darf daraus noch nicht auf die Unmöglichkeit eines solchen Nachweises geschlossen werden." (Kaufmann 1906, p. 535), quoted in (Hon 1995, p. 194).

with the experimental evidence. In other words, the rejection of Einstein's theory was a by-product of the direct refutation of Lorentz's theory.

Kaufmann's experimental results aroused much interest and received strictures as well as support from many distinguished physicists. These varied reactions are of considerable importance since they show how divergent the responses to the same set of experimental results could be.

The interest here is to respond directly to the motivating questions of this volume: What happens when different philosophical positions lead to different historical narratives? How do we assess competing accounts of the same historical case study? In the case of the Kaufmann experiment, the protagonists did, as it were, the work themselves; the actors, namely, Poincaré, Einstein, and Lorentz, responded to the experimental results in different narratives. Perhaps "narrative" is not the right term; rather, it is the response itself which is at stake. The baseline was there to respond to and several snapshots of this very baseline were taken. The baseline is mutual, for Kaufmann's published results were common knowledge, but the snapshot is idiosyncratic; it reflects the metaphysical and methodological commitment of the scientist. In the hand of the historian and philosopher of science the case study becomes a heuristic device which can uncover philosophical dispositions—the engine of scientific change.

3.3.1 Response I: Poincaré

Poincaré's heroes were those who sought a new theory of matter, that is, a theory which would explain the phenomenon of matter solely in terms of electrons immersed in ether: "Beyond the electrons and the ether," Poincaré proclaimed, "there is nothing" (Poincaré 1914, p. 209). Since Einstein set himself a completely different problem, he was not one of Poincaré's heroes. Einstein was concerned with principle theory and was not interested, at least in this domain of physics, in construction theory. It comes therefore as no surprise that Poincaré praised the theoretical work of Lorentz and Abraham and the experiments of Kaufmann, even when their results were not in agreement with his own ideas.

In 1905, Poincaré considered Kaufmann's results conclusive. He did not show any hint of skepticism that something might have gone amiss in the experiments. Three years later, when Einstein's relativity theory began to be understood as a comprehensive theory of principle, Poincaré still had much to say in praise of the results of Kaufmann:

> Abraham's calculations make us acquainted with the law in accordance with which the *fictitious* mass varies as a function of the velocity, and Kaufmann's experiment makes us acquainted with the law of variation of the total mass. A comparison of these two laws will therefore enable us to determine the proportion of the *actual* mass to the total mass. (Poincaré 1914, p. 206, italics in the original).

The result of Kaufmann's determination of this proportion was "most surprising: the actual mass is nil." And Poincaré concluded that

> we have thus been led to quite unexpected conceptions... What we call mass would seem to be nothing but an appearance, and all inertia to be of electromagnetic origin. (Poincaré 1914, pp. 206–207).

Poincaré considered the principle of relativity a law of nature which has been generalized from experience. He sided with Lorentz and by means of what is called Poincaré's stress gave a theoretical underpinning for the mechanism of the Lorentz contraction and the stability of the electron under such deformation.

As an exponent of conventionalism, it is difficult to understand why Poincaré set himself the trap which a realist would gladly lay before a conventionalist, namely, the pitfall of the *experimentum crucis*. Whereas in 1902 Poincaré had considered experiment a guide that helps us in our free choice among all possible conventions, in 1908 he considered its role to be that of an arbiter. "The method employed by Kaufmann," he remarked, "would... seem to give us the means of *deciding* experimentally between the two theories" (Poincaré 1914, p. 228, italics in the original). And in the concluding remarks to his account of the new mechanics, he stated that "further experiments will no doubt teach us what we must finally think of... [the new theories]. The root of the question is in Kaufmann's experiment and such as may be attempted in verification of it" (Poincaré 1914, p. 249). Why Poincaré, a noted conventionalist, conferred so much weight on a single experiment to the point of believing that it "revolutionizes at once Mechanics, Optics, and Astronomy," is at the root of the difficulty in comprehending Poincaré's responses to Kaufmann's experiment (Poincaré 1914, p. 286). Was it the case that he was conventionalist as a philosopher, but as a physicist he laid his trust in experiment? He was aware of this duality. "Have you not written," he invited his reader in 1905 to ask the author, that is, Poincaré himself,

> that the principles [of relativity theory], though of experimental origin, are now unassailable by experiment because they have become conventions? And now you have just told us that the most recent conquest of experiment put these principles in danger? (Poincaré 1946, p. 318).

Indeed, for Poincaré "principles are conventions and definitions in disguise. They are, however, deduced from experimental laws, and these laws have, so to speak, been erected into principles to which our mind attributes absolute value" (Poincaré 1952, p. 138). In other words, the principle of relativity and the principle of conservation of mass, for example,

> are results of experiment boldly generalized; but they seem to derive from their very generality a high degree of certainty. In fact, the more general they are, the more frequent are the opportunities to check them, and the verifications multiplying, taking the most varied, the most unexpected forms, end by no longer leaving place for doubt. (Poincaré 1946, p. 301).

But, then, new experimental results did compel one to doubt the absolute value which had been attributed to these principles. Specifically, in 1905 Poincaré himself questioned, "if there is no longer any mass, what becomes of Newton's law?... If the coefficient of inertia is not constant, can the attracting mass be? That is the question"

(Poincaré 1946, p. 312). Furthermore, in 1908 he acknowledged that Kaufmann's experiments "*have shown Abraham's theory to be right*. Accordingly, it would seem that the Principle of Relativity has not the exact value we have been tempted to give it" (Poincaré 1914, p. 228, italics in the original). Poincaré's answer is somewhat amusing: "Well, formerly I was right and today I am not wrong. Formerly I was right," he repeated and stressed that "what is now happening is a new proof of it" (Poincaré 1946, p. 318).

Poincaré's philosophy of science is not a rigid conventionalism. He appears to discern in science a spectrum of philosophies. At one extreme lies conventionalism which forms the philosophical foundations of geometry. And at the other extreme lies induction, the method of the physical sciences. Poincaré located the science of mechanics in between these two extremes. By contrast to geometry, in the physical sciences induction is the guiding method. It is here that "experiment is the source of truth." Here, experiment alone can teach us something new; it alone can give us certainty. He qualified, however, this strong view by remarking that all that experiment affirms, "is that under analogous circumstances analogous fact will be produced" (Poincaré 1952, pp. xxvi, 140–142). An experimental law is therefore always subject to revision.

"The principles of mechanics," Poincaré argued, "are presented to us under two different aspects. On the one hand, there are truths founded on experiment, and verified approximately as far as almost isolated systems are concerned; on the other hand," he continued, "there are postulates applicable to the whole of the universe and regarded as rigorously true" (Poincaré 1952, pp. 135–136). Thus, crucial experiments and conventions constitute the extremes of the spectrum and in between lies the science of mechanics. Poincaré considered Kaufmann's experiment an *experimentum crucis*, an arbiter among theories, and as such he accepted so to speak its verdict. Yet he was aware of its possible limitations, its approximations and possible errors. However, in his view, Kaufmann was a skillful experimenter who had taken "all suitable precautions and one cannot well see what objection can be brought" (Poincaré 1914, p. 229). Notwithstanding this confidence, Poincaré suspected that something had gone amiss regarding the measurement of the electric field. Was it uniform or not? "May it not be," he questioned, "that there is a sudden drop in the potential in the neighborhood of one of the armatures?" (Poincaré 1914, p. 229).

Nevertheless, Poincaré was reluctant to suspend judgment on an experiment he was suspicious of. In the final analysis, Poincaré submitted to Kaufmann's expertise on matters of experimentation and accepted the results. A theoretician for whom experimental results form the building blocks of knowledge–knowledge whose cement, so to speak, is the inductive method and to whose overall architecture conventionalism attends–Poincaré relied heavily on Kaufmann's stature as a leading experimenter and was prepared to concede, as late as 1908, that Abraham's theory had been shown to be right.

3.3.2 Response II: Einstein

In contrast to Poincaré who construed the principles of mechanics in general and the principle of relativity in particular to be essentially bold generalizations of a posteriori experience gleaned from several experiments, Einstein elevated the two principles of his relativity theory to a priori postulates which stipulate the features of the laws of nature so that there are no privileged observers. Though he spoke in his 1905 paper of unsuccessful experimental attempts to discover any motion of the earth relatively to the light medium, Einstein did not specify them. In the theory of relativity an empirical finding separated itself completely from its experimental origin and attained the status of a postulate which, together with another postulate, provided the axiomatic base for "a simple and consistent [einfachen und widerspruchsfreien] theory of electrodynamics of moving bodies" (Einstein 1905, p. 892). By comparison, for Poincaré the principles of mechanics always remained attached to their origins in experiments. Poincaré was thus prepared on both philosophical and experimental grounds for a refutation of the principle of relativity.

For Einstein a refutation of his theory would have meant more than just a rejection of a principle. It would mean discarding a comprehensive theory that, on the one hand, is simple and clear in terms of its logical deductions and, on the other hand, embraces a wide complex of phenomena. Einstein stressed that his was not a descriptive theory of the electron, or a constructive theory like that of Abraham; rather, it was a theory of principle, a theory which stipulates procedures for attaining physical knowledge. Although Einstein concluded his 1905 paper by deducing from the theory the properties of the motion of the electron that could be accessible to experiments, hence the theory is in principle refutable, he clearly intended his theory to be situated on some kind of meta-level. From this higher level, other theories and laws of nature may be looked upon and comprehensively correlated within a broad perspective.

What does it take to refute a theory of principle? Does Kaufmann's refutation of Einstein's relativity paper constitute such a case? Note that the results were available in 1905 but it seems that Einstein was not interested in testing his theory against Kaufmann's experimental results. But in 1906 Einstein was challenged when Kaufmann referred explicitly to the theory.

In a review article, published in 1907, "On the principle of relativity and the conclusions that follow from it," in a section entitled, "On the possibility of an experimental test of the theory of motion of the material point. Kaufmann's investigation," Einstein formulated his official response to Kaufmann's devastating results (Einstein 1907). Einstein began his response by acknowledging the validity of Kaufmann's experiment and indeed, like Poincaré, accepting Kaufmann's authority as a renowned experimenter. Einstein went into some details in describing the experiment and its apparatus.

> A prospect of comparison with experience of the results derived in the last section exists only where the moving electrically charged mass points possess velocities whose square is not negligible compared to c^2. This condition is satisfied in the cases of the faster cathode rays and the electron rays (β-rays) emitted by radioactive substances.
>
> There are three quantities for electron rays whose mutual relationships can be the subject of a more detailed experimental investigation, i.e., the generating potential or the kinetic energy of the rays, the deflectability by an electric field, and the deflectability by a magnetic field.
>
> [...]
>
> In the case of β-rays, only the quantities [of magnetic and electric deflections] are (in practice) accessible to observation. Mr. W. Kaufmann ascertained the relation between [the magnetic and electric deflections] for β-rays emitted by radium bromide granule with admirable care [bewunderungswürdiger Sorgfalt]. (Einstein 1907, pp. 436–437).

There follows then a description of the apparatus in some detail, and Einstein continues:

> In view of the difficulties involved in the experiment one would be inclined to consider the agreement as satisfactory [genügende]. However, the deviations are systematic and considerably beyond the limits of error of Kaufmann's experiment. That the calculations of Mr. Kaufmann are error-free [fehlerfrei] is shown by the fact that, using another method of calculation, Mr. Planck arrived at results that are in full agreement with those of Mr. Kaufmann. (Einstein 1907, p. 439).

Einstein acknowledged that Kaufmann's calculations are correct, but he noticed that the calculated results based on the relativity theory diverge systematically beyond the limits of experimental error. He therefore advised suspending judgment. He remarked:

> Only after a more diverse body of observations becomes available will it be possible to decide with confidence whether the systematic deviations are due to a not yet recognized source of errors or to the circumstance that the foundations of the theory of relativity do not correspond to the facts. (Ibid.)

This appears to be a reasonable response, much in line with traditional methodology: either there is an unrecognized error in the experiment or the theory is indeed wrong. In any event, one needs to wait: what one has thus far as experimental evidence does not suffice for a balanced judgment. Had Einstein stopped here, the response would be indeed traditional and not of much interest for the current discussion. But now comes the punch line:

> It should also be mentioned that Abraham's and Bucherer's theories of the motion of the electron yield curves that are significantly closer to the observed curve than the curve obtained from the theory of relativity. However, the probability that their theories are correct is rather small [ziemlich geringe Wahrscheinlichkeit], in my opinion, because their basic assumptions concerning the dimensions of the moving electron are not suggested by theoretical systems that encompass [umfassen] larger complexes of phenomena. (Ibid.)

What we have here is a clash between virtue and evidence. Einstein acknowledged openly that the theories of Abraham and Bucherer, that is, two classical theories of electrons, yield results which are significantly closer to the observed curve than the one obtained from his relativity theory. Einstein did not ignore this experimental evidence nor did he circumvent it. He stated the facts as they had appeared in the relevant publications. This is the baseline. He then appealed to probability; he remarked that the probability of these electron theories to be correct is rather small since the assumptions of these theories regarding the dimensions of the moving electron are not derived from systems which encompass larger complexes of phenomena. By implication, Einstein claimed–not explicitly–that his own proposed theoretical system has higher probability to be correct because, and this is the critical point, it encompasses larger complexes of phenomena. This is the snapshot. Factually, this is indeed the case; the foundations of Einstein's theory of relativity are categorically different from those of the classical ones. This is now well known, but not so clearly perceived at the time. Einstein's theory does indeed encompass a wide range of phenomena. Einstein concluded the subsection on Kaufmann's experiment with a condensed expression of a theoretical virtue.

Einstein elevated the principles of his relativity theory to a priori postulates which stipulate the features of the laws of nature so that there are no privileged observers. Given this axiomatic structure of the theory, what would an experimental refutation entail? It is not just the rejection of a principle; it means discarding a comprehensive theory that, on the one hand, is simple and clear in terms of its logical deductions and, on the other, embraces a wide complex of phenomena.

Einstein's strong belief that his theory transcends the status of a theory of matter and assumes the character of a theory of principle motivated his claim that Kaufmann's experiment was amiss. But his response was intuitive; he could not base his rejection of the evidence on either experimental or theoretical grounds. On the contrary, as we have seen, he was of the opinion that the measurements had been taken with "admirable care" and the calculations were "free of error". Einstein, it emerges, founded his case against these experimental evidence on theoretical virtue. What strengthened his confidence in his theory was the fact that according to the relativity theory the calculated values for the relations between the magnetic and electric deflection lay in a consistent manner above the observed curve; a fact that aroused his suspicion that a systematic error had vitiated Kaufmann's experiments. All the same, a theoretical virtue sufficed for Einstein to substantiate his intuitive rejection of the evidence.

The consistent methodological position of Einstein motivated, one is even inclined to say, compelled, Einstein to suspect an error whose source he could not determine; this suspicion gave, in his view, sufficient ground for rejecting the evidence.

According to Einstein a physical theory can be criticized from two perspectives: the degree of its "external confirmation" and the extent of its "inner perfection", that is, the degree of its "logical simplicity" (Einstein 1949, pp. 21–23). Here we have been concerned with the former demand which at first sight appears straightforward: "the theory must not contradict the evidence." However, Einstein immediately qualified this demand as he realized that its application had turned out to be quite delicate:

> it is often, perhaps even always, possible to adhere to a general theoretical foundation by securing the adaptation of the theory to the facts by means of artificial additional assumptions.

> In any case, however, this first point of view is concerned with the confirmation of the theoretical foundation by the available empirical facts. (Einstein 1949, pp. 21–23).

Since Einstein's suspicion that there was an error in Kaufmann's experiment and the consequent "blank" rejection of the experimental evidence have been vindicated, it is surprising that Einstein did not stress, or at least suggest, in his review of his intellectual autobiography, the possible occurrences of experimental errors. It is further surprising that he preferred rather to mention the possible adaptability of a theory to empirical facts and, further, to make the obvious point that the empirical knowledge which one can have at any historical juncture is limited. Compared to the possible occurrences of experimental errors, these two qualifications are indeed weak.

Be that as it may, one can follow Dirac and imagine Einstein responding to the evidence in the following vigorous words: "well, I have this beautiful theory, and I'm not going to give it up, whatever the experimenters find; let us just wait and see." As it happened, Einstein has so far proved right: "so it seems," Dirac surmised, "that one is very well justified in attaching more importance to the beauty of a theory and not allowing oneself to be too much disturbed by experimenters, who might very well be using faulty apparatus" (Dirac 1982, p. 83).

3.3.3 *Response III: Lorentz*

If Poincaré's and Einstein's reactions to Kaufmann's experimental results constitute the two possible extreme responses–both suspecting an error, the former, however, favoring the acceptance of the results and the latter categorically rejecting them–then Lorentz's reaction mediates or rather vacillates between the two poles. In 1924 Lorentz remarked that

> one of the lessons which the history of science teaches us is ... that we must not too soon be satisfied with what we have achieved. The way of scientific progress is not a straight line one which we can steadfastly pursue. We are continually seeking our course, now trying one path and then another, many times groping in the dark, and sometimes even retracing our steps. (Lorentz 1924, p. 608).

Although Lorentz illustrated this remark with the way physicists have interpreted the phenomenon of light throughout the ages, the development of his own view of the theory of electron in general and Kaufmann's experiment in particular offers an insightful case study of this characteristic of the march of science. Indeed, in his 1904 paper Lorentz accepted Kaufmann's results and considered them decisive, that is, he did not question their validity; rather, he demonstrated how his theory could account for the results: his formula for the transverse mass agreeing with the data at least as well as that of Abraham's formulae. According to Lorentz, Abraham's theoretical results were

> confirmed in a most remarkable way by Kaufmann's measurements of the deflection of radium-rays in electric and magnetic fields. Therefore, if there is not to be a most serious

> objection to the theory I have now proposed, it must be possible to show that those measurements agree with my values nearly as well as with those of Abraham. (Lorentz 1904, p. 826).

Lorentz, it appears, did not even contemplate, at least in this paper, the possibility that Kaufmann's experiment might be wrong. Moreover, he mentioned neither the small difference between the predictions of his theory and those of Abraham's theory nor the inaccuracy in Kaufmann's measurements reported by Abraham in 1903. For Lorentz it was apparently sufficient that a satisfactory agreement had been attained between the experimental results and his theory. He simply did not take the trouble to analyze the experiment itself and accepted its results, as he had accepted other experimental results, to serve as a test and a guide for his theory. Lorentz's reaction to Kaufmann's early results amounts in effect to the claim that there might have occurred an error of interpretation: Lorentz analyzed the data of Kaufmann's measurements within his own theory and showed how a satisfactory agreement with his own formulae could be obtained.

As a result of Lorentz's new interpretation, Kaufmann's experiment lost its conclusiveness. Kaufmann therefore redid the experiment in an attempt to force a decision primarily between the theory of Abraham and that of Lorentz. Although the new measurement did not vindicate Abraham's theory, they decisively refuted, according to Kaufmann, Lorentz's theory and thereby that of Einstein.

In 1906, in a series of talks at Columbia University, Lorentz presented his theory of electron. These talks, published in 1909 and, in a second edition, in 1915, demonstrate that Lorentz believed in an electromagnetic synthesis in which ether and charged particles are the fundamental concepts; matter is just superfluous. Kaufmann's final results in 1906 encouraged Lorentz to think that he was on the right track. Indeed, he cited Kaufmann's results with approval and remarked, "it will be best to admit Kaufmann's conclusion... that the negative electrons have no material mass at all." In Lorentz's view this conclusion "is certainly one of the most important results of modern physics" (Lorentz 1916, p. 43). But, of course, he was aware of the consequences:

> so far as we can judge at present, the facts are against our hypothesis.... Kaufmann has repeated his experiments with the utmost care and for the express purpose of testing my assumption. His new numbers agree within the limits of experimental errors with the formulae given by Abraham, but not so with [my equation for the transverse mass], so that they are decidedly unfavorable to the idea of contraction, such as I attempted to work out. (Lorentz 1916, pp. 212–213).

In March 1906 Lorentz wrote to Poincaré that

> unfortunately my hypothesis of the flattening of electrons is in contradiction with Kaufmann's new results, and I must abandon it. I am, therefore, at the end of my Latin. It seems to me impossible to establish a theory that demands the complete absence of an influence of translation on the phenomenon of electricity and optics. (Quoted in (Miller 1981, p. 334).)

Lorentz did not question the validity of Kaufmann's definitive results. Rather, he stressed that in speculating on the structure of the electron it should not be forgotten that

> there may be many possibilities not dreamt of at present; it may very well be that other internal forces serve to ensure the stability of the system, and perhaps, after all, we are wholly on the wrong track when we apply to the parts of an electron our ordinary notion of force. (Lorentz 1916, p. 215).

Lorentz adopted an agnostic position in order to pursue further his theory which, he acknowledged, had been refuted. In 1904, Lorentz demonstrated to his own satisfaction how his theory could account for Kaufmann's early set of results from the experiments conducted in 1902–1903; however, in 1906 Lorentz felt obliged to uphold the new results and, as a consequence, to relinquish his contraction hypothesis. He later expressed his idiosyncratic view in the following way: "each physicist can adopt the attitude which best accords with the way of thinking to which he is accustomed."[9]

3.3.4 Three Different Responses

Three theoretical physicists responded differently to the experimental results of Kaufmann regarding the constitution of the electron. Poincaré accepted Kaufmann's experimental results though he warned against a possible error. He suggested that an error might have occurred in the working of the apparatus. He, however, did not suspend his judgment and assented to the conclusion that classical theories of electron had been confirmed experimentally and, as a consequence, the relativity principle cannot serve as a foundation for physics. Einstein did not find any error; he suspected a systematic error but could not find its source. Indeed, he praised Kaufmann's expertise, but rejected the results all the same on methodological grounds. He could not give up the relativity principle. Lorentz, on his part, vacillated. Initially, he had thought that the interpretation was wrong; then, he found the results correct and had to relinquish his contraction hypothesis; but ultimately he continued to believe in his theory.

3.4 Conclusions

To return to the motivating questions: What happens when different philosophical positions lead to different historical narratives? How do we assess competing accounts of the same historical case study? Like scientists, historians and philosophers of science, too, rely essentially on two different kinds of background knowledge in order to call attention to the changes that have taken place in the edifice of scientific knowledge and to fathom the force of change. Two case studies were presented, theoretical and experimental, and in both cases the methodological approach consisted in differentiating between "baseline" and "snapshot." The distinction allows for treat-

[9]Quoted in Hirosige (1976), p. 70, see also Lorentz (1931), p. 210.

ing differently common knowledge, which is displayed in the public domain, the baseline, and idiosyncratic knowledge (if we can call it "knowledge") which reflects the personal perspective of the protagonist–this is the snapshot. The dynamic interaction between these two perspectives on knowledge offers a way of approaching the motivating questions and suggests a mode of grasping scientific change. What is illuminating in the cases at hand is the fact that the scientists themselves exhibited in their works the dynamics of "baseline" and "snapshot," in parallel to the practice of the historians and the philosophers of science.

References

Diderot, D., et al., (eds.). 1751–1765. *Encyclopédie ou dictionaire raisonné des sciences, des arts et des métiers*. Paris: Briasson. 17 vols.

Dirac, P.A.M. 1982. The early years of relativity. In *Albert Einstein: Historical and cultural perspectives*, ed. G. Holton, and Y. Elkana, 79–90. Princeton: Princeton University Press.

Dyck, W.V. 1904. Einleitender Bericht über das Unternehmen der Herausgabe der Encyklopädie der mathematischen Wissenschaften. In *Encyclopädie der mathematischen Wissenschaften mit Einschluss ihrer Anwendungen*, ed. W. Meyer, vol. 1, part. 1: Arithmetik und Algebra, v–xx. Leipzig: Teubner.

Einstein, A. 1905. Zur Elektrodynamik bewegter Körper. *Annalen der Physik* 17: 891–921. Reprinted in (Stachel, 1989, Doc. 23).

Einstein, A. 1907. Über das Relativitätsprinzip und die aus demselben gezogenen Folgerungen. *Jahrbuch der Radioaktivität und Elektronik* 4(4): 411–462. Reprinted in (Stachel, 1989, Doc. 47).

Einstein, A. 1949. Autobiographisches. In *Albert Einstein: Philosopher-Scientist*, ed. P.A. Schilpp, 1–96. La Salle (Illinois): Open Court.

Franklin, A. 1981. Millikan's published and unpublished data on oil drops. *Historical Studies in the Physical Sciences* 11: 185–201.

Hirosige, T. 1976. The ether problem, the mechanistic world view, and the origins of the theory of relativity. *Historical Studies in the Physical Sciences* 7: 3–82.

Hon, G. 1995. Is the identification of experimental error contextually dependent? The case of Kaufmann's experiment and Its varied reception. In *Scientific Practice: Theories and Stories of Doing Physics*, ed. J.Z. Buchwald, 170–223. Chicago: Chicago University Press.

Hon, G., and B.R. Goldstein. 2005. How Einstein made asymmetry disappear: Symmetry and Relativity in 1905. *Archive for History of Exact Sciences* 59: 437–544.

Hon, G., and B.R. Goldstein. 2009. Spotlight on: The nature of scientific change; in pursuit of conceptual change: The case of legendre and symmetry. *Centaurus* 51: 288–293.

Kaufmann, W. 1906. Über die Konstitution des Elektrons. *Annalen der Physik* 19: 487–453.

Lorentz, H.A. 1904. Electromagnetic phenomena in a system moving with any velocity smaller than that of light. Koninklijke Akademie van Wetenschappen te Amsterdam. *Section of Sciences. Proceedings* 6: 809–831.

Lorentz, H.A. 1916. *Theory of electrons and its applications to the phenomena of light and radiant heat*, 2nd ed. Leipzig: Teubner.

Lorentz, H.A. 1924. The radiation of light. *Nature* 113: 608–611.

Lorentz, H. A. 1931. *Lectures on theoretical physics. Lectures delivered at the University of Leiden in 1922, vol. 3*. London: McMillan.

Miller, A. I. 1981. *Albert Einstein's special theory of relativity: Emergence (1905) and early interpretation (1905–1911)*. Reading: Addison-Wesley.

Poincaré, H. 1914. *Science and method*. Trans. F. Maitland. London: Nelson.

Poincaré, H. 1946. *The foundations of science*. Trans. G.B. Halted, with an introduction by R.J. Royce (this is a collection of three works: Science and hypothesis, 9–197, The value of science, 201–355, Science and method, 359–546).

Poincaré, H. 1952. *Science and hypothesis*. Trans. J. Larmor. New York: Dover. Dover Reprint.

Stachel, J., ed. 1989. *The collected papers of Albert Einstein. Vol. 2. The swiss years: Writings, 1900–1909*. Princeton, N.J.: Princeton University Press.

Chapter 4
Two Modes of Reasoning with Case Studies

Wolfgang Pietsch

Abstract I distinguish a predictive and a conceptual mode of reasoning with case studies. These broadly correspond with two different kinds of analogical inference, one relying on common and differing properties, the other on structural similarity. The problem of generalizing from case studies is discussed for both. Regarding the predictive mode, eliminative induction provides a natural framework. In the conceptual mode, general rules are largely lacking not least due to a number of epistemological challenges like Raphael Scholl's underdetermination problem for HPS. In agreement with ideas of Richard Burian and Peter Galison, I argue that conceptual reasoning on the basis of case studies should not aim at grand universal schemes but rather at mesoscopic or middle-range theory. In the essay, I will repeatedly draw on insights from the social sciences, in which a much more extensive reflection on case study methodology exists compared with HPS.

4.1 Introduction

By examining historical episodes in view of specific conceptual questions, case studies provide the essential link between the history and the philosophy of science. They are thus rightly regarded as the central building block of an epistemology for an integrated HPS, in short &HPS. From this perspective, it is rather surprising that there exists little systematic literature on this topic. Presumably, the reasons lie in certain methodological prejudices of both fields. On the one hand, philosophy of science often relies on toy examples or caricatures rather than engaging with the complex and intertwined details of actual historical episodes. In history, on the other hand, the neglect of a systematic reflection seems to stem from a wide-spread hostility towards abstract theorizing. But obviously, as long as historical narratives are framed in a language, their recounting will require at least a certain amount of conceptualizing.

W. Pietsch (✉)
Munich Center for Technology in Society, Technical University Munich, Arcisstrasse 21, 80333 Munich, Germany
e-mail: pietsch@cvl-a.tum.de

T. Sauer and R. Scholl (eds.), *The Philosophy of Historical Case Studies*, Boston Studies in the Philosophy and History of Science 319,
DOI 10.1007/978-3-319-30229-4_4

In Sect. 4.2, a brief overview of some main characteristics of case studies will be given. Also, case studies will be contrasted with other research methodologies, in particular experiments and statistical approaches.

In Sect. 4.3, two modes of reasoning with case studies will be distinguished, an inductive mode aiming at prediction and an abstractive mode for concept development. I will argue that these two modes correspond to two different types of analogical inference, which have not always been kept apart: on the one hand, inferences on the basis of corresponding and differing properties, on the other hand, inferences on the basis of structural similarity.

It turns out that there are related epistemic worries about the reliability of analogical inference and about reasoning with case studies. A considerable number of researchers believe that both have only a heuristic value for suggesting plausible hypotheses. I will argue against this view in Sect. 4.4 and will show that the role of analogical and case-based reasoning is much richer. In particular, the problem how to generalize meaningfully from case studies will be addressed, both for the predictive and the conceptual mode. In line with most of the social science literature, I will argue that predictive inferences mainly rely on comparative reasoning based on the logic of eliminative induction in the tradition of Mill's methods. The situation is more complicated for the conceptual mode which is troubled by the ambiguities of concept formation. The crucial question, here, concerns the adequate level of generality for the conceptual framework.

In a brief outlook, Sect. 4.5 provides further evidence for the need of a thorough methodological assessment of case-based reasoning across the sciences. With the recent emergence of data-intensive science, the boundary between statistical approaches and case studies becomes increasingly blurred—requiring a combined methodology for both. Currently, this has an impact mainly on the predictive mode and less on the conceptual mode.

Section 4.6 will summarize the findings with respect to the role of case studies in integrated history and philosophy of science. Especially when examining questions of scientific methodology, well-developed case studies are an indispensable part of the inquiry.

4.2 The Nature of Case Studies

4.2.1 General Remarks

Notwithstanding their methodological importance, the nature and function of case studies is not a common topic in philosophy of science. In other fields, however, in particular the social sciences, there exists a sizeable and sophisticated literature on this issue, on which I will repeatedly draw in the following. While there are certainly substantial differences in the use of case studies compared with &HPS, there are also sufficient common grounds to merit a detailed look.

When defining the notion of case studies, the following cluster of criteria is usually mentioned (Yin 2009, p. 18; Gerring 2007, p. 17; George and Bennett 2005, pp. 17–19): (i) Case studies examine an episode in considerable detail and take into account the context in which it happened. Often, the examined phenomenon and its context are not clearly separable from each other. (ii) Relatedly, case studies take into account a large number of variables, while focusing on a single or at most a small number of instances. (iii) Often, the phenomenon of interest is examined from a variety of perspectives and with a variety of methods resulting in a heterogeneous data structure.[1] For example, case studies may involve empirical investigations, archival work, interviews, and surveys with open-ended questions. Triangulation is often used to verify that the different perspectives are actually coherent with each other. (iv) Importantly, case studies are always case studies for something—answering the question 'what is this a case of?', i.e. they relate the examined episode to a certain theoretical concept or a type of phenomenon.

Some people believe that case studies concern only a single, immutable instance, a snapshot of an event. But as for example John Gerring has pointed out, case studies always involve various kinds of within-case variation, for example temporal and spatial (Gerring 2007, p. 27). In general, the more detailed a case study is, the more variation will occur. Also, case studies never stand entirely on their own, but are always contrasted at least implicitly with similar or dissimilar cases in the background knowledge. If nothing else, already the use of both ordinary and scientific language establishes the link with comparable instances.

Case studies are generally categorized as a type of qualitative research. However, especially when dealing with complex episodes, they may very well incorporate various quantitative elements, e.g. gradual variation of certain parameters over time or between different entities. Thus, the qualitative-quantitative distinction is not really suitable to characterize this type of research.

Relatedly, case studies are often contrasted with statistical methods. Mostly, a certain hierarchy is thereby implied that reliable inferences can only result from statistical reasoning and that the conduct of case studies is merely of heuristic value. However, this viewpoint is strongly and in my view rightly contested by most proponents of case study research (e.g. Yin 2009, p. 6). As already mentioned, case studies often involve statistical arguments. Even more importantly, a systematic analysis of within-case variation can lead to reliable analogical inferences as pointed out in Sect. 4.4.2.

Finally, case studies are sometimes compared with experiments. Here, the latter in fact yield more reliable inferences because one can interact with and manipulate the phenomena while case studies mostly rely on observational data. Another difference is that experiments are usually carried out in a laboratory setting providing for a controlled and stable environment, while, as mentioned, the distinction between context and phenomenon is often helplessly blurred in case studies. However, the similarities between these types of research are more significant than the differences.

[1]The term is understood here in a very broad sense, e.g. a historical narrative could also constitute a data structure.

Both case studies and experiments examine in detail a specific instantiation of the phenomenon of interest. Crucially, case studies share with certain classes of experiments their exploratory nature and the comparative logic of eliminative induction, which will be described in more detail in Sect. 4.4.2.

A number of classification schemes have been proposed for further distinguishing different kinds of case studies, in particular (i) classification with respect to epistemological function, e.g. conceptual refinement, hypothesis generation, or causal analysis (see George and Bennett 2005, p. 75), or (ii) with respect to the representativeness of an episode.[2] As an example of the latter, Scholl and Räz (2013) offer a distinction between paradigm cases, hard cases that are difficult to make sense of, and important cases that are of intrinsic interest.[3]

For distinguishing different kinds of case studies, one can also have recourse to the comparative methodology underlying this type of research. 'Most similar' cases are as similar as possible in terms of circumstances, but the examined phenomenon fails to appear in one of the cases, allowing for the application of a weakened version of Mill's method of difference. 'Most different' cases are as different as possible, while the examined phenomenon still appears in both cases, allowing for the application of the method of agreement (Gerring 2007, pp. 89–90). In fact, already Francis Bacon in his well-known compilation of prerogative instances has listed a considerable number of useful types including 'solitary instances', which essentially correspond to 'most similar' and 'most different' cases, 'striking instances', which are paradigm or representative examples of a phenomenon, and the well-known 'instances of the fingerpost' or crucial instances (Bacon 1994).

4.2.2 Case Studies in Integrated History and Philosophy of Science

Very broadly, case studies provide the principal link between the factual level of the history of science and the conceptual level of the philosophy of science. One of the earliest attempts by James Conant, former president of Harvard University and mentor to Thomas Kuhn, to establish case-based reasoning in the history of science resulted in an influential book series *Harvard Case Histories in Experimental Science*. However, the idea that a few representative episodes can provide a complete picture for example of the nature of experimental science has rightly been criticized in the aftermath—maybe most forcefully by Peter Galison in his *Image and Logic* on the grounds that there can be no unifying conceptual scheme because there exists no universal experimental method (Galison 1997, Chap. 1). Such and similar qualms

---

[2] A further distinction concerns the type of variation that occurs in a case study (e.g. Gerring 2007, p. 28).

[3] See Scholl and Räz, in press (this volume). In the social-science literature, similar classification schemes can be found, e.g. Gerring (2007, pp. 89–90) suggests nine different types of cases: typical, diverse, extreme, deviant, influential, crucial, pathway, most-similar, and most-different.

were sometimes misinterpreted and have led to wide-spread skepticism about the usefulness of case studies among historians of science. However, Galison's point was the rejection of grand schemes and not an opposition towards any kind of conceptualization on the basis of case studies. Rather, he readily admits: "the very project of history must contend with the problem of how to make the particular stand for the general, and how to limit claims for such a stance" (Galison 1997, p. 60; see also Sect. 4.4.3).

And in fact there is a long list of very successful examples in the history of &HPS, for instance Kuhn's (1992) study of the Copernican revolution which played a crucial role in his development of general concepts like normal science or incommensurability, Hasok Chang's (2004) study of the development of a temperature scale that led to a general theoretical framework regarding the emergence of measurement techniques, or various analyses of Semmelweis' discovery of the importance of antiseptic procedures probing the adequacy of different kinds of scientific methodology (e.g. Hempel 1966, pp. 3–18; Scholl 2013).

4.3 The Predictive and the Conceptual Mode of Case-Based Reasoning

I will now distinguish two modes of reasoning with case studies, one aiming at the derivation of empirically verifiable predictions, the other at developing conceptual schemes that are useful to account for related phenomena. The first will henceforth be called the predictive mode and the second the conceptual or abstractive mode. From the explanations that will follow it should be obvious that the boundary between both modes is not entirely sharp.

The distinction is situated within a broadly Duhemian or Cartwrightian view of scientific theories as having a phenomenological and a theoretical level (Duhem 1954, Chap. 2; Cartwright 1983, pp. 1–20). The former consists of causal laws that are mostly experimentally established. In accordance with their causal nature, these phenomenological laws are primarily used to generate successful predictions and interventions. By contrast, the theoretical level–more exactly, there are often several layers of increasing generality–consists of abstract or theoretical laws that are construed on top of the phenomenological level. These general laws are developed from the causal laws in a process of abstraction and are mainly geared at unification and explanation. While the causal laws can be true or false in a straightforward sense in that predictions based on them turn out either right or wrong, this is not so for the theoretical laws. In fact, a wealth of literature has established a solid case for underdetermination regarding the laws on higher levels of abstraction such as physical axioms. Thus, truth is not a suitable criterion to evaluate the theoretical level. Rather, pragmatic criteria like simplicity or fruitfulness are more adequate. I am aware that many aspects of this epistemological view could be questioned and I cannot defend it here except by referring to the authority of philosophers of science like those mentioned above.

Basically, the predictive mode of reasoning remains at the phenomenological level. It employs case studies to make predictions about other instances that are sufficiently similar, aiming at true or probable inferences either in terms of causal laws or of singular statements about future or unknown events. Mostly, such inferences do not change the level of description, i.e. the vocabulary remains at the same level of abstractness. Inferences in the predictive mode are mainly evaluated in terms of empirical adequacy.

The predictive mode is relatively rare in &HPS. To illustrate it, let us therefore look at an example from medicine. More than thirty years ago the first cases of what was later recognized as acquired immune deficiency syndrome (AIDS) began to appear in the Western world. Certain cancers and pneumonias that previously occurred only in patients with congenital immunodeficiencies or who had received immunosuppressive chemotherapy, were suddenly reported in patients with no such background (MMWR 1981). Confronted with a hitherto unknown condition, medical personnel were facing the severe challenge to develop both diagnosis and treatment. In such situations, where in-depth theoretical knowledge and understanding is missing, case studies both of single patients and of small clusters of patients are an adequate methodology to explore the phenomenon and to make predictions about similar cases. For example, the doctors measured various parameters of the immune system such as lymphocyte counts or T-cell counts. Also, the social background of the patients was examined with the result that almost all of them were gay and/or drug abusers and that the rate of sexual intercourse between them was much higher than would have been expected by pure chance, eventually suggesting the venereal character of the disease. Treatments were tried without much success and a large number of patients died.

Apparently, the detailed case studies of patients and small patient clusters were used as a reference to make predictions about further cases. They provided doctors with some information about what to expect and how to treat patients with similar characteristics. For example, if someone without a history of immunologic weakness presented him-/herself with infections that normally do not occur in patients with a healthy immune system, the doctors could fairly reliably guess in what range the various immunologic parameters will be or they might check the sexual partners of the patient to find further diseased individuals.

This fits well with the characteristics for the predictive mode as depicted above. In particular, the case studies of individual patients can be used for direct predictions about other instances that are sufficiently similar. Also, there is little conceptualization going on at this early stage of the discovery of AIDS, but rather the predictions are mostly framed using terms and concepts that were already employed for the collection of data.

By contrast, the abstractive mode aims at conceptual schemes. Abstractive reasoning usually changes the level of description, i.e. it employs a case study for the development of abstract laws on the theoretical level. To this purpose, it generalizes by selectively focusing on those aspects that are deemed essential within a specific research context. According to the epistemological perspective sketched at the beginning of this section, conceptualizations should not be judged in terms of truth but rather usefulness.

The majority of case studies in &HPS belong to this mode. A paradigm example was already mentioned in the previous section, namely Thomas Kuhn's development of notions like normal science or incommensurability based on detailed historical work in particular on the Copernican revolution (Kuhn 1992). On the basis of this case study, Kuhn proposed his view of scientific revolutions as reconceptualizations of a certain scientific area, often with broader implications for scientific methodology but also for the general philosophical worldview. The study already contains some of the core ideas that were later elaborated in Kuhn's more famous book on *The Structure of Scientific Revolutions* (1996). Kuhn suggests that there may not be a final conceptualization for many fields of science and identifies non-cumulative elements in the process of scientific revolutions. In particular, while the range of phenomena that are accessible to explanation keeps growing, the nature of these explanations is subject to continuous changes during scientific revolutions (Kuhn 1992, pp. 261–265).

This case study, which is of a very different kind compared with the example discussed for the predictive mode, nevertheless fulfills the criteria that were stated at the beginning of Sect. 4.2.1. It illustrates well the characteristics of the conceptual mode. Obviously, concepts like a scientific revolution are situated on a much more theoretical level than the concrete historical events. Kuhn extracted the central features of such concepts by carefully abstracting from the complexities and contingencies of scientific practice. Propositions on the theoretical level, like "the paradigms before and after a scientific revolution are incommensurable", do not directly imply empirically verifiable statements. It therefore seems wrong to evaluate them in terms of truth. Instead, the crucial criterion is whether these concepts are useful for structuring the evidence in other historical episodes like for example the Darwinian or the chemical revolutions. While the notion of truth may apply when abstract concepts are used in concrete predictions on the phenomenological level, it generally cannot be employed to evaluate relations within the theoretical level. For example, Lamarckian concepts may fail to yield accurate predictions if applied to the question, why giraffes have acquired such long necks.[4] But this does not falsify Lamarckian ideas in general, it only shows a problem with a specific application of them.

Speaking of Kuhn, the use of case studies in &HPS bears much resemblance to that of exemplars in the natural sciences. Exemplars constitute representative cases that serve as role models for puzzle solving, they "are concrete problem solutions accepted by the group as, in a quite usual sense, paradigmatic" (Kuhn 1977, p. 298). Like case studies, they serve to develop a conceptual framework in order to analyze similar phenomena. The pendulum and the inclined plane are good examples, which in the hands of Galileo and others have contributed to establishing the conceptual foundations of classical mechanics (ibid., pp. 305–306). Another case in point is the harmonic oscillator whose principles can be understood by examining a simple

[4]The example was suggested by Raphael Scholl.

physical system of a point-mass and a spring, but which is applicable in a large variety of contexts far beyond physics.

Again, conceptual schemes based on exemplars are generally not used to predict but rather to adequately structure the perception of related phenomena. For example, the theory of harmonic oscillators does not itself allow for predictions in specific contexts of application, e.g. in a resonant circuit. Rather, a number of concrete electrical laws must be presupposed, which should be formulated in a way that the theory of harmonic oscillators can be applied. Thus, theoretical knowledge is used to structure the more concrete knowledge regarding a certain context of application resulting in phenomenological laws and predictions which can then be verified or falsified.

Darwin also relied on paradigmatic phenomena in his discovery and construction of the theory of evolution, maybe most famously the variation of species on different islands of the Galapagos archipelago. Again, the fundamental concepts of evolutionary theory like mutation and selection are quite useful in coherently structuring phenomena in the world of living things, but they usually do not generate specific predictions. For example, they fail to determine whether the human species will become extinct in the next century or why the dinosaurs vanished from the planet.

In spite of these similarities, there are also important differences between the use of exemplars in the natural sciences and the conceptual mode of case-based reasoning in &HPS. (i) The first concerns the very nature of the concepts. While those developed in the natural sciences mainly regard the ontology of the natural world like forces or genes, case studies in &HPS mostly aim at methodological or occasionally sociological concepts like underdetermination or scientific revolutions. (ii) Another important difference concerns the historical nature of the studied phenomena. The exemplars examined in the natural sciences are generally repeatable and therefore largely independent of a historical context. By contrast, the episodes studied in &HPS occurred only once, mostly in a distant past. Trivially, one cannot intervene or experiment in historical case studies and the relevant evidence is usually not directly accessible, but has to be transmitted over time using media like books or artefacts. (iii) Finally, the complexity of the examined phenomena differs. While the exemplars in the natural sciences can often be separated from context and examined in a laboratory setting, this is not possible for case studies in &HPS.

Remarkably, the distinction between the predictive and the conceptual mode is mirrored in two different versions of analogical inference, which have not always been clearly kept apart. In one type, which corresponds to the predictive mode, an assessment of similarity in terms of common and differing properties determines if a further property of one case will be instantiated in another case as well (e.g. Mill 1886, Chap. XX; Keynes 1921; or Carnap 1980, Sects. 16 and 17). The other type is analogy in terms of structural similarity which has been employed since the ancients in accordance with the original meaning of the term 'analogy', referring to likeness in relations. Such structural analogies can be found over and over in the history of science and they are mostly employed for the conceptual development of novel phenomena. For instance, James Clerk Maxwell famously emphasized their importance for his work. A good example is his use of hydrodynamical and mechanical analogies for the formulation of the field-theoretic approach to electrodynamics. In the context of &HPS, Kuhn's conceptual framework of scientific revolutions is an example of structural analogies between different historical episodes.

4.4 The Problem of Generalizing from Case Studies

4.4.1 The Problem Stated

Attempts to generalize from case studies are often dismissed as naïve, calling into question the role of case studies as a sound piece of scientific methodology. In the following, I will briefly survey a number of arguments in this regard indicating where they go wrong and then tackle the problem of generalizing in a more specific way for the two modes of reasoning that were identified in the previous section.

(i) A wide-spread argument concerns the claim that one supposedly cannot generalize from a single immutable instance. But this misconstrues the notion of case studies, which as we had seen always include a considerable amount of variation, for example in time or across subjects if the case study involves more than a single subject or entity (Gerring 2007, pp. 27–33). A simple example of within-case variation that can generate predictions concerns interventions in a controlled environment that lead to sudden and substantial changes of a phenomenon. If an otherwise healthy person drinks a potion brewed from a previously unknown herb and dies quickly afterwards, the death is highly likely a result of consuming the herb and it is not advised that other people drink from the potion. The logic of this reasoning can be formally reconstructed in terms of eliminative induction, more exactly by the method of difference. Besides within-case variation, case studies sometimes involve cross-case variation by drawing comparisons with related cases. Finally, the argument overlooks that case studies are generally evaluated with respect to a substantial amount of background knowledge, also allowing for comparative reasoning.

(ii) Relatedly, the Humean view still prevails that one can only generalize from a large number of instances and that the level of verification is somehow proportional to the number of cases studied. For example, Joseph Pitt in an influential article on *The Dilemma of Case Studies* makes this point: "it is unreasonable to generalize from one case or even two or three" (Pitt 2001, p. 373). This perspective is mistaken as the mentioned example of the poisonous potion shows. In a well-controlled environment, a causal connection can already be derived from two instances via Mill's method of difference, while a correlation even if established by a large number of instances may not mean anything. As argued by a number of methodologists in the tradition of Baconian eliminative induction,[5] today often referred to as Mill's methods, it is not the number of instances, but the variation between instances that counts. Thus, if

[5]The term 'eliminative induction' for Mill's methods has been used by several authors, in particular by Mill himself (1886, p. 256) and also by Mackie in his influential book *The Cement of the Universe*. Mackie writes: "In calling them eliminative methods Mill drew a rather forced analogy with the

there is sufficient variation within a single case, we can generalize to other cases with similar circumstances. Especially in the social science literature on case studies, it is generally taken for granted that comparative reasoning based on Mill's methods is well-suited for the analysis of case studies though usually difficult to implement in full rigor (e.g. George and Bennett 2005, Chap. 8; Hammersley et al. 2000).

(iii) Many scientists regard case studies as merely anecdotal, serving at best a heuristic function for suggesting novel concepts and hypotheses. Certainly, some narratives that have been called case studies in the past are nothing but mere story-telling. However, the problem of differentiating a meaningful case study from an anecdote just amounts to developing a sound methodology for case-based reasoning. What is required is an analysis of the different types of variation that may occur and how they can lead to reliable inferences both in the predictive and the conceptual mode. A general framework is needed that provides an outline how to research, write, and analyze case studies. While social scientists have made important attempts at this task (e.g. Thomas 2011; Gerring 2007; Yin 2009), in &HPS it still remains an underexplored question.

In the following, I will argue in further detail against the wide-spread view that case studies have a merely heuristic role in scientific method. I will again distinguish between the predictive and the conceptual mode.

4.4.2 Generalizing in the Predictive Mode

As already indicated, inferences from case studies in the predictive mode mainly rely on an assessment of corresponding and differing properties between two instances. Note that such inferences are often made on a case-to-case basis and mostly do not result in sweeping generalizations. In the usual terminology, coined by Keynes (1921), the properties that two instances share are called the positive analogy, the properties that differ the negative analogy. Furthermore, it is useful to distinguish the known positive or negative from the unknown analogy. On the basis of the known positive and negative analogy, P and N, respectively, a probability is determined that a property q which is present in one instance can be found in another instance as well. Many methodologists seem to believe that such analogical reasoning cannot be based on a systematic general framework (e.g. Norton 2011). But while I would agree that a definitive account has yet to be developed, I am quite optimistic that it is possible and will provide a broad outline below.

In the 20th century, two influential programs have tried to tackle the task, one developed by John Maynard Keynes in his *Treatise on Probability* (1921), the other by Rudolf Carnap in his work on inductive logic (Carnap 1980, Sects. 16–17). The latter but not the former has recently experienced some further developments (e.g.

(Footnote 5 continued)
elimination of terms in an algebraic equation. But we can use this name in a different sense: all these methods work by eliminating rival candidates for the role of cause" (Mackie 1980, p. 297).

by Kuipers 1984 and Romeijn 2006). While Carnap mostly treats very simple toy models, Keynes' attempts are closer to scientific practice and cover some real-world examples of case-to-case reasoning at least from a qualitative perspective. Several authors have suggested some general rules for analogical reasoning, for example (Keynes 1921, Chap. xix; Bartha 2013, Sect. 3.1): (i) The more extensive the positive analogy, the greater the probability of the analogical inference. (ii) The more extensive the negative analogy, the smaller the probability of the analogical inference. (iii) The more extensive the implication, the smaller the probability of the analogical inference. While these rules certainly fall short of a full-blown framework, they do provide some systematic access to reasoning with analogies beyond mere heuristics.

To illustrate this, consider again the poor individual who died after drinking the herbal potion, which in its framing is a simplified version of the AIDS-example discussed in Sect. 4.3. Now, another individual drinks from a similar potion and the question arises whether he will die or not. The positive analogy consists in all those properties that both cases have in common, e.g. that the drink contained the same herbs, that it was prepared in a similar way, or that the two individuals were both healthy men under forty. The negative analogy concerns all those properties that are different in both cases, e.g. the potion may have been served at different temperatures and only one of the men may be a pharmacist, while the other is a philosopher. For this example, the general rules stated above are very intuitive. In particular, the probability that the second man also dies should not decrease, if the potions were served at the same temperature. By contrast, the probability should not increase, if the mixture of herbs were slightly altered in one instance.[6] Finally, if the prediction is rendered more specific, e.g. that the death must occur within 48 h, the probability of the inference will in general be smaller.

In many epistemological treatments of induction, an ill-conceived focus on enumerative induction prevails. The idea of a mere repetition of instances suggests that these are all exactly alike. Thus, enumerative induction precludes from the beginning inferences based on only partial similarity. Herein lies one of the principal reasons why the inclusion of analogy in a general framework of induction is believed to be so difficult. By contrast, the long neglected eliminative induction in the tradition of Mill's methods constitutes a natural approach for reasoning from similarity and dissimilarity. At least in some accounts, it allows for inferences in the presence of a negative analogy by introducing a rule for determining the causal irrelevance of circumstances (see Pietsch 2014, Sect. 3c). Now, if the known and unknown negative analogy only concerns properties that are irrelevant to property q, then the analogical inference will be correct. By contrast, if at least one causally relevant property pertains to the negative analogy, then the analogical inference will fail.[7] In this way, eliminative induction provides a framework for treating predictive analogical inferences, while enumerative induction entirely fails to make sense of them.

---

[6]Note that this need not be the case for all properties. For example, if the second man were ill, the probability for his death would presumably increase. This underlines not so much the heuristic nature of analogical inferences but the need for a two-dimensional framework as outlined below.

[7]Additional complications may arise in the case of plural causation.

In principle, the sketched framework is deterministic, but probabilistic considerations can be integrated in a straight-forward manner: (i) when there are properties in the negative analogy of which one does not know whether they are relevant to q—as in the case of the temperature of the potion, which may for example destroy a poisonous substance; (ii) or when there are relevant properties in the unknown analogy, e.g. we may not know whether both men are healthy. In the first situation one needs to determine the probability that a property in the negative analogy is relevant or not, in the second situation the probability that a relevant property in the unknown analogy belongs to the negative or positive analogy. Of course, the combined situation can also occur of a property in the unknown analogy of which it is unknown if it is relevant. For determining these probabilities, the usual toolbox is available, in particular relative frequencies and symmetry considerations.

As a simple example, consider an analogical inference regarding two instances which differ only in irrelevant properties plus one property c of which it is not known whether it is causally relevant or not to phenomenon q. In one instance, q occurs in the presence of c and we are interested in the probability that in another instance, q will occur in the absence of c. Certainly, if we know the probability p that c is causally relevant for q, then the probability for the analogical inference, i.e. to observe q in the second instance, will be $(1-p)$. For example, a specific ingredient may be missing from the potion in the second instance, but otherwise the situation shall be identical both in terms of the individuals drinking the potion and the way it is prepared and served. Now, if we can determine the probability that this additional ingredient is causally relevant, then we know how likely the death of the second person is, given the death of the first. Of course, substantial work is required how to determine and interpret these probabilities.

Or consider a situation, where all properties that are relevant to q belong to the positive analogy except for one c in the unknown analogy. Given the probability p that c is in the positive analogy, i.e. $(1-p)$ that it is in the negative analogy, then the probability for the analogical inference of q will be p. Here, one option to determine p explicitly refers to the known positive and negative analogy, examining if there is any causal, statistical or deductive connection between properties in P and N with c. For example, all properties in P and N might be independent of c except one x in the positive analogy. Now, if it is known that x entails c with probability p, then we would have p for the analogical inference.

As an example, consider again the two men taking a sip from the herbal drink. The situations shall be alike in all relevant circumstances except that this time there is uncertainty, whether a certain crucial ingredient c of the potion is present in the second instance. The uncertainty determines how likely the second person is going to die. Some evidence may suggest the respective probability. For example, the color x of the potion may be known for both instances while being statistically correlated with the presence of the crucial ingredient c.

Thus, causal or deductive connections between properties constitute an important criterion to determine the probability for an analogical inference. That the internal causal structure between properties is important for analogical reasoning has been emphasized in particular by Hesse (1966, p. 59) and more recently by Bartha (2010).[8] A second criterion concerns the number of properties in the positive analogy compared with the negative and the unknown analogy, as is evident from the three rules that were given in the beginning of this section. Of course, there are enormous difficulties with adequately determining these ratios, e.g. detecting the independence of properties as well as their respective evidential weight. As also Norton (2011) emphasizes, the recognition of both criteria lies at the basis of the important two-dimensional model of analogical reasoning, where the horizontal relation determines the amount of similarity between instances in terms of properties and the vertical relation determines the nature of the connection between the properties.

In summary, eliminative induction provides a framework for reasoning by analogy or similarity in the predictive mode. Thus, the epistemic uncertainty connected with many analogical inferences stems not from the absence of a common logic but rather from the fact that the available evidence rarely is good enough for a quantitative assessment. The whole distinction between (enumerative) induction and analogy was ill-conceived from the beginning. Every induction is based at least partly on reasoning by similarity, because there are no instances that are alike in all respects. Eliminative induction can account for this.

The analysis of mechanisms is a further central element in the predictive mode of reasoning with case studies, where broadly understood mechanisms trace the relation between input and output quantities by looking at the detailed arrangement of parts and features. The study of mechanisms traces the causal processes in a case study and thereby further corroborates any causal relations between input and output quantities that may have been identified by eliminative induction. In the social sciences, the analysis of mechanisms is often referred to as process tracing, on which there exists abundant literature (e.g. George and Bennett 2005, Chap. 10). With respect to the two-dimensional model sketched above, mechanistic reasoning concerns the vertical relations, how the various properties within a case are connected.

In the philosophy of medicine, a current debate tries to understand the relation between comparative and mechanistic evidence (Russo and Williamson 2007). Very briefly, my viewpoint is that there is no real opposition, both can again be understood from the perspective of eliminative induction. Comparative reasoning establishes causal factors for a specific phenomenon, while mechanistic reasoning corroborates the relevance of these factors by analyzing more fine-grained connections. Well-established mechanisms provide additional confirmation for the causal relevance of factors by making the link to other bodies of evidence, i.e. essentially by unification. For example, explicating the chemical mechanism of a certain medication links up the causal effects of the medication with a well-confirmed system of fundamental chemical laws.

---

[8]"The validity of [an argument by analogy] will depend, first, on the extent of the positive analogy compared with the negative [...] and, second, on the relation between the new property and the properties already known to be parts of the positive or negative analogy, respectively" (Hesse cited in Norton 2011, p. 9).

While the predictive mode of reasoning with case studies is ubiquitous in the natural sciences, it is not so prominent in the history of science. This is mainly due to the immensely complex causal structures with which historians deal that seldom allow the identification of sufficiently similar cases. However, when developing a case study, historians often rely on more or less predictive analogies with respect to other episodes in their background knowledge, to determine where to look for interesting material and which research questions to ask.

4.4.3 Generalizing in the Conceptual Mode

As also shown by the role of exemplars in the natural sciences, case-based reasoning is often employed for concept development rather than for predictions. This observation is closely related to an interesting distinction made by some social scientists between eliminative and analytic induction (e.g. Hammersley et al. 2000). The basic idea in the latter is that one constantly needs to reformulate the hypothesis and redefine the phenomena when examining various instances in order to arrive at universal hypotheses—drawing attention to the often neglected fact that induction always goes hand in hand with conceptual refinement. Compared with eliminative induction, which mainly proceeds by parameter variation as discussed in the previous section, concept development is a much more complex and varied process. In fact, no general rules and principles seem to exist and presumably, it involves considerable creativity and intuition.[9] Historical studies suggest that the social and psychological context within which scientists work plays a considerable role if only by providing a set of possible analogies.

Obviously, concept development can occur at different levels of coarse-graining. It can concern observational terms, but also more abstract generalizations. In the following, the focus lies mainly on the latter, henceforth referred to as a process of abstraction. The basic idea is that the conceptual scheme developed for one case can be transferred to another case—just as in the example of Maxwell's use of hydrodynamic analogies for developing electromagnetic field theory. With sufficient ingenuity, it is often possible to develop any conceptual analogy to a certain extent, but of course, not all frameworks are equally fruitful.

In abstraction, there are always different ways how to conceptualize, how to tell a story. In the natural sciences, this is known as the underdetermination of scientific theories. Raphael Scholl has recently pointed out an analogous predicament for &HPS illustrating his claims with the example of Semmelweis' discovery of the origin of childbed fever which can be reconstructed according to different methodologies, e.g. hypothetico-deductive or causal (Scholl 2013, 2015). Both in the sciences and in &HPS, the pluralism of interpretations is closely connected with the fact that

---

[9]For a good overview from a philosophy-of-science perspective consult Nersessian (2008), who stresses the role of analogy, imagery, and thought experiments. See also Hempel's (1952) classic treatise on the subject.

in an abstractive mode some characteristics of a phenomenon are highlighted and others neglected depending on research interests. The resulting framework has to be evaluated in terms of pragmatic criteria, most importantly simplicity of description and fruitfulness for further research.

Besides underdetermination, a further difficulty in the conceptual mode concerns the theory-dependence of whether a case study is particularly representative or deviant. For example, the pendulum, the inclined plane, or the trajectory of a projectile are paradigmatic cases for classical mechanics, while being hard cases in Aristotelian physics with its concept of motion directed towards a natural place. By contrast, burning fire or falling rain were representative phenomena for Aristotelian dynamics but are largely uninteresting in classical mechanics. As an example from &HPS, various methodological paradigms like hypothetico-deductivism or causal analysis each have their paradigmatic and hard cases.

It is beyond doubt that one can generalize conceptually from case studies. Rather, the crucial question is how general such a conceptual level should be. In retrospect, attempts at very high-level theories in history, such as the Marxist conception of historical law, have proved of dubious merit at best. They explain everything and nothing. A similar failure of high-level ideologies can be observed in the social sciences. Presumably, the main reason lies in the complexity and contextuality of the phenomena in both fields. By contrast, high-level conceptualizations are possible in physics, supposedly because it deals with much simpler phenomena.

Thus, a number of scholars have argued that historians should not aim at very general theories but rather at "mesoscopic" concepts that are well-adapted to specific contexts and can be subject to change over time as science evolves. For example, Peter Galison favors a study of "history claiming a scope intermediate between the macroscopic (universalizing) history that would make the cloud chamber illustrative of all instruments in all times and places and the microscopic (nominalistic) history that would make Wilson's cloud chamber no more than one instrument among the barnloads of objects that populated the Cavendish Laboratory during this century." (Galison 1997, p. 61).[10] Richard Burian makes a related suggestion: "Case studies cannot and should not be expected to yield universal methodologies or epistemologies. Rather, they yield local or, better, *regional* standards, and fallible ones at that." (Burian 2001, p. 400). In a similar vein, Chang (2011, pp. 110–111) has argued for replacing the term 'case study' by the notion of 'episode' for much the same reasons. For Chang an episode denotes the variation on a theme and not a mere instantiation of a general concept. Thus concepts have no static meaning, but have to be contextualized, while still allowing for inferences from one episode to the other.

Remarkably, a comparable suggestion was made by Merton (1949), arguing for the importance of middle-range theory in the social sciences: "Middle-range theory is principally used in sociology to guide empirical inquiry. It is intermediate to gen-

[10]Galison argues for a "sited, not typical, history", the aim of which is "to evoke the mesoscopic periods of laboratory history, not a universal method of experimentation" (Galison 1997, p. 63). I largely agree but would add that scientific method nevertheless possesses a universal logical core, which has to be contextualized when analyzing specific episodes.

eral theories of social systems which are too remote from particular classes of social behavior, organization, and change to account for what is observed and to those detailed orderly descriptions of particulars that are not generalized at all. Middle-range theory involves abstractions, of course, but they are close enough to observed data to be incorporated in propositions that permit empirical testing. Middle-range theories deal with delimited aspects of social phenomena, as is indicated by their labels" (Merton 1949, p. 531). Of course, coherence between different mesoscopic, regional, or middle-range concepts should still be sought. However, in view of the complexity of the phenomena both in history and the social sciences, one should abandon the search for extremely general high-level frameworks as in physics. Nevertheless, concept development remains an essential element of any scientific endeavor because trivially without concepts, one cannot account for anything.

The problem of generalizing is thus very different for the conceptual and the predictive modes. One important aspect concerns the plurality of interpretations in the conceptual mode that is much less prominent when dealing with case-based predictions. Even though predictive statements may be phrased in different ways, their truth-value does not change. Another aspect concerns the criteria for evaluating analogical reasoning, which are of largely pragmatic nature in the conceptual mode, e.g. referring to simplicity and fruitfulness, while in the predictive mode empirical adequacy is the dominant factor. In a way, conceptualizations cannot turn out wrong, they just cease to be useful. While in the predictive mode, one is mainly interested in the reliability of inferences, the crucial question in the conceptual mode concerns the adequate level of generality such that the concepts are universal enough to be applicable in various contexts but not so general that they cease to be meaningful.

4.5 The Vanishing Boundary Between Case and Statistical Studies

The basic moral of the previous section is that detailed case studies are as important as cross-case comparisons for both reliable predictions and adequate concept development. It is mainly due to the limitations of the human cognitive apparatus that thus far one always had to make a choice between two complementary options how to deal with complexity: either by looking at a few cases in considerable detail or by looking at a large number of instances rather superficially. But in the end, the boundary between statistical reasoning and case studies is rather artificial. The best basis for both prediction and concept development would certainly involve acquaintance with a large number of cases in their full complexity.

Remarkably, there are some indications that the boundary between case studies and statistical reasoning is indeed beginning to dissolve due to the recent emergence of data-intensive science resulting from advances in information technology (from a philosophy-of-science perspective e.g. Leonelli (2012), Pietsch (2015)). Although it is too early to really judge the impact of these developments, data-intensive methods are sometimes able to analyze a large number of cases in considerable detail.

An example concerns recent US election campaigns, in particular Obama's bid for office in 2012. Both Republicans and Democrats possess large data bases, in which for each citizen who is eligible to vote various data is gathered, e.g. on political preferences, demographic data and sometimes consumer data. In addition, some individuals volunteer information about how they voted in previous elections and on their current commitments. This data can then be used to make predictions about other individuals who have not disclosed their voting behavior. The algorithms providing for such predictions mostly rely on reasoning by analogy, they basically search for sufficiently similar individuals. Certainly, the characterization of voters in terms of a large number of parameters is still very crude. In this manner, it is hard to account for qualitative data or for the nature of connections between different properties. Still, data-intensive science currently constitutes the most promising approach to combine the analysis of within-case complexity with comparisons across a large number of cases.

At the moment, data-intensive science is still chiefly about predictions. Thus, its impact is felt especially in fields like the social sciences or medicine, where the predictive mode of reasoning with case studies is widely used. But I see no in-principle reasons why a data-based automation of concept development should not be possible. However, a much deeper theoretical understanding of the basic principles of concept development will be required. Also, much more complex knowledge architectures would need to be constructed than are currently available. After all, extensive background knowledge is required for the pragmatic evaluation of conceptual frameworks.

This means that natural scientists dealing with exemplars and philosophers of science reasoning on the basis of case studies will be fairly safe against any threat from information technologies taking over their work any time soon. However, the emergence of data-intensive science shows that it is high time to develop a common epistemological framework for case studies and statistical approaches as well as for analogy and induction.

4.6 Some Final Comments

Two modes of reasoning with case studies were sketched, one predictive, the other conceptual. While the latter provides the natural link between history and philosophy of science, the former plays only a minor role in &HPS, mainly in the day-to-day work of the historian, when deciding which sources to consult and what questions to ask by comparison with related cases in the background knowledge.

In the conceptual mode, case studies serve a number of purposes. They can ground methodological considerations in actual scientific practice. Historical case studies can provide a corrective to contemporary philosophical debates, they can probe and challenge philosophical theory. Case studies can also play a role in the discovery of novel theoretical phenomena and in forging new concepts to adequately account for these. A philosophy of science that cannot make a connection to scientific practice, both historical and contemporary, has no use whatsoever. Often philosophers have been much too quick in developing grand conceptual schemes on the basis of caricature-like toy examples instead of genuine history. On the other hand, a historical analysis that does not allow for methodological insights and generalizations is meaningless since there is nothing to learn from it. Thus, suggestions emerging from a direct confrontation of detailed history with larger philosophical claims, e.g. concerning the importance of mesoscopic theory, should be taken very seriously.

Acknowledgments I am much grateful to the editors of this volume, Tilman Sauer and Raphael Scholl, for helpful comments on the manuscript and also for organizing the inspiring workshop in Bern and contributing so many interesting ideas to the subject themselves. I also thank Christian Joas, Désirée Schauz, and Elsbeth Bösl for helpful discussions as well as Karin Zachmann for pointing me to the insightful discussion by Peter Galison.

References

Bacon, F. 1994. Novum Organon. Chicago, IL: Open Court (Original edition: London, 1620).

Bartha, P. 2010. *By parallel reasoning: The construction and evaluation of analogical arguments*. New York: Oxford University Press.

Bartha, P. 2013. Analogy and analogical reasoning. The Stanford encyclopedia of philosophy (Fall 2013 Edition). http://plato.stanford.edu/archives/fall2013/entries/reasoning-analogy.

Burian, R. 2001. The dilemma of case studies resolved: The virtues of using case studies in the history and philosophy of science. *Perspective on Science* 9(4): 383–404.

Carnap, R. 1980. A basic system of inductive logic. In *Studies in inductive logic and probability*, ed. R. Jeffrey, 7–155. Berkeley, CA: University of California Press.

Cartwright, N. 1983. *How the laws of physics lie*. Oxford: Oxford University Press.

Chang, H. 2004. *Inventing temperature: Measurement and scientific progress*. Oxford: Oxford University Press.

Chang, H. 2011. Beyond case-studies: History as philosophy. In *Integrating history and philosophy of science: Problems and prospects*, ed. S. Mauskopf, and T. Schmaltz, 109–124. Dordrecht: Springer.

Duhem, P. 1954. *The aim and structure of physical theory*. Princeton, NJ: Princeton University Press.

Galison, P. 1997. *Image and logic. A material culture of microphysics*. Chicago: University of Chicago Press.

George, A., and A. Bennett. 2005. *Case studies and theory development in the social sciences*. Cambridge, MA: MIT Press.

Gerring, J. 2007. *Case study research. Principles and practices*. Cambridge: Cambridge University Press.

Hammersley, M., P. Foster, and R. Gromm. 2000. Case study and theory. In *Case study method: Key issues, key texts*, ed. R. Gromm, M. Hammerseley, and P. Foster, 234–258. London: Sage.

Hempel, C. 1952. *Fundamentals of concept formation in empirical science*. Chicago, IL: Chicago University Press.

Hempel, C. 1966. *Philosophy of natural science*. Upper Saddle River, NJ: Prentice Hall.

Hesse, M. 1966. *Models and analogies in science*. Notre Dame, IN: Notre Dame University Press.

Keynes, J.M. 1921. *A treatise on probability*. London: Macmillan.

Kuhn, T.S. 1977. *The essential tension*. Chicago, IL: Chicago University Press.

Kuhn, T.S. 1992. *The Copernican revolution*. Cambridge, MA: Harvard University Press.
Kuhn, T.S. 1996. *The structure of scientific revolutions*, 3rd ed. Chicago: University of Chicago Press.
Kuipers, T. 1984. Two types of inductive analogy by similarity. *Erkenntnis* 21: 63–87.
Leonelli, S. 2012. Introduction: data-driven research in the biological and biomedical sciences. *Studies in the History and Philosophy of the Biological and Biomedical Sciences* 43(1–3)
Mackie, J.L. 1980. *The cement of the universe*. Oxford: Clarendon Press.
Merton, R. 1949. On sociological theories of the middle range. In *Classical sociological theory*, ed. C. Calhoun, J. Gerteis, J. Moody, S. Pfaff, and I. Virl, 531–542. Oxford: Wiley-Blackwell.
Mill, J.S. 1886. *A system of logic, ratiocinative and inductive*. London: Longmans, Green and Co.
MMWR. 1981. Pneumocystis Pneumonia-Los Angeles. *Morbidity and Mortality Weekly Report* 30(21): 1–3.
Nersessian, N. 2008. *Creating scientific concepts*. Cambridge, MA: MIT press.
Norton, J.D. 2011. Analogy. Draft chapter of a book on the material theory of induction, http://www.pitt.edu/~jdnorton/papers/material_theory/Analogy.pdf.
Pietsch, W. 2014. The nature of causal evidence based on eliminative induction. *Topoi* 33(2): 421–435.
Pietsch, W. 2015. Aspects of theory-ladenness in data-intensive science. *Philosophy of Science* 82(5): 905–916.
Pitt, J. 2001. The dilemma of case studies: Toward a Heraclitian philosophy of science. *Perspective on Science* 9(4): 373–382.
Romeijn, J. 2006. Analogical predictions for explicit similarity. *Erkenntnis* 64(2): 253–280.
Russo, F., and J. Williamson. 2007. Interpreting causality in the health sciences. *International Studies in the Philosophy of Science* 21(2): 157–170.
Scholl, R. 2013. Causal inference, mechanisms, and the Semmelweis case. *Studies in the History and Philosophy of Science* 44(1): 66–76.
Scholl, R. 2015. Inference to the best explanation in the catch-22: How much autonomy for Mill's method of difference? *European Journal for Philosophy of Science* 5(1): 89–110.
Scholl, R., and T. Räz. 2013. Modeling causal structures: Volterra's struggle and Darwin's success. *European Journal for Philosophy of Science* 3(1): 115–132.
Scholl, R., and Räz, T. in press. Towards a methodology for integrated history and philosophy of science. In *The philosophy of historical case studies. Boston studies in the philosophy and history of science*, ed. T. Sauer, and R. Scholl.
Thomas, G. 2011. *How to do your case study. A guide for students and researchers*. Los Angeles: Sage.
Yin, R. 2009. *Case study research, design, and methods*. Los Angeles: Sage.

Chapter 5
Towards a Methodology for Integrated History and Philosophy of Science

Raphael Scholl and Tim Räz

A glaring asymmetry, obvious at this meeting, is that historians dress better than philosophers – historians always being interested in the details, sartorial and otherwise, while philosophers seem concerned only with dressing in general.
Richards (1992)

Abstract We respond to two kinds of skepticism about integrated history and philosophy of science: foundational and methodological. Foundational skeptics doubt that the history and the philosophy of science have much to gain from each other in principle. We therefore discuss some of the unique rewards of work at the intersection of the two disciplines. By contrast, methodological skeptics already believe that the two disciplines should be related to each other, but they doubt that this can be done successfully. Their worries are captured by the so-called dilemma of case studies: On one horn of the dilemma, we begin our integrative enterprise with philosophy and proceed from there to history, in which case we may well be selecting our historical cases so as to fit our preconceived philosophical theses. On the other horn, we begin with history and proceed to philosophical reflection, in which case we are prone to unwarranted generalization from particulars. Against worries about selection bias, we argue that we routinely need to make explicit the criteria for choosing particular historical cases to investigate particular philosophical theses. It then becomes possible to ask whether or not the selection criteria were biased. Against worries about unwarranted generalization, we stress the iterative nature of the process by which historical data and philosophical concepts are brought into alignment. The

R. Scholl (✉)
Department of History and Philosophy of Science, University of Cambridge, Cambridge, UK
e-mail: raphael.scholl@gmail.com

T. Räz
FB Philosophie, University of Konstanz, 78457 Konstanz, Germany
e-mail: tim.raez@gmail.com

T. Sauer and R. Scholl (eds.), *The Philosophy of Historical Case Studies*, Boston Studies in the Philosophy and History of Science 319,
DOI 10.1007/978-3-319-30229-4_5

skeptics' doubts are fueled by an outdated model of outright confirmation versus outright falsification of philosophical concepts. A more appropriate model is one of stepwise and piecemeal improvement.

5.1 Introduction

Integrated history and philosophy of science must steer clear of two opposite methodological pitfalls. One is the unwarranted generalization of grand philosophical theses from a handful of historical case studies. The other is the outright imposition of philosophical concepts upon cases that serve as mere Potemkin villages. The task is difficult, and many have claimed it not to be worth the trouble. Some philosophers will argue that their discipline seeks a kind of systematic knowledge that does not need to be sensitive to the detailed study of cases from actual scientific practice. Conversely, some historians of science will argue that historical understanding is possible without the abstract concepts that contemporary philosophy of science has to offer.

Thus, the present contribution has a foundational and a methodological purpose. Section 5.2 articulates our view of what we can find at the intersection of history and philosophy of science (HPS) that each discipline on its own cannot offer. We expect some version of our "fundamental motivation" for an integrated HPS to meet with broad consensus—not only among those who self-identify as historians and philosophers of science, but also among the many philosophers of science who embrace case studies (from contemporary or historical science) as a routine part of their work.

Our remaining discussion takes up the greater difficulty: It outlines some best methodological practices for relating history to philosophy (and vice versa) in case studies. While a robust grounding in both historical and philosophical methods is indispensable, we believe that the integrated HPS project requires an additional set of skills that currently receive little discussion. We will begin with the so-called dilemma of case studies, which encapsulates the two methodological challenges we introduced above (Sect. 5.3). First, how can we apply philosophical concepts to historical cases without selection bias—that is, without choosing our historical cases such that they are merely convenient illustrations of preconceived philosophical theses? We will outline how historical cases can play a *probative* role in philosophical investigations by allowing us to test and refine our best concepts about science (Sect. 5.4). Second, how should the confrontation of philosophical concepts and historical theses proceed? We will suggest a model that emphasizes iterative, piecemeal conceptual engineering instead of generalizations from particular cases (Sect. 5.5).

The view of integrated HPS we will be presenting is not new. We see our contribution as an attempt at explicating a method already widely used by many practitioners in the field. Our hope is that other scholars will accept this invitation to join us in a renewed explicit reflection on the methodology of integrated HPS.

5.2 The Fundamental Motivation for an Integrated History and Philosophy of Science

We recognize a plurality of different approaches within HPS.[1] However, for the purposes of the present contribution our interests mainly belong to a particular view of the history-philosophy interaction. On this view, philosophy of science provides us with a set of concepts and worthwhile questions about the history of science—questions that are particularly pertinent to science's core epistemological concerns. Conversely, the history of science grounds philosophical reflection in actual science—in brief, it enables a naturalistic approach to the questions of philosophy of science concerning such issues as confirmation, explanation, reduction, discovery, long-term theory change, and so on. Our flavor of integrated HPS derives its value from this unique interaction: it provides the historian with theoretically rich concepts and questions, and it grounds its philosophical concepts in actual scientific practice.[2]

5.2.1 *Questions for History*

Historical sources are powerful: they can bring into focus vague questions and diffuse notions about science. But it is also a well understood fact that the sources fail to speak for themselves. The sources speak because we ask specific, carefully crafted questions of them. It is at first a curiosity to learn that Darwin, in his *Beagle* notebooks, used various spellings of words like "occasion", "coral" and "Pacific"—but the curiosity becomes a revealing fact once we use these spelling variants to date Darwin's conversion to transformationism after the Galapagos (Sulloway 1982). If the sources sometimes seem to almost force themselves on us, it is because we read them with well prepared minds.

The focus of most current historians of science is on questions about the cultural, social and material context of scientific inquiry. Three decades ago, this shift was welcome as a corrective to an earlier tradition which was overly focused on great ideas and great men, or theories and theoreticians. We now have a much richer understanding of science as a human activity than in the past.

However, other types of questions have been neglected or even rejected. In particular, the core philosophical concerns about science have not in recent years received much historical elucidation: discovery, justification, representation, explanation, and

[1] Schickore (2011) gives a good overview of the debates about integrated history and philosophy of science since the middle of the 20th century, and she cites many key works. Howard (2011) offers a more long-term history of the relation of the two fields and discusses some of the fundamental reasons for their separation in the 20th century. A good snapshot of the state of the field in the new millennium can be found in Arabatzis and Schickore (2012).

[2] While we focus on one flavor of HPS, we see others as complementary and equally valuable. For instance, a key project is to trace the origin and growth of modern concepts, theories and questions (Lennox 2001; Schickore 2011).

so on. It is time to reverse this trend. To echo an earlier generation of scholars, the historical neglect of philosophical issues is an opportunity missed and a responsibility avoided (cf. Laudan et al. 1986, p. 152). By reintegrating philosophical concerns with historical scholarship, we will gain richer histories.[3] The penalty for foregoing this reintegration is an inadequate picture of the scientific enterprise—a picture just as inadequate as one that pays no attention to cultural context or institutional frameworks.

A serious historical investigation of philosophical questions still promises to teach us much about how scientists conceive of new hypotheses, and more broadly about what strategies they employ for solving empirical problems. We will be able to study in greater depth how actual scientific communities have debated different types of empirical evidence (not just rhetorically, but epistemologically) and how they have adjusted their judgments in accordance with it. We will learn whether individual scientific disciplines grow essentially cumulatively or by sharp discontinuities—which remains, in many ways, as open a question as it was when Kuhn put it on the scene in the 1960s. Similarly, history will guide us beyond the traditional philosophical focus on individual scientists and the relationship between their theories and their data. Instead, it will enable us to study how shared epistemic goals are reached (or not) by collaboration and competition both within and between multiple research groups. For the most part, historically deep and philosophically informative historical studies of these and similar issues remain to be written.

In summary, philosophical issues are a rich resource for asking historical questions that are particularly pertinent to science, but this approach remains under-appreciated both in theory and in practice. In consequence, we miss the chance of a deeper understanding of what makes science special as an epistemic enterprise. However, we do not envision the use of philosophical questions in historical scholarship as a one-way interaction, as we will discuss further in the next subsection and in the remainder of this paper: While philosophical questions should be asked of history, we also believe that history will allow us to refine our philosophical questions and indeed to answer them in unexpected and uniquely informative ways.

5.2.2 *Naturalism for Philosophy*

Fifty years ago, philosophical skepticism about integrated history and philosophy of science centered around the issue of normativity: If the goals of philosophy of science are normative, then what does a descriptive project like the history of science have to do with it? For one may study Galielei's or Darwin's methodology to one's heart's content, but this will not shed any light on the normative question of whether their methods are justified. When we ask whether science *should* proceed in one

[3]An indication of this is the tradition of "historical epistemology" (see e.g. Rheinberger 1997; Daston and Galison 2007; and the special issue of *Erkenntnis* edited by Feest and Sturm 2011).

way or another, the answer will not come from a descriptive historical study but from a normative philosophical argument. In effect, history was short-circuited out of philosophical discussions.

The normativity objection found its most famous expression in Ronald Giere's "marriage of convenience" paper, long a cornerstone of philosophical skepticism about integrated HPS. Giere wrote:

> If one grants that epistemology is normative, it follows that one cannot get an epistemology out of the history of science — unless one provides a philosophical account which explains how norms are based on facts. This ought to be a central problem for historically oriented philosophers of science, but few seem willing even to acknowledge the question, let alone attack it head on (Giere 1973, p. 290).

From the normative perspective, Giere's challenge is perhaps as close as one can get to an ironclad argument against the philosophical relevance of the history of science. On this view, history can only play a heuristic role by allowing us to identify problems and outline possible solutions that then require proper philosophical analysis. At most, normative philosophy of science needs a link to the history of science in order to be a philosophy *of science*, rather than of some logically possible state of affairs. Or as Hanson (1962) put it, without history the pure philosopher's "analytical skill may be admirable, but it does not take us anywhere" (p. 586).

However, in the decade after the "marriage of convenience" paper, Giere changed his mind and began to give a much more crucial role to history. He now argued that he had misconstrued the issue; that all known normative approaches had stalled; and that naturalism offered the best prospect for a successful philosophy of science (Giere 1985). In a recent paper, he summarized his change of mind:

> I came to the conclusion that the philosophy of science should be transformed into something like the theory of science. That is, philosophers should be in the business of constructing a theoretical account of how science works. Philosophical claims about science would then have the status of empirical theories. In short, the philosophy of science should be naturalized. This means, among other things, giving up pretensions to finding autonomous standards for the practice of science (Giere 2011, pp. 60–61).

In a naturalized philosophy of science, history of science ceases to be a heuristic crutch.[4] Instead, it turns into the indispensable empirical basis for a theoretical enterprise that is best understood in analogy to other theoretical enterprises in the natural and social sciences. Much as ecologists model interspecific competition, or as macroeconomists model the effects of monetary policy, philosophers of science model scientific confirmation, explanation, theory choice, and so on. By reconstruing the HPS project along the lines of the empirical sciences, the naturalistic turn deflects the normativity objection.

[4]We prefer not to draw a strong distinction between historical and contemporary scientific practice as an object of study. What would have been "contemporary" science to Giere in 1985 is "historical" now, but the theoretical questions we ask about our cases remain largely the same. The only difference are in the methods of study: How recent an episode is will partly dictate whether our tools will include archival studies, oral histories, laboratory notebooks, or questionnaires, not to mention "embedding" oneself in a research group. Depending on method, of course, some questions will be easier to answer than others.

Even though the naturalized project begins with and emphasizes description, it shares many of the goals of the old normative project. HPS does not turn into a purely descriptive project because we understand it to have two tasks: description *and* justification (see also Lipton 2004). That is, we wish to give an adequate description of past and present science, but we also wish to understand how the epistemic successes and failures of science can be accounted for. However, the traditional normative project took the descriptive task to be fairly trivial, while many naturalists would argue that adequate description may be the harder of the two tasks.

The naturalist's approach to the task of justification offers at least two advantages. First, a close engagement with the past and present of scientific practice can serve as an accelerator. Even if it were possible to do normative philosophy of science from first principles (such as logic or probability calculus), the project is likely to advance more quickly if existing practice is taken as a guide. If we wish to understand epistemology, we should begin with the most successful epistemological enterprise that we know. A second, much stronger naturalistic argument holds that many issues in the philosophy of science cannot be tackled from first principles at all. This is because scientific practices such as induction or explanation may ultimately be grounded in facts about the world (on induction, see Norton 2003). For example, the justification for biologists' interest in mechanistic explanations (Machamer et al. 2000; Bechtel 2006) probably does not derive from any formal philosophical property of such explanations. More likely, biologists have learned in the course of research that mechanistic explanations are adequate to many parts of their area of inquiry. What counts as a "normatively" adequate explanatory standard in this case has a necessary empirical and historical dimension: it concerns what is the case in the world and how we have learned about it. Thus, while the old normative project aimed for some sort of extra-empirical justification for the methods of science, strong naturalists expect the task of justification to be continuous with empirical science itself.

Where the methodology of integrated HPS is concerned, a commitment to some sort of naturalism is the beginning and not the end of the discussion. It is far from trivial to see how the combined goals of descriptive adequacy and philosophical insight can be achieved in practice. Therefore, in the remainder of this contribution we will discuss what is now perhaps the most pressing methodological problem in the discipline: best practices for relating history and philosophy of science to each other.[5] We will begin our discussion at the skeptical extreme: with the "dilemma of case studies", which suggests that the project of integrated history and philosophy of science—whether construed naturalistically or not—may be doomed in principle.

[5]Pinnick and Gale (2000) commented that "despite the possibility of doing so, philosophers have not pursued a method of case-study design" (p. 116). They also observed that disciplinary consensus about method coincides with progress.

5.3 The Dilemma of Case Studies

If we wish to use historical cases in order to test, refine and expand our best concepts about science, we are faced with two main problems: selection bias and inappropriate generalization from individual cases. In an issue of *Perspectives on Science* devoted to the legacy of Thomas Kuhn, Pitt (2001) labeled these twin dangers as "the dilemma of case studies". If we approach our project "top down" (proceeding from philosophy to history) then this leads into the first horn of the dilemma:

> [I]f the case is selected because it exemplifies the philosophical point being articulated, then it is not clear that the philosophical claims have been supported, because it could be argued that the historical data was manipulated to fit the point (p. 373).

Yet it is no solution simply to stick closer to the facts of history, since proceeding "bottom up" (form history to philosophy) only leads into the second horn of the dilemma:

> [I]f one starts with a case study, it is not clear where to go from there – for it is unreasonable to generalize from one case or even two or three (p. 373).

Pitt's dilemma seems to us to capture the core worries of both philosophers and historians who are presently skeptical about the project of integrated HPS. Many historians are wary of philosophically motivated work since the philosophy might well dictate which historical cases are chosen and how they are interpreted—very much in line with the first horn of Pitt's dilemma. Conversely, many philosophers (even those with a broadly naturalistic outlook) worry that any conclusions drawn on the basis of historical case studies are ultimately unwarranted generalizations—and this mirrors the second horn of Pitt's dilemma.

A number of practitioners of integrated history and philosophy of science have responded to Pitt's challenge (Burian 2001, 2002; Schickore 2011; Chang 2011). Chang in particular has articulated conceptual moves that may allow us to break free from the dilemma of case studies. He argues that the we should not think of philosophical concepts as *general* and historical facts as *particular*—for this would indeed lead into fruitless debates about how many white swans are needed to show that all swans are white. Instead, philosophy provides *abstract* concepts which are instantiated to various degrees by *concrete* historical cases. Chang compares this—using an admittedly imperfect metaphor—to the relationship between the setting of a TV series and its episodes:

> When we have an episode of *The Simpsons*, or *Buffy the Vampire Slayer*, or what have you, the episode is not really a case or an example of whatever the general idea of the show might be. Rather, the episode is a concrete instantiation of the general concepts (the characters, the setting, the type of events to be expected, etc.), and each episode also contributes to the articulation of the general concepts (Chang 2011, pp. 110–111).

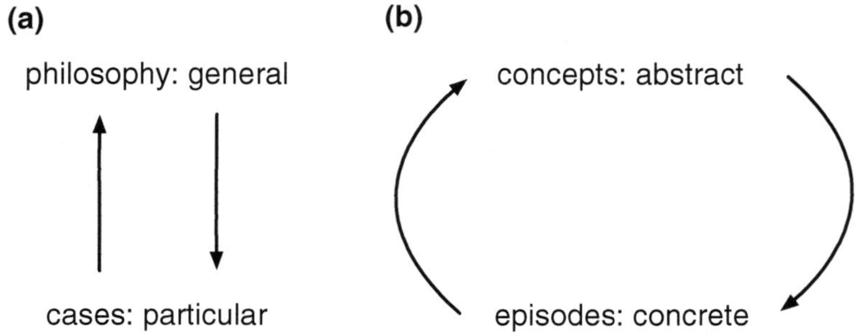

Fig. 5.1 **a** The traditional view of a contrast between general philosophical concepts and particular historical cases, which are related to each other either *top-down* or *bottom-up*. The usual problems present themselves: How can we move *top-down* without selection bias? How can we move *bottom-up* without unwarranted generalizations? **b** A schema of Chang's alternative view. Instead of contrasting the general with the particular, it contrasts abstract concepts with concrete instances in historical episodes. Instead of conceiving of either a *bottom-up* or a *top-down* confrontation of concepts and episodes, Chang proposes an iterative, cyclical movement between concepts and episodes. The entry into the cycle can occur either with concepts or with episodes: It does not require us to decide that either concepts or episodes are primary

Once the relationship between history and philosophy of science is construed in this way, we stop thinking in terms of working "top down" or "bottom up". Instead, we start thinking of a cyclical process: Just as abstract concepts help us to understand concrete episodes, so concrete episodes help us to further elaborate our conceptual tools. On Chang's view, doing HPS consists in a repeated cycling between the concrete and the abstract (Fig. 5.1).[6]

Chang's cyclical model is a useful metaphor for the interaction between history and philosophy of science and fits better with the actual practice of HPS than Pitt's top-down/bottom-up model. However, it requires further elaboration if it is to provide the basis for a methodology of HPS.[7]

In the remainder of this contribution, we will discuss the particulars of both the downward arrow (from concepts to episodes) and the upward arrow (from episodes to concept), as well as their cyclical interaction. In Sect. 5.4, we will discuss criteria

---

[6]While we adopt Chang's general framework, we do not think that much hinges on whether we speak of "episodes" or "cases", and we will continue to use both terms.

[7]Schickore (2011) has argued that the history-philosophy relationship should not be understood in terms of a confrontational model, in analogy to the empirical sciences, but in terms of hermeneutics, or "the art of gradually reconciling provisional analytic concepts with a provisional reading of the historical record" (p. 459). However, we believe that the confrontational and the hermeneutic models can be reconciled. Certainly the confrontational model must be conceived, as we discuss, in cyclical and iterative terms. But this is no surprise, since the empirical sciences—on which the confrontational model is based—are similarly iterative in theory testing. Moreover, HPS is in part concerned with the beliefs and desires of human actors, the traditional domain of interpretive, hermeneutic approaches. But this has ample room in the confrontational model, which understands the study of human beliefs and motives in terms of empirical theses about cognitive states (how ever difficult these may be to ascertain).

for moving from concepts to episodes without incurring the risk of selection bias. In Sect. 5.5, we will discuss the handling of agreements and conflicts between concepts and episodes, which we call the dynamics of confrontation. In particular, we will discuss the cyclical movement between concepts and episodes which allows us to achieve a certain kind of generality for our concepts about science.

5.4 The Selection of Case Studies

Many of the best works in integrated HPS explicitly discuss the merits of their chosen cases: they tell us why a particular case will not merely illustrate but investigate the worth of particular philosophical claims, or why conclusions reached for one case are likely to extend to others. However, for the most part the reasons for the choice of case studies remain implicit, and there is little discussion of a general methodology by which the selection of case studies should proceed. The aim of this section is to develop the outlines of such a general methodology. Like many systematic investigations, we begin with typology: There are a number of different purposes that case studies typically serve, and the issues of selection bias and generalization must be understood in relation to these purposes.

5.4.1 Hard Cases

The basic idea of hard cases is to seek out challenges: instead of selecting cases that illustrate a philosophical thesis particularly well, we prefer those that are difficult to accommodate and that therefore put a thesis to the test. To use an engineering example: If we build a self-driving car that can navigate the busy and complex traffic of Beijing without accident, then it can probably handle the more serene streets of Zurich as well. Hard cases demonstrate the power of a principle, and they show that the same principle can plausibly handle a host of similar but less difficult cases.

Analogies in the sciences are easy to find. Consider for example evolution by natural selection, for which the Giraffe's neck is a traditional and well-worn illustration. The illustration is excellent as a didactic tool: it is easy to understand and contrasts well with the alternative of evolution by use and disuse. However, those who are skeptical of the power of natural selection will not find the case particularly compelling: neck length may be a fairly trivial trait, and it would be easy to accept its origin by natural selection while denying that selection can produce more intricate and complex traits. It is no surprise, then, that evolutionists from Darwin onward have been particularly interested in hard cases for natural selection such as the human eye. If a trait as complex as the eye can be explained by natural selection, then selection has passed a high bar; and this success immediately makes plausible that more trivial cases can be explained in the same way.

To illustrate the point using an example from our own work, Scholl and Nickelsen (2015) investigated the genesis of Peter Mitchell's mechanism of oxidative phosphorylation. This is the main process by which mitochondria transform the energy in foodstuffs into a chemical compound called ATP, which cells, tissues and organs then use to drive their various processes. Mitchell received a 1978 Nobel Prize for the formulation of the mechanism, and it has long counted as one of the most spectacularly original contributions to 20th century biology. Leslie Orgel once wrote that "[n]ot since Darwin and Wallace has biology come up with an idea as counterintuitive as those of, say, Einstein, Heisenberg and Schrödinger" (Orgel 1999, p. 17). In our study we were able to show that the genesis of the theory can be explained using concepts from two recent strands in the philosophy of scientific hypothesis generation. One is interested in how the unknown causes of phenomena are sought (Graßhoff and May 1995; Lipton 2004); another is interested in how new mechanistic hypotheses are generated based on known entities and interactions (Darden 2006). The first strand allowed us to see that Mitchell's process of hypothesis generation, however spectacular the result, occurred in a well-defined space of possible causal hypotheses. The second strand allowed us to see how this well-defined space of possible hypotheses was investigated by generating "how possibly" mechanisms.

Our study has special probative force because it deals with a hard case of scientific discovery. No one would claim that the mechanism of oxidative phosphorylation was a trivial extension of existing biochemical knowledge: It was a theory of acknowledged novelty and originality. If its genesis is intelligible in terms of a number of basic heuristics, then the power of those heuristics is credibly demonstrated. Moreover, that the hard case could be accommodated provides some warrant for the speculation that many further but less difficult cases of scientific discovery are amenable to similar analyses.

There is a general recognition that hard cases can be particularly telling. To pick just one example, many of the philosophically most interesting theses in *Inventing Temperature* (2004) rely on Chang first convincing his readers that the measurement of temperature is, against expectation, a hard case in the history of measurement techniques. Chang's notion of epistemic iteration becomes compelling precisely when we realize that it can illuminate a particularly difficult epistemic advance.

5.4.2 Paradigm Cases

Many cases play a special role in HPS because they have become paradigms of some aspect of science. In a sense, these cases function like model organisms in biology: we study them not only as particulars, but as more or less typical instances of some aspect of science. Like model organisms, paradigm cases offer the advantage of pre-built resources. The relevant historical documents and the historical context are usually reasonably well understood, so that new conceptual studies can proceed rapidly.

A good example of a paradigm case is Semmelweis's discovery of the cause of puerperal fever around the middle of the 19th century. The case was introduced to the philosophy of science in Carl G. Hempel's *Philosophy of Natural Science* (1966), where it served to illustrate aspects of the confirmation of theory by data. The case was later revisited by Lipton (2004), who challenged Hempel's hypothetico-deductive reconstruction of Semmelweis's procedure and outlined an alternative in terms of inference to the best explanation (IBE). Lipton argued that a number of aspects of Semmelweis's investigation—including its context of discovery, the rejection of alternative hypotheses, and the confirmation of accepted hypotheses—remain obscure on a hypothetico-deductive reconstruction but become intelligible in the framework of IBE. Importantly, Lipton was able to draw on rich existing material concerning Semmelweis's discovery such as, for example, K. Codell Carter's translation of Semmelweis's main work (Semmelweis 1983). The translation, in turn, was partly produced in order to facilitate the use of the Semmelweis case in a course in philosophy of science. The discussion of the Semmelweis case has continued in recent years: Gillies (2005) has argued that a Kuhnian perspective is necessary for understanding the reception of Semmelweis's work; Bird (2010) sees Semmelweis as an instance of inference to the *only* explanation; and one of us has argued that Mill's four methods of experimental inquiry play an important role in the confirmation of Semmelweis's data (Scholl 2013)—a fact which was previously overlooked because Carter's translation omitted Semmelweis's copious numerical tables, which seemed irrelevant from a Hempelian perspective. For the most part, these authors are not primarily interested in Semmelweis *qua* Semmelweis: the topic of interest is confirmation, of which Semmelweis is taken to be a representative instance.

Whether a case deserves the status of a paradigm is itself open to debate. For example, Tulodziecki (2013) has recently argued that the discussion of Semmelweis proceeds from the false assumption that Semmelweis was an excellent reasoner. She discusses a number of flaws in Semmelweis's arguments which indicate that the case is not, after all, a representative instance of successful scientific reasoning. In our view, such explicit arguments for and against the representativeness of a case are required when using paradigm cases.

Importantly, it remains an empirical question whether concepts can be transferred from the paradigm to other cases. The fact that paradigms are considered typical instances gives us reason for some optimism that many concepts, once developed and refined, can be transferred from them to other cases—but whether this is in fact the case must be checked in further detailed studies. This again mirrors the use of model organisms, where we also have the expectation but no guarantee of transferability to other organisms.

Many classical works of HPS use paradigm cases. Take for instance Shapin and Schaffer's *Leviathan and the Air-Pump* (1985). The authors' discussion of experimental knowledge is powerful precisely because the air-pump is emblematic of experimental science. What is true for the air-pump is plausibly true for countless other experiments. While we would disagree with many of Shapin and Schaffer's specific claims, from a methodological point of view the air-pump is a properly deployed paradigm case.

5.4.3 Big Cases

The most traditional and straightforward reason for choosing a case study is that it concerns a big scientific achievement. It may be an achievement that served as a scientific template for many further works; it may be the foundation for a large branch of present-day science; it may have yielded an understanding of a fundamental aspect of nature. In many of the most interesting cases, such as the works of a Newton or a Darwin, all of the above will apply.

Unlike paradigm cases, big cases cannot be expected to generalize particularly well. We often assume that big cases are also in some way typical of an aspect of science, and we may therefore be tempted to generalize from them in the same way as we do from paradigm cases (Sect. 5.4.2 above). But of course typicality is something that cannot be assumed. It is possible that Newton's efforts to confirm his theories were quite atypical of how most confirmation in science happens; it may be that Darwin's standards for what is an acceptable explanation were atypical of scientific explanations at most times; and so on. That a big case is also typical of some aspect of science must be explicitly argued for (or at least stated as a premise)—and then these cases generalize in virtue of being paradigms.

Similarly, we should not be misled into thinking that all big cases are hard cases. Certainly influential scientists like Newton and Darwin solved hard empirical problems. But that does not mean that their work always qualifies as hard cases in the sense of Sect. 5.4.1: Whether something is a hard case in our sense depends on the philosophical thesis under consideration. The genesis of Darwin's theory may have been particularly conceptually challenging, which makes it a hard case for those who argue that scientific discovery is explicable in terms of basic heuristics. At the same time, however, other aspect of Darwin's work may not constitute a hard test of relevant philosophical ideas—maybe there is little to be learned from finding that Darwin's concept of explanation conformed to the notion that good explanations are mechanistic. Our modest point is simply that whether big cases are also hard cases in the sense discussed here depends on the philosophical thesis under test.

Even though there is no reason to think that lessons learned from big cases are necessarily transferable to other cases, we do not think that selection bias is a major concern. A scholar may choose a big case specifically because it bears out his or her philosophical *idée fixe*—but this would nevertheless teach us something interesting about a case we already consider to be important, provided that the concepts actually apply. Simply put, it is inherently fascinating to understand the particulars of an important episode. Moreover, we certainly wish to know whether our best philosophical concepts can illuminate the epistemic advances we find most important. It is not only allowed but even necessary, sooner or later, to apply our conceptual tools to the big cases. The only mistake would be to think that inherent fascination is a substitute for carrying on with the broader program of HPS: Ultimately, the range of applicability of concepts must be checked using diverse cases. To know whether a Newton or a Darwin is typical of science at his time, or typical of key concepts from the philosophy of science, is part of understanding the episode.

Big cases are particularly prone to the underdetermination problem of HPS (see the introduction to this volume by Sauer and Scholl): the same historical episode is usually told again and again in different philosophical terms, which raises concerns that philosophical concepts hinder rather than help our understanding of science. Most influential scientists have had multiple careers in the literature: as good inductivists, as resourceful hypothetico-deductivists, as epistemically cautious Popperian falsificationists, perhaps as methodological anarchists, and finally as contingent products of mostly social forces. Not all of these accounts can be true, but deciding among them is hindered by the fact that the key questions often concern cognitive processes of past scientists—to which we have little access. The best defense against the mindless retelling of big cases according to prevailing philosophical fashion is not, however, to retreat to some form of historical positivism, but to take the cyclical model of HPS seriously: We must consider a wide range of cases from the history of science, use them to improve our conceptual tools, and deploy these tools to understand episodes at different levels of importance. We will have more to say about how cases are used to evaluate and refine concepts in Sect. 5.5.

5.4.4 Randomized Cases

There already exists a widely accepted method for avoiding selection bias in the sciences: randomization. It is at least conceivable that case studies big and small could be chosen randomly from a database and submitted to philosophical analysis. If one had a particular hypothesis about, say, the steps by which model-building proceeds, it might be possible to ask the database for random instances of model-based science in order to check the applicability of the hypothesis. While we do not believe that this should (or could) replace historical judgment in the choice of case studies, it could be a valuable complement to the way in which historical scholarship traditionally proceeds.

Before the randomization of case studies becomes feasible, both practical and conceptual problems need to be addressed. On the practical side, no suitable database of case studies from the history or the philosophy of science currently exists. However, it is a reason for optimism that such a database would be desirable for any number of purposes apart from randomization. For instance, a database of case studies could restore unity to a field that has lately focused on historical micro-studies rather than grand narratives. On the conceptual side, the organization of the database would be a challenging issue. How are case studies to be individuated and classified in a way that is historically adequate, reasonably theory-neutral and useful for data retrieval? At minimum, something akin to Pubmed's Medical Subject Headings (MeSH) vocabulary would be required. Importantly, long-term institutional backing would be a prerequisite for the credibility of such a project.

5.4.5 First Sketch of a Typology

We have distinguished between four types of case studies. Each has its own conceptual relationship to key concerns such as selection bias and generalizability.

HARD CASES are chosen to be difficult for the philosophical concepts under study to handle. What counts as a hard case will thus vary depending on the philosophical concepts we are interested in. If a philosophical principle survives contact with a hard case, this speaks to its power. Hard cases circumvent selection bias by seeking challenges rather than convenient illustrations. They allow us to draw more forceful conclusions than individual cases normally do since they give us reason to think that the tested philosophical principle is powerful enough to handle less difficult cases as well.

PARADIGM CASES are the model organisms of HPS. We use them in teaching and research as typical instances of particular aspects of science. Because they are already accepted as typical, and because the relevant historical sources are usually easily available, paradigm cases are efficient tools for making new points and for revising existing concepts. Importantly, whether a case qualifies as paradigmatic is usually itself a point of debate. And whether concepts that apply to the paradigm case can be extended to further cases remains, as in the case of model organisms, an empirical question.

BIG CASES concern influential scientific achievements. They have particular appeal because of their conceptual or historical centrality to the scientific enterprise. However, big cases must not be assumed to be typical of some aspect of science without further argument; nor are they necessarily hard cases, since this depends on the philosophical concepts under study. Finally, big cases are particularly attractive targets for retellings according to prevailing philosophical fashions. This impulse must be resisted by committing in earnest to the cyclical model of HPS.

RANDOMIZATION of case studies to counteract selection bias is currently little more than a neat idea, but we think that it is coherent in principle. A reason to pursue the idea is that a database of case studies would be a useful tool with many additional uses.

Our typology is not intended as something static and upfront. Cases do not present themselves to us with a label identifying them as "paradigm" or "hard" cases. Whether we understand a case as paradigmatic, or as hard with respect to a philosophical thesis, and so on, should be allowed to develop as our studies progress. New historical evidence—the cyclical revisions we discuss in the next section—may well alter our assessment of a case study's status. Moreover, our typology is unlikely to be complete: we look forward to the critical discussion of additional types with their own particular functions. Our main point is modest: As we develop research projects, we should pay greater attention to the ways in which our case studies relate to the philosophical theses we wish to demonstrate, criticize or test.

5.5 Dynamics of Confrontation

Once philosophical theses are confronted with historical cases, agreements and disagreements between them are found. The temptation is to think of these as simple instances of the confirmation or rejection of general hypotheses by particular facts. But this temptation must be resisted on pain of two related philosophical sins: the universal sin and the existential sin. When we commit the universal sin, we assume that what is true for our case study is true for all of science. When we commit the existential sin, we assume that a single counterexample can serve to reject a philosophical approach. Both procedures are misguided, but how can we do better?

An example from our own research (Scholl and Räz 2013) will serve to illustrate some of the moves we think are appropriate for iteratively evaluating and revising philosophical concepts and for interpreting historical episodes. We are not holding up our paper as particularly significant, much less do we think the paper should compel universal assent. Its usefulness as an example rests to a large extent on its ordinariness: The goal is to illustrate a methodological approach, but this is largely independent of the acceptance of the philosophical and historical theses defended.

The starting point of our project was a paper by Weisberg (2007), who had proposed a distinction between different types of theoretical practices in science: modeling and "abstract direct representation" (ADR). In order to illustrate modeling, Weisberg used Volterra's work on predator-prey dynamics; to illustrate ADR, he turned to Darwin's work on the origin of coral atolls and Mendeleev's work on the periodic table of the elements. While we agreed that theorizing is heterogeneous, we were interested in how the scientists themselves understood the subcategories of their methodological practices, and so turned to the historical sources.

5.5.1 Agreement

To investigate Volterra's own views on method, we studied Volterra and D'Ancona (1935), which presents a monograph-length exposition of the predator-prey model. Perhaps as a reaction to the critics of earlier publications of the basic model (Volterra 1926a, b, 1928), the monograph contains a methodological discussion as a preface. These methodological reflections support the idea that there is a contrast between more "direct" and more "indirect" theoretical practices. Volterra and d'Ancona place their own work on predator-prey dynamics in the latter subcategory. They label their own indirect approach as "deductive", but we argue (with Weisberg) that it should be understood in terms of our present-day notion of modeling. Thus, we found an agreement between our philosophical thesis and the historical case: a distinction between modeling and more "direct" theoretical practices was suggested by the scientists themselves.

The finding of an agreement is good news for our philosophical thesis (that model-based science is "indirect" theoretical practice that should be distinguished from more "direct" practices) and for our understanding of the historical episode. Yet it is only a first step: We must immediately ask whether the agreement is the result of selection bias such that the historical episode is, in truth, a mere illustration (as opposed to a study) of a pet philosophical thesis. Is the case hard for our thesis? Is it paradigmatic of some aspect of science? (We would argue that the predator-prey model is indeed a paradigmatic case of model-based science.) Do we have reasons for believing that the thesis, although borne out by the case, might apply only locally? In general, agreement is nothing but a sign that we must now investigate the range of applicability of our thesis (see Sect. 5.5.5).

5.5.2 *Conflict*

Next, we reexamined one of Weisberg's two examples of ADR: Charles Darwin's explanation of the origin and structure of coral reefs and atolls in the Pacific (Darwin 1842). Weisberg understood this as a case of ADR, because "at all times, Darwin was talking about the actual atolls in the Pacific" (2007, p. 228). We disagreed with this assessment. Darwin had to examine the effects of the interaction between known processes: the subsidence of islands and the growth of corals, processes which he investigated in some detail during his journey on the *Beagle*. However, it was impossible for him to observe this interaction directly because it extends over centuries. Darwin thus had to resort to a mental model of the process and its consequences: This allowed for the indirect investigation of the genesis and structure of coral atolls that is typical of model-based science. We thus reclassified Darwin's explanation as a case of modeling, extending the category to an additional, less obvious case of non-mathematical modeling.

We must be careful about the conclusions drawn from this conflict. The first instinct is to conclude that the thesis of ADR has now been rejected, as practiced in the popular philosophical game of counterexamples. However, the appropriate thing to do is simply to reassign Darwin to the modeling category without making judgments about the existence or range of applicability of ADR, which the reevaluation of an individual case does not permit.

5.5.3 *Incompleteness*

After determining agreements and conflicts between existing concepts and cases, we may discover that there are some issues for which we do not have any conceptual tools at all. This sort of incompleteness may of course indicate that our existing concepts need to be expanded or generalized by strenuous tinkering, but that strikes us as a problematical "winner takes all" mentality. Incompleteness may just as well indicate that a new concept needs to be developed in addition to those we already have.

In the example in hand, Volterra's methodological discussion confirmed to us the existence of an incompleteness in Weisberg's scheme of theoretical practices. Volterra and d'Ancona preferred to contrast model-based science with a more "direct" theoretical investigation in terms not of ADR but of experimental causal inference. The authors write that for their own investigation they would have *preferred* an experimental approach, which would have allowed for direct causal inferences in the system under scrutiny. However, this was unfeasible for practical reasons having to do with the size and time scales of ecological systems. In light of this, we added the subcategory of causal inference as an additional theoretical practice. While the historical case did not suggest the *existence* of this type of theorizing (since philosophers have studied it for a long time), the sources certainly urged that causal inference as a theoretical practice is a relevant *contrast* to model-based science.

Instances such as this one illustrate the *generative* role of historical cases. By turning to the historical sources, we were able to recognize a natural subcategory—one of considerable philosophical interest—that needed to be added to Weisberg's initial range of theoretical practices.

5.5.4 *Redundancy*

The complement to incompleteness is redundancy, where we may find that we have developed elaborate philosophical concepts that are not applicable to *any* part of science. There is a philosophical prejudice in such cases for thinking that the concepts are fine but that the search for good examples goes on. We would suspect, however, that a failure to find good instances of a concept is an indication that the concept is either not well developed or wholly misguided.

In the present example, the historical cases led us to suggest a redundancy. We were convinced that Darwin's work on coral atolls did not fit in the subcategory of ADR, while other scholars had already argued that Mendeleev's discovery of the periodic table should be understood in different terms as well.[8] We took this as an indication that the entire subcategory of ADR may be superfluous (unless, of course, additional instances can be presented).

[8]See Scerri (2012) for a critique of Weisberg's interpretation of Mendeleev's work.

5.5.5 Ranges of Applicability

In all the steps discussed above, it is useful to keep thinking about the range of applicability of a thesis. An agreement may indicate not that a philosophical approach is correct overall, but only that it captures a subset of cases well, while other concepts may apply to other subsets. Similarly, a supposed counterexample may indicate not that a thesis is false, but only that our case is outside its range of applicability. Thus, we concluded that Volterra's work on predator-prey dynamics did indeed fit in Weisberg's subcategory of model-based science, while Darwin (as explained in Sect. 5.5.2) had to be removed from the ADR subcategory and moved to modeling. An examination of the Mendeleev case led us to remove that case from the subcategory of ADR as well.[9] Finally, we immediately recognized cases that fit in the newly added subcategory of causal inference, as suggested by Volterra's own methodological discussions. For instance, Semmelweis's investigation of the cause of puerperal fever (Scholl 2013) has long been discussed as a paradigm case of causal inference. Finding multiple instances of a single subcategory can be taken as an indication that a natural and substantial subcategory has been found.

5.5.6 Summary

Just as the same historical episode can simultaneously be a hard case, a paradigm case and an big case (depending on the philosophical concepts under study), confrontation will rarely involve just one of the possibilities listed above: as we investigate the relationship between historical cases and philosophical theses, we will usually make several or all of the moves discussed. Some concepts will be in agreement with the historical case, while many will be in various degrees of conflict; some of these conflicts will be resolved by adjusting ranges of applicability, while others will require us to consider incompleteness or redundancy. As discussed above, we see the process of confrontation as cyclical, where a combination of the moves outlined above will occur over multiple iterations. For the case discussed here, we summarize the cyclical revisions in Fig. 5.2.

[9]For the time being we refrained from assigning Mendeleev to any of the other subcategories, although Scerri (2012) suggested "classification"—which we should presumably count among theoretical practices.

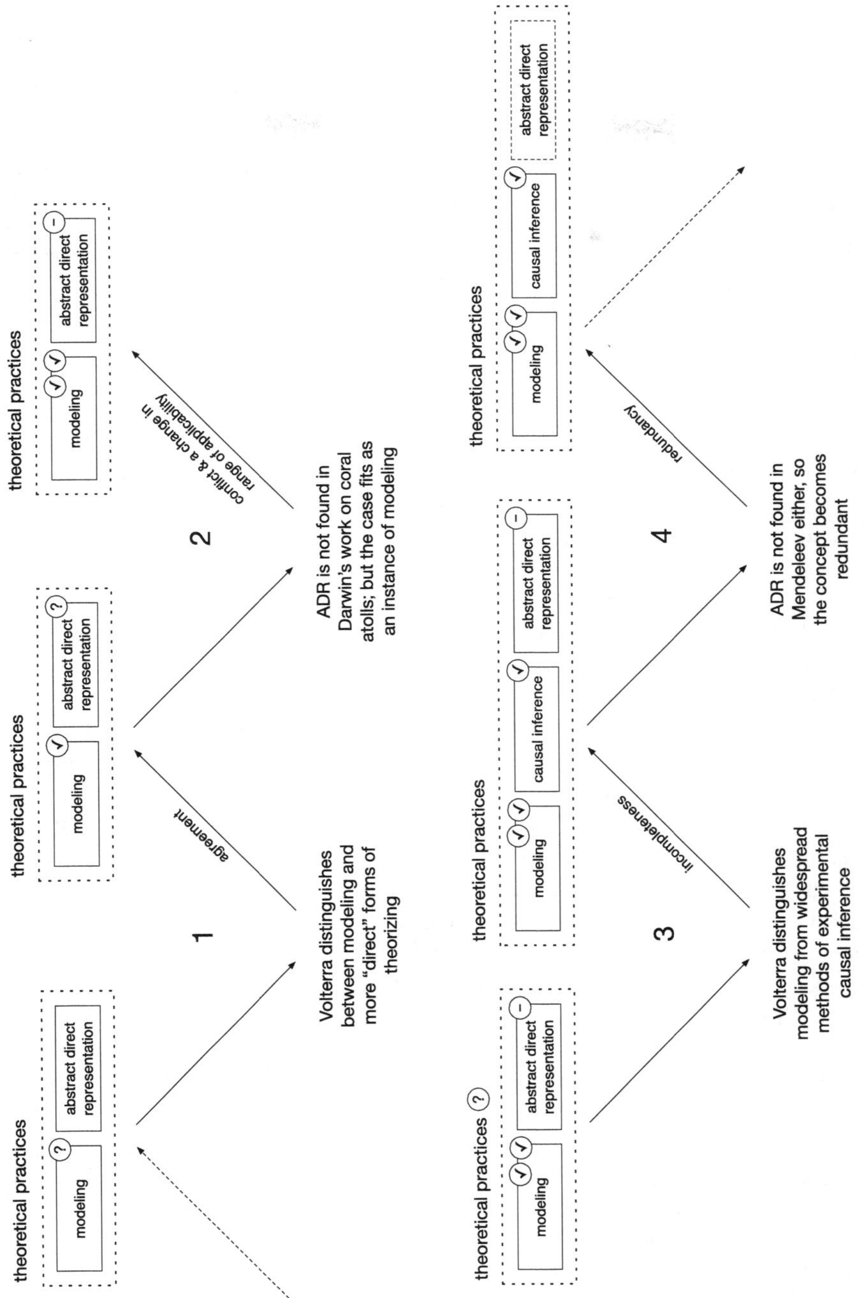
theoretical practices
modeling
abstract direct representation
1
agreement
Volterra distinguishes between modeling and more "direct" forms of theorizing
theoretical practices
modeling
abstract direct representation
2
conflict & a change in range of applicability
ADR is not found in Darwin's work on coral atolls; but the case fits as an instance of modeling
theoretical practices
modeling
abstract direct representation
theoretical practices
modeling
abstract direct representation
3
incompleteness
Volterra distinguishes modeling from widespread methods of experimental causal inference
theoretical practices
modeling
causal inference
abstract direct representation
4
redundancy
ADR is not found in Mendeleev either, so the concept becomes redundant
theoretical practices
modeling
causal inference
abstract direct representation

◀ **Fig. 5.2** Cyclical HPS and the exploration of theoretical practices. We begin with Weisberg's notion that two kinds of theoretical practices should be distinguished: modeling and abstract direct representation (ADR). In cycle 1, we find an **agreement** between concepts and case: Volterra, in his work on predator-prey dynamics, recognizes a type of theorizing that corresponds to modeling. In cycle 2, we test whether Darwin, in his work on coral atolls, follows a theoretical practice that corresponds to Weisberg's ADR. We find a **conflict** between concepts and case. Based on textual evidence, we adjust the modeling subcategory's **range of applicability** to include Darwin's work on coral atolls, giving modeling a second instance. Next, in cycle 3, we recognize an **incompleteness**: Volterra's chosen contrast for model-based science is experimental causal inference rather than ADR. Finally, cycle 4 suggests that the concept of ADR is **redundant** since it does not apply to Weisberg's second suggested instance either: Mendeleev's work on the periodic table

5.6 Conclusions

We have discussed a fundamental motivation for integrated history and philosophy of science, and for case-study-based philosophy of science more generally. Philosophical concepts are a useful resource for asking questions about the historical or contemporary practice of science: They provide particularly pertinent questions that relate to science's core epistemic project. Conversely, historical and contemporary science provides the empirical basis for a naturalized philosophy of science, which should be properly conceived as a theory of science in analogy to theory construction in other empirical disciplines.

Moreover, we have examined the dilemma of case studies as a methodological challenge for integrated HPS. First, how can we minimize the danger of selection bias in choosing our case studies? How can we choose case studies that play a probative instead of merely illustrative role? We have suggested that an initial remedy is the explicit discussion of criteria for choosing a case study, and of the function that the case study will play vis-à-vis the philosophical concepts under study. In particular, we have outlined four main types of case studies: Hard cases, which are chosen such that they challenge rather than illustrate the relevant philosophical concepts; paradigm cases, which are taken to be typical of some aspect of science; big cases, which concern particularly influential or otherwise important scientific findings; and randomized cases, where selection bias is minimized by leaving selection to chance.

Second, how should our conceptual engineering proceed when we are confronting philosophical concepts with historical cases? We have argued that the main mistake is to look for concepts that are too general, and to be content with their facile confirmation or rejection. A disciplined pluralism appears to pay dividends: We must ask about all the main categories of concepts about science (such as confirmation, explanation, and many more) whether they can be understood in terms of a number of subconcepts. Much of the work will consist in describing, elaborating and applying these subconcepts. In this procedure, individual cases do not usually confirm or disconfirm broad categories of concepts (such as "mechanistic explanation"). Instead, agreements and conflicts between concepts and cases prompt us to assess the concepts' ranges of applicability. If our existing concepts prove inadequate to the understanding of historical cases, this incompleteness will challenge us to create new concepts. In this way, historical cases are more than just a testbed for philosophy. They also play a crucial generative role.

In 1992, David Hull asked philosophers of science for renewed ambition:

> Although grand theories about the nature of science are currently out of fashion, I think that we need to rehabilitate them. We need to construct theories about science the way that scientists construct theories about fluids, gene flow and continental drift (Hull 1992, p. 473).

We agree with Hull's ambition and with his naturalistic approach, but his notion of "grand theories" requires clarification. In line with the above discussion, we should not expect to find *the* grand theory of science, akin to Popper deriving all aspects of proper science from *modus tollens*. Rather, we should expect theories of the kind of we find in most special sciences: a series of overlapping models (to use Giere's term) whose interaction and integration with each other are themselves important questions for study. In brief, grand theories are not necessarily unified theories, and pluralists need not fear them.

Many important issues are left unanswered by our discussion. Our types of case studies are likely not exhaustive: We may well find a range of further types, with distinct functions in HPS, that we have not yet considered. And similarly, much of what we have said about the dynamics of confronting concepts and cases remains to be made more precise and to be extended by further procedures. Moreover, there are a number of important issues that we have only touched upon. For instance, case studies usually leave a certain room for interpretation. Different scholars may see different and contradictory concepts instantiated in the same episode, even when there is no malicious intent to misrepresent. The problem is defanged somewhat by explicit criteria for choosing case studies, and by an awareness of the conceptual engineering involved in confronting concepts with cases. Nevertheless, the underdetermination problem of integrated HPS and its relationship to methodology deserves much further discussion (see also Franklin and Collins, this volume; Kinzel, this volume; and Räz, this volume).

To conclude, integrated history and philosophy of science promises a unique understanding of the scientific enterprise. However, we again agree with Hasok Chang, who writes:

> I believe that the neglect to clarify the nature of the history-philosophy relationship in case-studies has contributed decisively to a widespread disillusionment with the whole HPS enterprise (2011, p. 109).

We expect that a renewed methodological discussion can bring about the needed clarification. The effort promises to pay off on both the philosophical and the historical front: we stand to gain better general theories of science and a more adequate understanding of its history.

Acknowledgments For helpful comments and debates, which have shaped this contribution significantly, we thank first and foremost the participants of the workshop "The philosophy of historical case studies" at the University of Bern (November 21–22, 2013). In addition, we are indebted to Michael Bycroft, Allan Franklin and Jutta Schickore; the participants of the Fifth Conference on Integrated History and Philosophy of Science (&HPS5) at the University of Vienna (June 26–28,

2014); the Visiting and Postdoctoral Fellows at the University of Pittsburgh's Center for Philosophy of Science (2014–2015); and the members of the Lake Geneva Biology Interest Group (lgBIG). Raphael Scholl was supported by a grant from the Swiss National Science Foundation (grant number P300P1_154590).

References

Arabatzis, T., and J. Schickore. 2012. Ways of integrating history and philosophy of science. *Perspectives on Science* 20(4): 395–408.

Bechtel, W. 2006. *Discovering cell mechanisms: The creation of modern cell biology*. Cambridge: Cambridge University Press.

Bird, A. 2010. Eliminative abduction: Examples from medicine. *Studies in History and Philosophy of Science* 41: 345–352.

Burian, R.M. 2001. The dilemma of case studies resolved: The virtues of using case studies in the history and philosophy of science. *Perspectives on Science* 9(4): 383–404.

Burian, R.M. 2002. Comments on the precarious relationship between history and philosophy of science. *Perspectives on Science* 10(4): 398–407.

Chang, H. 2004. *Inventing temperature: Measurement and scientific progress*. Oxford: Oxford University Press.

Chang, H. 2011. Beyond case-studies: History as philosophy. In *Integrating history and philosophy of science*, ed. S. Mauskopf, and T. Schmaltz, 109–124. Netherlands: Springer.

Darden, L. 2006. *Reasoning in biological discoveries*. New York: Cambridge University Press.

Darwin, C. 1842. *The structure and distribution of coral reefs*. London: Smith, Elder and Co.

Daston, L., and P. Galison. 2007. *Objectivity*. New York: Zone Books.

Feest, U., and T. Sturm. 2011. What (good) is historical epistemology? Editors' introduction. *Erkenntnis* 75(3): 285–302.

Franklin, A., and H. Collins. in press. Two kinds of case study and a new agreement. In *The philosophy of historical case studies, boston studies in the philosophy and history of science*, ed. T. Sauer, and R. Scholl.

Giere, R.N. 1973. History and philosophy of science: Intimate relationship or marriage of convenience?

Giere, R.N. 1985. Philosophy of science naturalized. *Philosophy of Science* 52(3): 331–356.

Giere, R.N. 2011. History and philosophy of science: Thirty-five years later. In *Integrating history and philosophy of science*, ed. S. Mauskopf, and T. Schmaltz, 59–65. Netherlands: Springer.

Gillies, D. 2005. Hempelian and Kuhnian approaches in the philosophy of medicine: The Semmelweis case. *Studies in History and Philosophy of Biological and Biomedical Sciences* 36(1): 159–181.

Graßhoff, G., and M. May. 1995. Methodische Analyse wissenschaftlichen Entdeckens. *Kognitionswissenschaft* 5: 51–67.

Hanson, N.R. 1962. The irrelevance of history of science to philosophy of science to philosophy of science. *The Journal of Philosophy* 59(21): 574–586.

Hempel, C.G. 1966. *Philosophy of natural science*. Englewood Cliffs: Prentice Hall.

Howard, D. 2011. Philosophy of science and the history of science. In *The continuum companion to the philosophy of science*, ed. S. French, and J. Saatsi, 55–71. London: Continuum.

Hull, D. 1992. *Testing philosophical claims about science*, 468–475. In PSA: Proceedings of the Biennial Meeting of the Philosophy of Science Association.

Kinzel, K. in press. Pluralism in historiography: A case study of case studies. In *The philosophy of historical case studies. Boston studies in the philosophy and history of science*, ed. T. Sauer, and R. Scholl.

Laudan, L., A. Donovan, R. Laudan, P. Barker, H. Brown, J. Leplin, P. Thagard, and S. Wykstra. 1986. Scientific change: Philosophical models and historical research. *Synthese* 69(2): 141–223.
Lennox, J.G. 2001. History and philosophy of science: A phylogenetic approach. *História, Ciências, Saúde-Manguinhos* 8(3): 655–669.
Lipton, P. 2004. *Inference to the best explanation*. London: Routledge.
Machamer, P., L. Darden, and C.F. Craver. 2000. Thinking about mechanisms. *Philosophy of science* 67(1): 1–25.
Norton, J.D. 2003. A material theory of induction. *Philosophy of Science* 70(4): 647–670.
Orgel, L. 1999. Are you serious, Dr Mitchell? *Nature* 402(6757): 17–17.
Pinnick, C., and G. Gale. 2000. Philosophy of science and history of science: A troubling interaction. *Journal for General Philosophy of Science* 31(1): 109–125.
Pitt, J.C. 2001. The dilemma of case studies: Toward a heraclitian philosophy of science. *Perspectives on Science* 9(4): 373–382.
Räz, T. in press. Gone till november: A disagreement in Einstein scholarship. In *The philosophy of historical case studies. Boston studies in the philosophy and history of science*, ed. T. Sauer, and R. Scholl.
Rheinberger, H.-J. 1997. *Toward a history of epistemic things: Synthesizing proteins in the test tube*. California: Stanford University Press.
Richards, R.J. 1992. *Arguments in a sartorial mode, or the asymmetries of history and philosophy of science*, 482–489. In PSA: Proceedings of the Biennial Meeting of the Philosophy of Science Association.
Sauer, T., and R. Scholl. in press. Introduction. In *The Philosophy of Historical Case Studies. Boston Studies in the Philosophy and History of Science*.
Scerri, E.R. 2012. A critique of Weisberg's view on the periodic table and some speculations on the nature of classifications. *Foundations of Chemistry* 14(3): 275–284.
Schickore, J. 2011. More thoughts on HPS: Another 20 years later. *Perspectives on Science* 19(4): 453–481.
Scholl, R. 2013. Causal inference, mechanisms, and the Semmelweis case. *Studies in History and Philosophy of Science Part A* 44(1): 66–76.
Scholl, R., and T. Räz. 2013. Modeling causal structures: Volterra's struggle and Darwin's success. *European Journal for Philosophy of Science* 3(1): 115–132.
Scholl, R., and K. Nickelsen. 2015. Discovery of causal mechanisms: Oxidative phosphorylation and the Calvin-Benson-cycle. *History and Philosophy of the Life Sciences* 37(2): 180–209.
Semmelweis, I.P. 1983. *The etiology, concept, and prophylaxis of childbed fever*. Trans. K.C. Carter. Madison: University of Wisconsin Press.
Shapin, S., and S. Schaffer. 1985. *Leviathan and the air-pump: Hobbes, boyle, and the experimental life*. Princeton: Princeton University Press.
Sulloway, F.J. 1982. Darwin's conversion: The beagle voyage and its aftermath. *Journal of the History of Biology* 15(3): 325–396.
Tulodziecki, D. 2013. Shattering the myth of Semmelweis. *Philosophy of Science* 80(5): 1065–1075.
Volterra, V. 1926a. Fluctuations in the abundance of a species considered mathematically. *Nature* 118(2972): 558–560.
Volterra, V. 1926b. Variazioni e fluttuazioni del numero d'individui in specie animali conviventi. *Memorie della R. Accademia dei Lincei* 6(2): 31–113.
Volterra, V. 1928. Variations and fluctuations of the number of individuals in animal species living together. *Journal du Conseil/Conseil Permanent International pour l'Exploration de la Mer* 3(1): 3–51.
Volterra, V., and U. D'Ancona. 1935. *Les associations biologiques au point de vue mathématique*. Paris: Hermann.
Weisberg, M. 2007. Who is a modeler? *The British Journal for the Philosophy of Science* 58(2): 207–33.

Part II
Controversies Reconsidered

Chapter 6
Two Kinds of Case Study and a New Agreement

Allan Franklin and Harry Collins

Words, words. They're all we have to go on.
Stoppard (1967)

Abstract The debate between Collins and Franklin over the demise of the credibility of Joseph Weber's gravitational wave claims has been treated as an iconic case of conflict over rival interpretations of the history of science (see, for example, Kinzel, this volume). Collins conducted contemporaneous interviews with the scientists and argued that the existence of the experimenter's regress meant that scientists who generated results that conflicted with Weber were not forced to claim that he was wrong—a possible interpretation was that the critics' experiments were less sound than Weber's. Collins argued that the crucial intervention was made by a scientist whose rhetoric encouraged everyone to interpret Weber's results, rather than their own, as flawed. Franklin drew largely on published sources and claimed that the accumulation of negative results was the inevitable outcome of rational processes. Collins and Franklin still disagree strongly about method and interpretation but the interesting thing discussed here is that, for them, the violence has gone out of the debate. In the early days they found themselves insulting each other but nowadays they find themselves cooperating in joint enterprises. This change reflects a change in the history of science: nowadays it is impossible to believe that there is no social component involved in the acceptance of scientific results so the disagreement between Franklin and Collins is no longer over deep epistemological principle but over methodological approach and their views concerning the intentions of different historical actors. This is the stuff of normal disagreement between historians rather

A. Franklin (✉)
Department of Physics, University of Colorado, Boulder CO 80309-0390, USA
e-mail: allan.franklin@colorado.edu

H. Collins
Distinguished Research Professor of Sociology, Cardiff University, Cardiff, UK
e-mail: CollinsHM@cf.ac.uk

T. Sauer and R. Scholl (eds.), *The Philosophy of Historical Case Studies*,
Boston Studies in the Philosophy and History of Science 319,
DOI 10.1007/978-3-319-30229-4_6

than mutual incomprehension born of incommensurable approaches. The change in the tenor of the debate is a consequence of the fact that a revolution in historiography has taken place.

6.1 Introduction

Franklin: When two scholars[1] offer different accounts and interpretations of the same episode, is it possible to decide which is correct? One of the best known examples of such different accounts…

Collins: Or so it is often said!

…are Collins's and Franklin's accounts of the early experiments that attempted to detect gravitational waves, in particular Joseph Weber's experiments. In the early 1970s, Weber claimed a first detection of gravitational radiation (Collins 1975, 1981a, 1992, Chap. 4, 2004; Collins and Pinch 1998, Chap. 5; Franklin 1994, 1998, 2002, Chap. 2). Collins's and Franklin's studies illustrate two different approaches.

Collins: But how conflicting are the accounts of Collins and Franklin? We will explore some of ways they differ as we work through this chapter but in retrospect there is something strange about the whole debate. A couple of decades or so back, I wrote the nastiest things about Franklin I have ever put in print and he wrote some very unpleasant things about me. Looking back from this vantage point, however, it is hard to see why. Let me say that the thaw in relations was initiated by gracious and generous gestures from Franklin, who told me, during the 2001 history of science meeting in Denver, Colorado, that he liked the talk I gave there on the Laser Interferometer Gravitational-Wave Observatory and, subsequently, having read *Gravity's Shadow*, he told me my work on gravitational wave physics was good. I doubt that I would have been capable of such gestures and I thank him.

Subsequently, Allan gave me invaluable help as I sorted out my statistical arguments for *Gravity's Ghost* and he tells me that my queries and comments gave him the idea of writing *Shifting Standards: Experiments in Particle Physics in the Twentieth Century*. So, while, to reiterate, we disagree over some of the early Weber events, we don't any longer seem to disagree in such a way to as to give rise to insults in print. What has changed aside from Allan's generous gesture?

The current exercise, which I initially thought would involve me simply inserting a few clarificatory comments here and there into Allan's text, has grown in significance. As I started to add comments I could not stop wondering about the transformation in the nature of our argument. I think the explanation of what has happened, reflects our very discussion of the early Weber days (see below): it has to do with the difference

[1] This paper was written by Franklin who then invited Collins to be co-author. Collins agreed and made some small sub-edits and technical corrections accepted by Franklin. Where Collins thought that certain differences were revealing for the purposes of the exercise he added comments. Franklin and Collins reach a new agreement, and comment on remaining differences, in Sect. 6.5.

between the contemporary versus the retrospective view. In this case, the difference between perspectives is exceptionally strong because we have lived, I believe, through a scientific revolution in science studies.[2] A few decades ago the only way science was supposed to have worked was with theorists putting up ideas and experimentalists proving them or otherwise. If anyone said there was a social acceptance component to what counts as proving or disproving, that person was said to be crazy. I am going to make a deliberately provocative statement: as far as the possibility of bringing the social into the nature of science is concerned, the world has changed and the presence of a social component to experimental credibility is now treated as a matter of course. (This will be further illustrated, below, by the joint work of me and Allan.) There is, therefore, no further pressing need for Allan to treat me as crazy nor need for me to call him an idiot for not being able to see what is in front of his face. What is left is matters of emphasis and disagreement over certain episodes and certain methods—all fairly normal stuff.[3]

Franklin: For the past forty-plus years Collins has immersed himself within the experimental gravity wave community and has conducted numerous interviews with the participants. It is fair to say that he has developed interactional expertise, "the ability to master the language of a specialist domain in the absence of practical competence" (Collins and Evans 2007, p. 14; Giles 2006).[4] In his study Franklin uses primarily published sources; papers, published letters, and conference proceedings.[5]

6.2 The Underlying Positions

Franklin: Before we begin a detailed discussion of the episode it is worth pointing out some of our general agreements and disagreements. Both of us are in agreement that science provides us with knowledge of the world. Thus Collins states: "For all its fallibility, science is the best institution for generating knowledge about the natural world that we have" (Collins 1992, p. 165). More recently he has remarked that "… one cannot take away integrity in the search for evidence and honesty in declaring one's results and still have science; one cannot take away a willingness to listen to anyone's scientific theories and findings irrespective of race, creed, or social eccentricity and still have science; one cannot take away the readiness to expose one's findings to criticism and debate and still have science; one cannot take away the idea that the best theories will be able to specify the means by which they could be shown

[2] Elsewhere, Evans and I have described 'three waves' of science studies, the crucial transition from Wave 1 to Wave 2 taking place in the early 1970s (Collins and Evans 2002, 2007).

[3] I exclude the humanities types, still fighting their anti-science corner in the two-cultures debate, trying to justify a radical post-modernism.

[4] This is in contrast to *contributory expertise*, the ability to participate and contribute to the science.

[5] In some of his other studies Franklin was a participant.

to be wrong and still have science; one cannot take away the idea that a lone voice might be right while all the rest are wrong and still have science; *one cannot take away the idea that good experimentation or theorization usually demand high levels of craft skills and still have science*; and one cannot take away the idea that, in virtue of their experience, some are more capable than others at both producing scientific knowledge and at criticizing it and still have science. These features of science are 'essential,' not derivative" (Collins 2013, p. 156, emphasis added).

Collins also advocates methodological relativism, the position that the sociologist of scientific knowledge should behave as if "… the natural world has a small or nonexistent role in the construction of scientific knowledge" (Collins 1981b, p. 3). Collins does not believe that experimental results can resolve issues of controversy in science, or of confirmation or refutation.

Collins: It's a bit more complicated. From the early 1970 s until 1981 I think I might well have been doing something that fits Allan's description—I certainly thought I was doing something deep and philosophical born in the new freedoms of thought and action made possible by the '1960s.' But toward the end of that period it became clear to me that I could not 'prove' the kind of philosophical point I had in mind through empirical case-studies. I explained my new position—methodological relativism—in a paper published in 1981 (Collins 1981c). This is the position I have held since. It is that to do good sociology of scientific knowledge it is vital not to short circuit the analysis by explaining the emergence of what people count as the truth by the fact that it is the truth; if you do that you can stop the analysis whenever you like and that makes for bad work. Therefore, in the course of the analysis of what comes to count as the truth of the matter you have to assume there is no truth of the matter. It is just as if you were trying to explain why Catholics think the bread and the wine transubstantiate into the body and blood of Christ: you would not say 'they believe it because it does change,' or at least you would not say it if you were a sociologist as opposed to a priest. One applies the same principle to science: one does not say, 'scientists came to believe in relativity because it's true': that short circuits the whole social analysis project. Therefore, to do good sociological analysis you have to assume the world has no effect on scientists' beliefs about the world. That's methodological relativism.

While on the topic of changes over time, I started my gravitational wave project in 1972 and it is still ongoing. Later in this paper Allan is going to describe my approach to the study as emphasizing interviews rather than study of the published literature. There is some validity to this as regards my earliest work—I did not search the published literature as assiduously as I would have done if I had nothing else to go on—but my large book, *Gravity's Shadow*, which was published in 2004, uses every resource I could get my hands on including all the published materials I could read plus private correspondence painfully extracted from Joe Weber.

Franklin: Collins bases his attitude to experimental results on what he calls the "experimenter's regress." In discussing the question of the observation of gravity waves he asks, "*What is the correct outcome?*… What the correct outcome is depends on whether there are, or are not, gravity waves hitting the earth in detectable fluxes. To find this out we must build a good gravity wave detector and have a look. But we won't know if we have built a good detector until we have tried it and obtained the correct outcome. But we don't know what the correct outcome is until … and so on *ad infinitum*. This circle can be called the 'experimenter's regress'" (Collins 1985; Collins and Pinch 1998, p. 98). Collins states that when the appropriate range of outcomes is known at the outset this provides a universally agreed criterion of experimental quality and the regress can be broken. Where such a criterion does not exist other means must be found to break the regress, which must be independent of the experimental result itself.

In Collins' view the regress is eventually broken by negotiation within the appropriate scientific community, a process influenced by factors such as the career, social, and cognitive interests of the scientists, their reputations and that of their institutions, and the perceived utility for future work, but one that is not decided by what we might call epistemological criteria, or reasoned judgment. Thus, Collins concludes that his regress raises serious questions concerning both experimental evidence and its use in the evaluation of scientific hypotheses and theories.

Collins: I don't think I say that 'epistemological criteria' and reasoned judgment cannot decide controversies, I think I say that on their own they cannot decide them if the controversy is deep and the parties are determined.

Franklin: Franklin, on the other hand, advocates an essential role for experimental evidence in the production of scientific knowledge. "Science is a social construction because it is constructed by the scientific community. But, … it is constructed from experimental evidence, rational discussion and criticism, and the inventiveness of scientists" (Franklin 1990, p. 197). Franklin argues that one can decide what is a correct experimental result independent of that result by applying what he calls the "epistemology of experiment." This is a set of strategies that scientists legitimately use to argue for the correctness of their experimental results. These include:

1. Experimental checks and calibration, in which the experimental apparatus reproduces known phenomena;
2. Reproducing artifacts that are known in advance to be present;
3. Elimination of plausible sources of error and alternative explanations of the result (the Sherlock Holmes strategy)[6];
4. Using the results themselves to argue for their validity. In this case one argues that there is no plausible malfunction of the apparatus, or background effect, that would explain the observations;
5. Using an independently well-corroborated theory of the phenomena to explain the results;
6. Using an apparatus based on a well-corroborated theory;
7. Using statistical arguments;

[6] As Holmes remarked to Watson, "How often have I told you that when you have eliminated the impossible, whatever remains, however improbable, must be the truth" (Conan Doyle 1967).

8. Manipulation, in which the experimenter manipulates the object under observation and predicts what they would observe if the apparatus was working properly. Observing the predicted effect strengthens belief in both the proper operation of the experimental apparatus and in the correctness of the observation;
9. The strengthening of one's belief in an observation by independent confirmation;
10. Using "blind" analysis, a strategy for avoiding possible experimenter bias, by setting the selection criteria for "good" data independent of the final result. For details, see Franklin (2007, pp. 220–225, 2002, Chap. 6).

Franklin suggests that this set of strategies is also neither exclusive nor exhaustive. No single strategy, or group of strategies, is necessary to argue for the correctness of an experimental result. Nevertheless, the use of such strategies is, he believes, necessary to establish the credibility of a result. We shall discuss below how these very different views are applied to the episode of gravity wave detection.

Collins: I applaud this list of guidelines for strengthening the credibility of experiment. That my position is complex can be seen from my more scientific work. At the time of writing, I am just beginning the fourth year of a € 2.26M research project, based on the idea of interactional expertise, which involves doing a new kind of social survey by carrying out hundreds of imitation games (Turing Tests played with humans) on many different topics (mostly not to do with science) in many different countries. I recently led a new research application which included the following sentiment: "We have found that, with samples of 200, differences in pass rates of 10 % are statistically significant but we have chosen the more demanding criterion of replicability, as our 'gold standard'." A careful reading of Collins (e.g. 1985) shows that I have always been a defender of replication as a criterion of the soundness of experimental findings even as I was trying to show that it did not work as the standard account would indicate: "replicability is a perfectly appropriate criterion for distinguishing the true from the false; replicability is the scientifically institutionalized equivalent of the stability of perception" (Collins 1992, p. 130). The way it works is that establishing the replicability of a result is co-extensive with resolving the experimenter's regress and this means is it co-extensive with deciding what counts as a competent experiment in the experimental area under dispute. Someone who wants to prove something with repeated experiments has to (a) show that the results can be seen as continuing to come out the same way and (b) has to establish that the experiments were competently performed. In areas of deep dispute, showing the later will require more than experimental skill though establishing it will have no effect unless the 'epistemological criteria' are also met—replications have to be seen to 'work.' My oft discussed TEA-laser case shows what happens when there is no deep dispute; there is no need to establish the competence of the experiment because it is un-controversially read-off the outcome. The so-called 'epistemological criteria' are necessary for establishing the existence of a new phenomenon (as Allan says) but they are not a sufficient criterion where dispute runs deep. We should already half-know this from Duhem and Quine's pointing out that sub-hypotheses could always be used to explain away mismatches between theory and data but what I think I did was (a) to show that something similar happens when an experimental outcome is

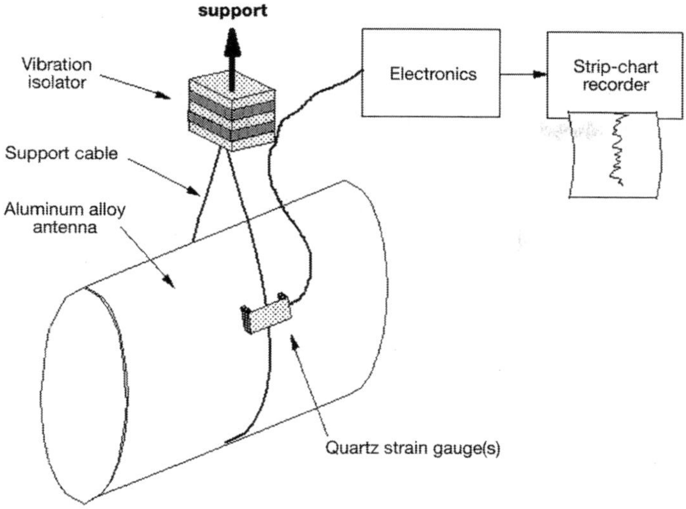

Fig. 6.1 A Weber-type gravity wave detector. From Levine (2004, p. 46)

compared to a conflicting one and (b) to show how this affects the unfolding of the day-to-day life of science in disputed areas. That the point is general can be seen by trying it on other episodes—e.g. the history of Michelson-Morley-type experiments. (The Duhem-Quine accounting of the experimenter's regress is a 'chronological lie,' of course—see the mention of Medawar, below—I certainly did not have Duhem-Quine in mind when I was 'discovering' the experimenter's regress—see Collins 2009 for a more 'true to life' account of how it happened.)

6.3 An Agreed upon History of the Early Gravity Wave Experiments

Franklin: Beginning in the late 1960s and extending into the 1970s Joseph Weber, using an experimental apparatus of his own design, which would become the standard apparatus for all of the early experiments, claimed to have observed gravity waves (Fig. 6.1). Weber used a massive aluminum alloy bar,[7] or antenna, which was supposed to oscillate when struck by gravitational radiation.[8] The oscillation was to

[7]This device is often referred to as a Weber bar.

[8]At this time, gravity waves were predicted by Einstein's General Theory of Relativity. Just as an accelerated electrically charged particle will produce electromagnetic radiation (light, radio waves, etc.), so should an accelerated mass produce gravitational radiation (gravity waves). Such radiation can be detected by the oscillations produced in a large mass when it is struck by gravity waves. Because the gravitational force is far weaker than the electromagnetic force, a large mass must be accelerated to produce a detectable gravity wave signal. (The ratio of the gravitational force between

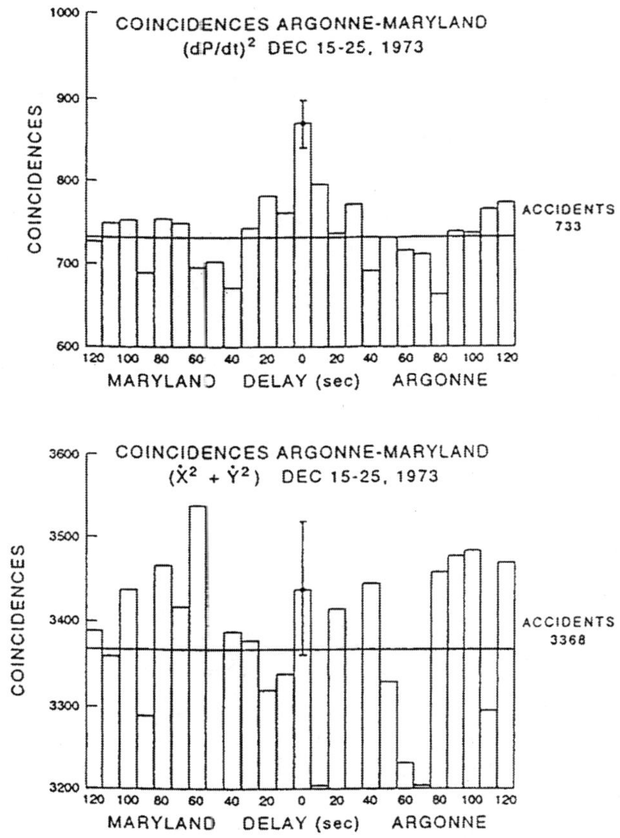

Fig. 6.2 Weber's time-delay data for the Maryland-Argonne collaboration for the period 15–25 December, 1973. The *top graph* was obtained using the nonlinear algorithm preferred by Weber, whereas the *bottom graph* used the linear algorithm. The zero-delay peak is seen only with the nonlinear algorithm. From Shaviv and Rosen (1975, p. 250)

be detected by observing the amplified signal from piezo electric crystals, or other strain gauges, attached to the antenna. The amplified signal was then sent to either a chart recorder or digitized and sent to a computer. The signals were expected to be quite small (the gravitational force is quite weak in comparison to electromagnetic force) and the bar had to be well insulated from other sources of noise such as electrical, magnetic, thermal, acoustic, and seismic forces. Because the bar was at

(Footnote 8 continued)
the electron and the proton in the hydrogen atom compared to the electrical force between them is 4.38×10^{-40}, a small number indeed.) The difficulty of detecting a weak signal is at the heart of this episode. There had been an earlier controversy about whether General Relativity did, in fact, predict gravitational radiation. For an excellent history of the theory of gravitational radiation, see Kennefick (2007); for a very interesting analysis of the detection of gravitational waves via the decay of a binary star system, for which a Nobel Prize was awarded, see Kennefick (2014).

a temperature different from absolute zero, thermal noise could not be avoided, so Weber set a threshold for pulse acceptance that was in excess of the size expected from most of the pulses caused by thermal noise.[9] In his early papers Weber made no discovery claim concerning gravity waves and merely suggested that the observed coincidences might be due to gravitational radiation. In 1969, after observing coincidences between two widely-separated detectors, Weber claimed to have detected approximately seven pulses/day due to gravitational radiation. A sample of Weber's data is shown in Fig. 6.2.

Because Weber's reported rate was far greater than that expected from the most plausible calculations of cosmic events (by many orders of magnitude), his early claims were met with skepticism. During the late 1960s and early 1970s, however, Weber introduced several modifications and improvements that increased the credibility of his results. He claimed that above threshold peaks had been observed simultaneously in two detectors separated by one thousand miles. It was extremely unlikely that such coincidences were due to random thermal fluctuations. In addition, he reported a 24 h periodicity in his peaks, a sidereal correlation that indicated a single source for the radiation, located near the center of our galaxy. Weber also added a delay to the signal from one of the antennas and found that the excess coincidences disappeared, as they should if the signals were real. These results increased the plausibility of his claims sufficiently so that by 1975 six other experimental groups had constructed apparatuses and begun attempted replications of Weber's experiment.[10] All of these attempted replications found no evidence for gravity waves.[11]

---

[9]Given any such threshold there is a finite probability that a noise pulse will be larger than that threshold. The point is to show that there are pulses in excess of those expected statistically.

[10]In a later commentary on these early experiments, James Levine, who collaborated with Richard Garwin on one of these experiments, stated that it was this sidereal effect that was most important in persuading him, and others, to attempt the replications (Levine 2004). Levine's commentary was not available when Collins and Franklin wrote their initial accounts.

[11]Collins, in some early work, offered two arguments concerning the difficulty, if not the virtual impossibility of replication.

Collins: I don't understand this remark. I am engaged in analyzing the process and meaning of replication, not saying it is impossible. Some of the problems of what it means to replicate are discussed in Collins (1992, ch 2).

The first is philosophical. What does it mean to replicate an experiment? In what way is the replication similar to the original experiment? Franklin suggests that a rough and ready answer is that the replication measures the same physical quantity. Whether or not it, in fact, does so can, he believes, be argued for on reasonable grounds, as discussed earlier. Collins' second argument is pragmatic. This is the fact that in practice it is often difficult to get an experimental apparatus, even one known to be similar to another, to work properly. Collins illustrates this with his account of Harrison's attempts to construct two versions of a TEA leaser (Transverse Excited Atmospheric) (Collins 1985, pp. 51–78). Despite the fact that Harrison had previous experience with such lasers, and had excellent contacts with experts in the field, he had great difficulty in building the lasers. Hence, the difficulty of replication. Ultimately Harrison made the laser work after a series of adjustments. As Collins explains, "…in the case of the TEA laser the circle was readily broken. The ability of the laser to vaporize concrete, or whatever, comprised a universally agreed criterion of experimental quality. There was never any doubt that the laser ought to be able to work and never any doubt about when one was working and when it was not" (Collins 1985, p. 84).

6.4 The Accounts Diverge

Franklin: By 1975 it was generally agreed that the flux of gravity waves claimed by Weber did not exist and that Weber's experiment, including his analysis procedures, was inadequate. Certainly the six failed replications played a major role in this. At this point Collins invokes the "experimenter's regress" and argues that theses attempted replications were not as persuasive as they might seem.

Collins: Two things are going on here. The first is the extraordinary care we must take to understand how persuasive or otherwise the counter-experiments seemed in the early 1970s, when the dispute was still live; it is quite different when one looks back from a deeply entrenched consensual position. (It may be relevant that I started fieldwork on gravitational waves in 1972, when the controversy was still live, whereas Allan's analysis began after it was over). In *Gravity's Shadow* (Collins 2004, Chap. 5) I use the metaphor of a steep conical island with Joe Weber trying to maintain his grip on the dry land of belief in his results while the waters of skepticism rise around him. At the end of the book, looking back from the vantage point of what we know now, I write:

"When I now read, as I have just read, the correspondence between Joe Weber, Dick Garwin, and others, such as Dave Douglass, I read it knowing how things turned out. I read this correspondence as through a template that allows me to focus on where Weber went wrong and hides all those places where he went right. The pattern of the template is Weber desperately struggling to hide his mistakes and shift his position; he wriggles and struggles to maintain his foothold on the island. Knowing that he's going to drown, I see his feet and hands grasping and slipping where once I saw them clinging and climbing. The difference between grasping and slipping and clinging and climbing is almost nothing—it is just what you are primed to see" (Collins 2004, p. 210).

Those 6 counter-experiments are pretty convincing to us but they were not as convincing before 1975 because of the experimenter's regress.

The second thing that is going on is a difference between my overall approach and Allan's. I am always asking, would someone determined to believe in the reality of Joe Weber's claims *be forced* to reject them by 'this' or 'that'? Among the 'this's' and 'that's' are the counter-experiments. I argue that if you were determined in that way, the experiments would not prove to be decisive though, of course, they would still be important evidence. Joe Weber would have been much, much happier if others' experiments had supported his own but the other experimental results did not, and could not, force him or his allies—of which there were a few (see Collins 2004)—to give up.

Franklin: The decision to reject Weber's conclusion rested on what was a good gravity wave detector and who was a competent experimenter. Collins supports his view with quotations taken from interviews with experimenters critical of Weber's work. Several experimenters commented about problems with some of the other experimental apparatuses and their reported results. Comments about Experiment

W[12] include, "*Scientist a*: ... that's why the W thing, though it's very complicated, has certain attributes so that if they see something, it's a little more believable ... They've really put some thought into it" (Collins 1992, p. 84). Scientist b on the other hand stated that, "They hope to get very high sensitivity but I don't believe them frankly. There are more subtle ways round it than brute force" (p. 84). Scientist c is more critical, "I think that the group at ... W ... are just out of their minds" (p. 84). Scientists also commented on the possible significance of differences in the detectors. "iii...it's very difficult to make a carbon copy. You can make a near one, but if it turns out that what's critical in the way he glued his transducers, and he forgets to tell you that the technician always puts a copy of Physical Review on top of them for weight, well, it could make all the difference" (p. 86). Weber also felt that the differences between detectors was crucial and that the other detectors were less effective than his. "Well, I think it is very unfortunate because I did these experiments and I published all relevant information the technology,[13] and it seemed to me that one other person should repeat my experiments with my technology, and then having done it as well as I could do it they should do it better ... It is an international disgrace that the experiment hasn't been repeated by anyone with that sensitivity" (p. 86).[14]

As noted earlier, James Levine had remarked that the sidereal effect was the most important piece of evidence that convinced him to attempt a replication of Weber's experiment. Collins quotes an anonymous scientist who agreed with Levine and stated that, "The sidereal correlation to me is the only thing of that whole bunch of stuff that makes me stand up and worry about it. ...If that sidereal correlation disappears then you can take that whole ... experiment and stuff it some place" (p. 87). Collins remarks that Weber's use of a computer had added to the credibility of his results for some, but not all of the scientists. "You know he's claimed to have people write computer programs for him 'hands off.' I don't know what that means ... One thing that me and a lot of people are unhappy about, is the way he's analysed the data, and the fact that he's done it in a computer program doesn't make that much difference" (p. 87).[15]

Collins also cites a "list of 'non-scientific reasons that scientists offered for their belief or disbelief in the result of Weber's and others' work reveals the lack of an 'objective criterion of excellence. This list comprised: (1) Faith in a scientist's experimental capabilities and honesty, based on previous working partnership; (2) Personality and intelligence of experimenters ; (3) Reputation of running a huge lab; (4) Whether the scientist worked in industry or academia; (5) Previous history of failures[16]; (6) 'Inside information'; (7) Style and presentation of results; (8) Psycho-

[12]In this publication Collins maintains the anonymity of both the institutions and the experimenters. In later work, as we shall see he identifies one of the experimenters. Scientist Q is Richard Garwin.

[13]This point will be important in one of the criticisms made of Weber's results and one that is cited by Franklin.

[14]Weber's critics would disagree with that comment.

[15]Franklin's discussion of some of the problems with Weber's data analysis is given below.

[16]As we have seen Weber's experiments on gravity waves were regarded as a failure. Weber later made a very speculative hypothesis concerning coherent neutrino scattering. For a discussion of how this hypothesis was treated, see Franklin (2010).

logical approach to experiment; (9) Size and prestige of university of origin; (10) Integration into various scientific networks; (11) Nationality" (Collins 1992, p. 87).

Thus, Collins argues, on the basis of these interviews, that the six negative results obtained by Weber's critics did not have sufficient weight to destroy the credibility of Weber's results. He suggests, however, that they did raise questions about those results. He also argues that opinion crystallized against Weber because of the results and the presentation of those results by Richard Garwin. Collins states that prior to Garwin's work, critics were more tentative in their rejection of Weber's results and had been willing to explore other possible explanations of those results. After Garwin's publications their comments were more negative. Garwin was quite clear in his publication that he believed Weber's results were wrong. He stated that his results were in substantial disagreement with those reported by Weber. Other critics expressed reservations about Garwin's work. "…as far as the scientific community in general is concerned, it's probably [Garwin's] publication that generally clinched the attitude. But in fact the experiment they did was trivial—it was a tiny thing … But the thing was, the way they wrote it up … Everybody else was awfully tentative about it … It was all a bit hesitant … And then [Garwin] comes along with this toy. But it's the way he writes it up you see" (Collins 1992, p. 92). Another critic stated, "[Garwin's paper] was very clever because its analysis was actually very convincing to other people and that was the first time that anybody had worked out in a simple way just what the thermal noise from the bar should be … It was done in a very clear manner and they sort of convinced everybody" (Collins 1992, p. 92). Collins concludes that "The growing weight of negative reports, all of which were indecisive in themselves, were crystallized, as it were, by [Garwin]. Henceforward, only experiments yielding negative results were included in the envelope of serious contributions to the debate" (Collins 1992, p. 92).

Franklin questions the role of Garwin as the crystallizer of the opposition to Weber. As discussed below, other scientists, at the time, presented similar arguments against Weber's results. At the GR7 Conference (Shaviv and Rosen 1975), Garwin's experiment was mentioned only briefly, and although the arguments about Weber's errors and analysis were made, they were not attributed to the absent Garwin.[17]

Franklin takes the six negative results more seriously than does Collins. He believes that sufficient arguments were given for the credibility of these results. He argues that these six results, combined with several problems found with Weber's experiment and with its analysis procedures demonstrated that the scientific community was both reasonable and justified in their rejection of Weber's results.

Collins: Here Allan introduces terminology that seems to me entirely superfluous—'reasonable and justified.' I have done hundreds and hundreds of interviews with scientists of holding fiercely competing views and, the more dishonest or stupid critics of parapsychological research aside, I have never come across anyone whose arguments were not reasonable and justified. The concepts of 'rational,' 'reasonable'

[17]The panel discussion on gravitational waves covers 56 pages, 243–298, in Shaviv and Rosen (1975). Tyson's discussion of Garwin's experiment occupies one short paragraph (approximately one quarter of a page) on p. 290.

and 'justified' are idle wheels in the history of science (see also Collins 1981c). I think, by the way, that this is one of the few places where there is a deep disagreement between us—I think using terms like 'reasonable' and 'rational' really is a waste of time because it is almost impossible to find anything that you can be sure is not reasonable and rational. What is worth noticing, once more, is that the deep disagreement between us that appeared to be beyond question when we were insulting each other in print a few decades ago has dissolved. Nowadays it seems obvious to everyone that acceptance of an idea is, at least in part, a process of social acceptance so that all that is left to argue about is the *relative contribution* of the social and the 'epistemic' whereas once it was a matter of whether there was any social at all. And the 'relative contribution' argument is not something that is going to cause people to insult each other in print. So, just as we need to be very careful about analyzing the way Joe Weber lost his credibility by reading backwards from where we are now, we have to be careful about analyzing the disagreement between Allan and me by reading backwards from where we are now.

Franklin: One important difficulty with Weber's experiment was his failure to successfully calibrate his experimental apparatus. Calibration is the use of a surrogate signal to standardize an instrument. If an apparatus reproduces known phenomena, then we legitimately strengthen our belief that the apparatus is working properly and that the experimental results produced with that apparatus are credible and reliable. If calibration fails, then we do not trust the experimental results produced with that apparatus. Thus, if your optical spectrometer reproduces the known Balmer series in hydrogen, you have reason to believe that it is a reliable instrument. If it fails to do so, then it is not an adequate spectrometer. Collins states that calibration cannot provide grounds for belief that an experimental apparatus is working properly. "The use of calibration depends on the assumption of near identity of effect between the surrogate signal and the unknown signal that is to be measured (detected) with the instrument" (Collins 1992, p. 105). Franklin (1997) argues that in most cases calibration is unproblematic because the adequacy of the surrogate signal is clear. In the case of gravity waves, however, both Collins and Franklin agree that there is no standard laboratory source of gravity waves that one can use to calibrate a gravity wave antenna, and that calibration is more problematic in this case. In this episode, Weber's critics injected pulses of acoustic energy into their antennas and found that they could observe them (Fig. 6.3). Weber was unable to detect such signals with his experiment and admitted that the six other experimental groups could not only detect such pulses, but did so with an efficiency twenty times greater than that of his own apparatus. Under ordinary circumstances Weber's calibration failure would be sufficient grounds for rejecting his results. The detection of gravity waves is not, however, an ordinary case. In this episode scientists were searching for a hitherto unobserved phenomenon with a new type of apparatus. Thus, Weber could argue that the waveforms of the potential gravitational waves accounted for the difference in calibration performance. He could have argued that it something more mysterious. Calibration was important, but not decisive.

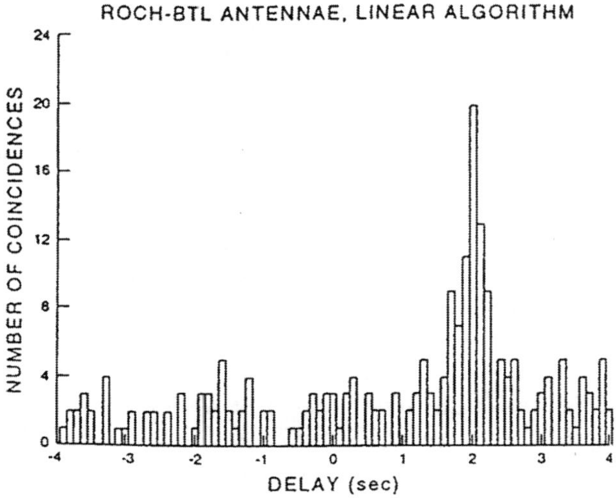

Fig. 6.3 A plot showing the calibration pulses for the Rochester-Bell Laboratory collaboration. The peak due to the calibration pulses is clearly seen. This was also a data run so the peak is displaced 2 s so as not obscure any possible signal. From Shaviv and Rosen (1975, p. 285)

One crucial difference between the analysis procedures used by Weber and by his critics concerned the algorithm used to analyze the signals emerging from the gravity wave antenna. A gravity wave antenna operating at a finite temperature is always producing thermal noise. If a gravity wave strikes the antenna the two signals are added together, producing a change in both the amplitude and the phase of the output signal. Weber used a non-linear algorithm that was sensitive only to the amplitude of the signal, whereas his critics used a linear algorithm that was sensitive to both the amplitude and the phase. In this episode there was remarkable cooperation between Weber and his critics. They exchanged both analysis programs and data tapes. The critics used Weber's preferred non-linear algorithm on the calibration data and could find no calibration signal (Fig. 6.4). (In this case the calibration signal was inserted during a data run, but with a 2-s offset so as not to obscure any real signal, which would appear at zero time delay. Neither a signal nor the calibration pulses are seen.)

Weber responded, correctly, that the calibration pulses used by his critics were short pulses, of the type they expected for gravity waves and for which the linear algorithm was better. He stated that real gravity wave pulses were longer, for which the non-linear algorithm was better. Weber's critics responded by analyzing their data with both algorithms. They found no gravity wave signal with either algorithm (Figs. 6.4 and 6.5 show the data for the non-linear and linear algorithms, respectively). If Weber was correct a signal should have appeared when the critics' data was analyzed with the non-linear algorithm. It didn't. The critics' results were robust against changes in the analysis procedures. In addition, Ronald Drever reported that he had looked at the sensitivity of his apparatus with arbitrary waveforms and pulse lengths and found no effect with either algorithm (Shaviv and Rosen 1975, pp. 265–

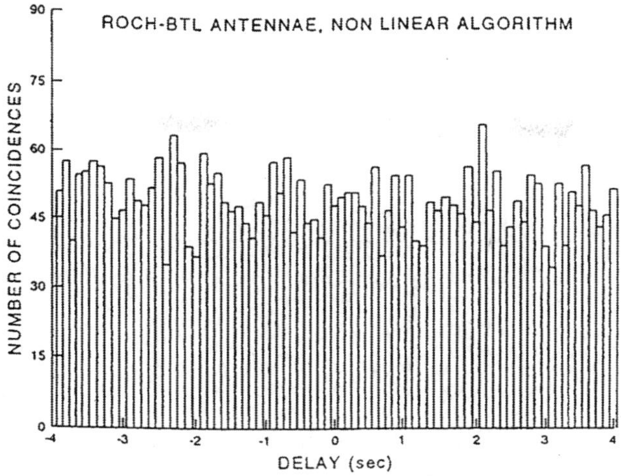

Fig. 6.4 A time-delay plot for the Rochester-Bell Laboratory collaboration, using the nonlinear algorithm. No sign of a zero-delay peak is seen. From Shaviv and Rosen (1975, p. 284)

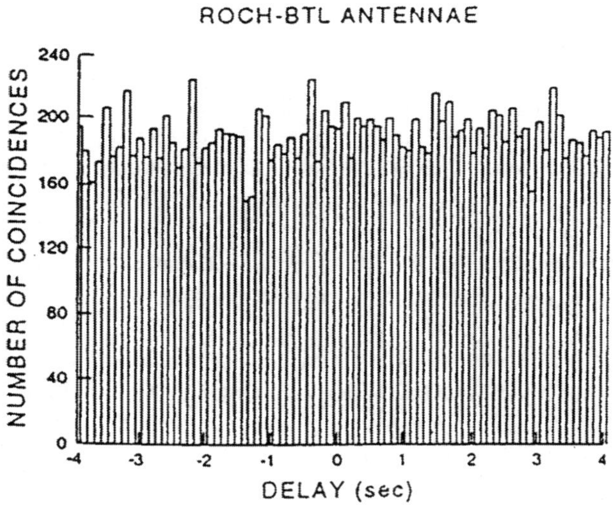

Fig. 6.5 A time-delay plot for the Rochester-Bell Laboratory collaboration, using the linear algorithm. No sign of a zero-delay peak is seen. From Shaviv and Rosen (1975, p. 285)

276). Nevertheless Weber preferred the non-linear algorithm. His reason for this was that the non-linear algorithm provided a more significant signal than did the linear algorithm. This is shown in Fig. 6.6. Weber remarked, "Clearly these results are inconsistent with the generally accepted idea that [the linear algorithm] should be the better algorithm" (Shaviv and Rosen 1975, pp. 251–252).

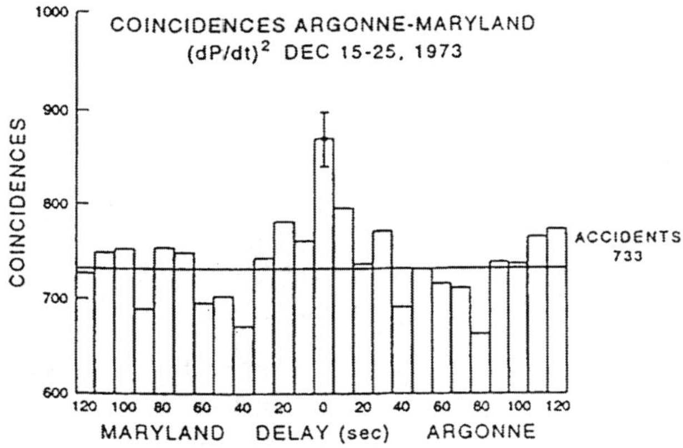

Fig. 6.6 Weber's time-delay data for the Maryland-Argonne collaboration for the period 15–25 December 1973. The data were analyzed with the nonlinear algorithm. A peak at zero time delay is clearly seen. From Shaviv and Rosen (1975, p. 250)

How then did Weber obtain his positive result when his critics, using his own analysis procedures, could not? It was suggested that Weber had varied his threshold cuts, to maximize his signal, whereas his critics used a constant threshold. Tony Tyson, one of Weber's critics remarked, "I should point out that there is a very important difference in essence in the way in which many of us approach this subject and the way Weber approaches it. We have taken the attitude that, since these are integrating calorimeter type experiments which are not too sensitive to the nature of pulses put in, we simply maximize the sensitivity and use the algorithms which we found maximized the signal to noise ratio, as I showed you. Whereas Weber's approach is, he says, as follows. He really does not know what is happening, and *therefore he or his programmer is twisting all the adjustments in the experiment more or less continuously, at every instant in time locally maximizing the excess at zero time delay.* I want to point out that there is a potentially serious possibility for error in this approach. No longer can you just speak about Poisson statistics. *You are biasing yourself to zero time delay, by continuously modifying the experiment on as short a time scale as possible (about 4 days), to maximize the number of events detected at zero time delay.* We are taking the opposite approach, which is to calibrate the antennas with all possible known sources of excitation, see what the result is, and maximize our probability of detection. Then we go through all of the data with that one algorithm and integrate all of them. Weber made the following comment before and I quote out of context: 'Results pile up.' I agree with Joe (Weber). But I think you have to analyze all of the data with one well understood algorithm (Shaviv and Rosen 1975, p. 293, emphasis added).

Richard Garwin agreed, and pointed out that he and James Levine had used a computer simulation to demonstrate that varying the threshold could produce a positive result. This "delay histogram" was obtained by partitioning the computer generated data into 40 segments. For each segment, "single events" were defined in each "chan-

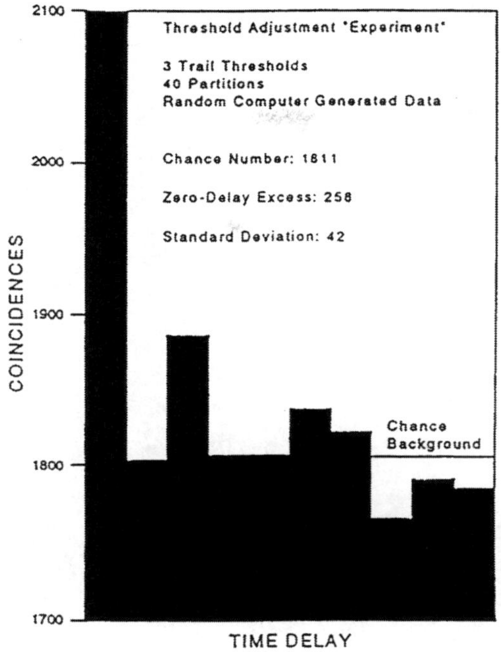

Fig. 6.7 The results of selecting thresholds that maximized the zero-delay signal for Levine's computer simulation. From Garwin (1974, p. 10)

nel" by assuming one of three thresholds a, b, or c. That combination of thresholds was chosen for each segment which gave the maximum "zero delay coincidence" rate for that segment. The result was 40 segments selected from one of nine "experiments." The 40 segments are summarized in Fig. 6.7, which shows a "six standard deviation" zero delay excess (Garwin 1974, pp.9–10).

Weber also cited evidence provided by Kafka as supporting a positive gravity wave result. Kafka did not agree. This was because the evidence resulted from performing an analysis using different data segments and different thresholds on real data. Only one graph showed a positive result, indicating, in fact, that such selectivity could produce a positive result. Kafka's results are shown in Fig. 6.8. Note that the positive effect is seen in only the bottom graph. "The very last picture (Fig. 6.8) is the one in which Joe Weber thinks we have discovered something, too. This is for 16 days out of 150. There is a 3.6 σ [standard deviation] peak at zero time delay, but you must not be too impressed by that. It is one out of 13 pieces for which the evaluation was done, and I looked at least at 7 pairs of thresholds. Taking into account selection we can estimate the probability to find such a peak accidentally to be of the order of 1 %" (Shaviv and Rosen 1975, p. 265).

Weber denied the charges. "The computer varies the thresholds to get a computer printout which is for 31 different thresholds. The data shown are not the results of looking over a lot of possibilities and selecting the most attractive ones. We obtain a result that is more than three standard deviations for an extended period for a wide range of thresholds. I think it is very important to take the point of view that the

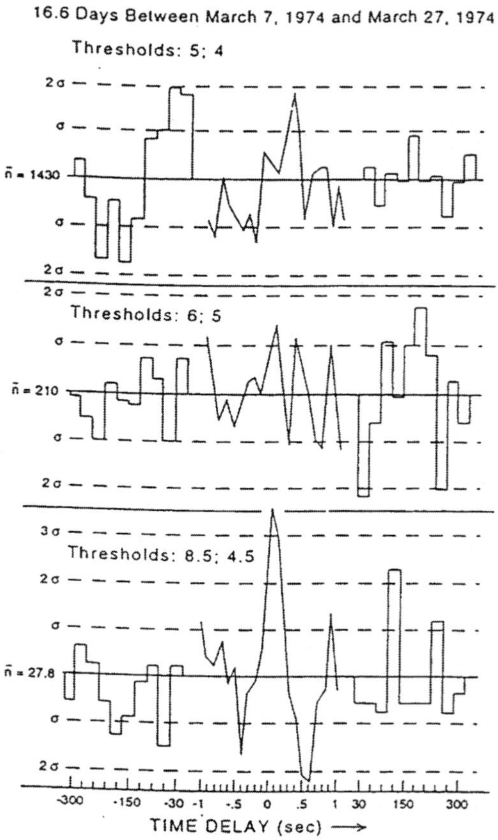

Fig. 6.8 Kafka's results from real data using various thresholds. A clear peak is seen at zero time delay in the bottom graph. From Shaviv and Rosen (1975, p. 266)

histogram itself is the final judge of what the sensitivity is" (Shaviv and Rosen 1975, pp. 293–294). Weber did not, however, specify his method of data selection for his histogram. In particular, he did not state that all of the results presented in a particular histogram had the same threshold.[18]

As noted earlier, there was considerable cooperation among the various groups. They exchanged both data tapes and analysis programs. "There has been a great deal of intercommunication here. Much of the data has been analyzed by other people. Several of us have analyzed each other's data using either our own algorithm or each other's algorithms" (Tyson in Shaviv and Rosen 1975, p. 293). This led to the first of several questions about possible serious errors in Weber's analysis of his data.

[18]There is some anecdotal evidence that supports the view that Weber tuned his analysis procedures to maximize the signal. Collins suggests that Weber might have been influenced by Weber's experience on a submarine chaser during World War II. In those circumstances a false positive results only in a few wasted depth charges, whereas missing a positive signal would have had fatal consequences. Collins quotes an unnamed physicist who stated, "Joe would come into the laboratory—he'd twist all the knobs until he finally got a signal. And then he'd take data. And then he would

David Douglass first pointed out that there was an error in one of Weber's computer programs.

> The nature of the error was such that any above threshold event in antenna A that occurred in the last or the first 0.1 s time bin of a 1000 bin record is erroneously taken by the computer program as in coincidence with the next above threshold event in channel B, and is ascribed to the time of the later event. Douglass showed that in a four day tape available to him and included in the data of (Weber et al. 1973), nearly all of the so called 'real' coincidences of 1–5 June (within the 22 April to 5 June 1973 data) were created individually by this simple programming error. Thus not only some phenomenon besides gravity waves *could*, but in fact *did* cause the zero delay excess coincidence rate (Garwin 1974, p. 9).

Weber admitted the error, but did not agree with the conclusion.

> This histogram is for the very controversial tape 217. A copy of this tape was sent to Professor David Douglass at the University of Rochester. Douglass discovered a program error and incorrect values in the unpublished list of coincidences. Without further processing of the tape, he (Douglass) reached the incorrect conclusion that the zero delay excess was one per day. This incorrect information was widely disseminated by him and Dr. R. L. Garwin of the IBM Thomas J. Watson Research Laboratory. After all corrections are applied, the zero delay excess is 8 per day. Subsequently, Douglass reported a zero delay excess of 6 per day for that tape (Weber in Shaviv and Rosen 1975, p. 247).

Although Weber reported that his corrected result had been confirmed by scientists at other laboratories and that copies of the documents had been sent to editors and workers in the field, Franklin found no corroboration of any of Weber's claims in the published literature. At the very least, this error raised doubts about the correctness of Weber's results. There was also a rather odd result reported by Weber.

> First, Weber has revealed at international meetings (Warsaw, 1973, etc.) that he had detected a 2.6 standard deviation excess in coincidence rate between a Maryland antenna [Weber's apparatus] and the antenna of David Douglass at the University of Rochester. Coincidence excess was located not at zero time delay but at "1.2 s," corresponding to a 1 s intentional offset in the Rochester clock and a 150 ms clock error. At CCR 5, Douglass revealed, and Weber agreed, that the Maryland Group had mistakenly assumed that the two antennas used the same time reference, whereas one was on Eastern Daylight Time and the other on Greenwich Mean Time. Therefore, the "significant" 2.6 standard deviation excess referred to gravity waves that took 4 h, zero minutes and 1.2 s to travel between Maryland and Rochester (Garwin 1974, p. 9).

(Footnote 18 continued)
analyze the data: he would define what he would call a threshold. And he'd try different values for the thresholds. He would have algorithms for a signal—maybe you square the amplitude, maybe you multiply things … he would have twelve different ways of creating something. And then thresholding it twenty different ways. And then go over the same data set. And in the end, out of these thousands of combinations there would be a peak that would appear and he would say, "Aha—we've found something." And [someone] knowing statistics from nuclear physics would say, "Joe—this is not a Gaussian process—this is not normal—when you say there's a three-standard-deviation effect, that's not right, because you've gone through the data so many times." And Joe would say, "But—What do you mean? When I was working, trying to find a radar signal in the Second World War, anything was legal, we could try any trick so long as we could grab a signal" (Collins 2004, pp. 394–395). This is not an eyewitness account but a widely held view, here expressed by one of Weber's critics.

Weber answered that he had never claimed that the 2.6 standard deviation effect he had reported was a positive result. By producing a positive result where none was expected, Weber had, however, certainly cast doubt on his analysis procedures. As Collins remarked Weber was, "effectively conjuring a signal out of what should have been pure noise" (Collins 1992, p. 90).

Levine and Garwin (1974) and Garwin (1974) raised yet another doubt about Weber's results. This was the question of whether or not Weber's apparatus could have produced his claimed positive results. Here again, the evidence came from a computer simulation.

> Figure 6.9b shows the 'real coincidences' confined to a single 0.1 s bin in the time delay histogram. James L. Levine and I observed that the Maryland Group used a 1.6 Hz bandwidth "two stage Butterworth filter." We suspected that mechanical excitations of the antenna (whether caused by gravity waves or not) as a consequence of the 1.6 Hz bandwidth would not produce coincident events limited to a single 0.1 s time bin. Levine has simulated the Maryland apparatus and computer algorithms to the best of the information available in (Weber et al. 1973) and has shown that the time delay histogram for coincident pulses giving each antenna 0.3 kT is by no means confined to a single bin, but has the shape shown in Fig. 6.9a (Garwin 1974, p. 9).

One further problem for Weber's results was the disappearance of the sidereal effect. (Recall that several critics had regarded that effect as the major reason for their belief in the plausibility of Weber's results.) In addition, critics pointed out that the sidereal effect should have had a 12-h, not a 24-h, period. The earth does not shield the detector from gravity waves.

Thus, we have two very different accounts of the same episode, the early experiments on gravity waves.

Collins: Though they don't seem *very* different to me—not nowadays.[19]

Franklin: In Collins's view all of the negative evidence provided by Weber's critics was insufficient for the rejection of Weber's claim to have observed gravity waves. "Under these circumstances it is not obvious how the credibility of the high flux case [Weber's result] fell so low. In fact, it was not the single uncriticized experiment[20] that was decisive: scientists rarely mentioned this in discussion. Obviously the sheer weight of negative opinion was a factor, but given the tractability, as it were, of all the negative evidence, it did not have to add up so decisively. There was a way of assembling the evidence, noting the flaws in each grain, such that outright rejection of the high flux claim was not the necessary inference" (Collins 1992, p. 91). Collins attributes the resolution of the controversy to the criticism of Weber offered by Richard Garwin. "Thus, without the actions of Garwin and his group it is hard to see how the gravity wave controversy would have been brought to a close. That such a contribution was needed is, once more, a consequence of the experimenter's regress" (Collins and Pinch 1998, p. 106).

[19]Franklin shares this view.

[20]Collins pointed out that only one of the six attempted replications of Weber's experiment was not criticized by other practitioners.

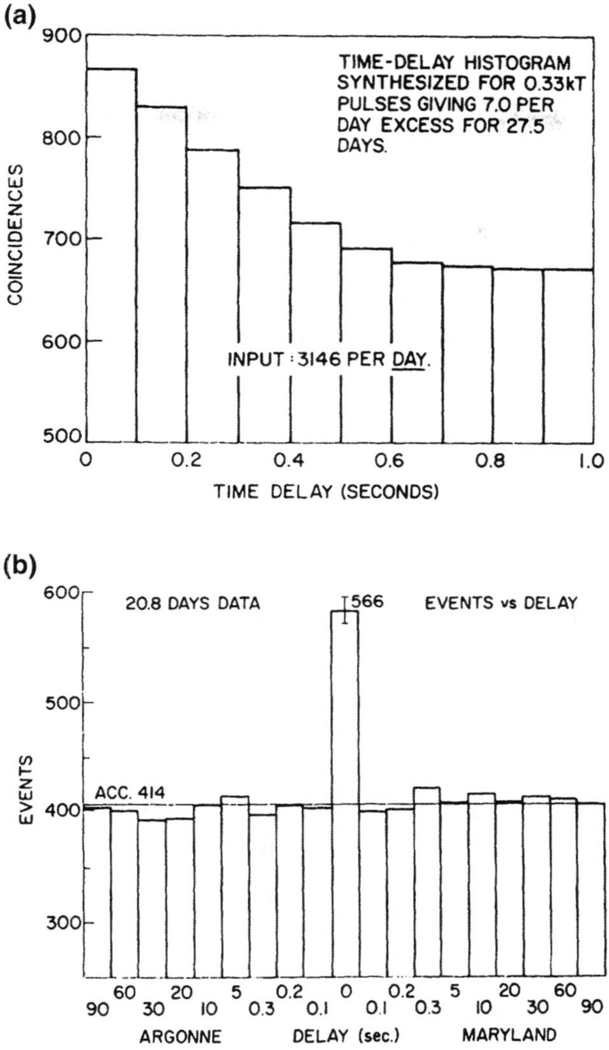

Fig. 6.9 **a** Computer-simulation result obtained by Levine for signals passing through Weber's electronics. **b** Weber's reported result. The difference is clear. From Levine and Garwin (1974, p. 796)

Franklin, on the other hand believes that the credible negative results of the critics' experiments, bolstered by their use of the epistemology of experiment, combined with the problems of Weber's experiment and its analysis procedures, made the rejection of Weber's results reasonable, if not rational. In his view the decision was based on valid experimental evidence and on reasoned and critical discussion.

Collins: This is something we do disagree about though, again, it does not seem to me to something that would cause one to insult one another in print unless Allan still wants to say that the notions of 'valid experimental evidence' and 'reasoned and critical discussion' have no component of social credibility—something which I simply cannot get my head round since what seems reasonable and what seems a reasonable experiment is always a matter of social context—where social includes the agreements found within scientific communities; this is the world we live in on this side of the scientific revolution in science studies. Indeed, Allan and I have done joint work on precisely the nature of these social agreements as they have changed over the history of high-energy physics—social agreements over what level of statistical significance is required to justify a discovery claim (it changed from 3 sigma in the 1960s to 5 sigma now and I can report that at least one statistical analyst has opined that it should be 7 sigma for gravitational waves).

6.5 Convergence on a New Agreement

Franklin and Collins: Franklin agrees that the change from 3 sigma to 5 sigma as a criterion for discovery in high-energy physics was a social change—it was a change in the way the community of physicists agreed to act. But Franklin suggests that this change was brought about primarily through reasoning and empirical observation—it was observed that too many mistakes were being made under the 3 sigma criterion. On the other hand, Franklin concedes that, given that most sciences still use only 2 sigma as the criterion for discovery, reason and observation on their own are soon exhausted as an explanatory resource. So Franklin is ready to accept that social change within the scientific community is part of the explanation of scientific change and scientific difference is, in part, social difference. On the other hand, Franklin insists that in the physics he studies there is no visible influence of outside social forces affecting the science except during pathological periods such as the Nazi regime. He is ready to concede that this separation between science and society is less clear in the case of sciences such as economics.

Collins agrees that in the physics he studies there is no obvious influence from the wider society on the conclusions that physicists reach though there are clear influences on the way the science is conducted. Collins is more interested in stressing the, in-principle, possibility of outside influence—a possibility opened up by the interpretative flexibility of scientific results—the experimenter's regress etc. This is because Collins wants to use physics as a 'hard case' for understanding science as a whole and as the basis for understanding sciences that are less isolated from outside forces. But Collins also believes it is a defining criterion of science that scientists must strive to eliminate outside influences even if they cannot remove all of them—notably those they do not understand. In respect of the gravitational wave physics field, as far as he can see, the scientists have been successful in eliminating nearly all outside pressure on the substance of their findings.

Collins: So we can now assume that the nasty battle is over. We can now disagree, in a nicely un-insulting way, about (a) the contribution of Garwin and his team to the crystallization of the loss of Weber's credibility and (b) whether that contribution is better shown by restricting the use of evidence to the published record or by using all the evidence one can get one's hands on.

The first thing to note is that Garwin and his team thought they were trying to bring about what the sociologist would call 'closure.' I quote from pages 164–5 of *Gravity's Shadow*: "… At that point it was not doing physics any longer. It's not clear that it was ever physics, but it certainly wasn't by then. If we were looking for gravity waves, we would have adopted an entirely different approach [e.g., an experiment of sufficient sensitivity to find the theoretically predicted radiation] … there's just no point in building a detector of the [type] … that Weber has. You're just not going to detect anything [with such a detector—you know that both on theoretical grounds and from knowing how Weber handles his data], and so there is no point in building one, other than the fact that there's someone out there publishing results in *Physical Review Letters*. … *It was pretty clear that [another named group] were never going to come out with a firm conclusion* … so we just went ahead and did it … we knew perfectly well what was going on, and it was just a question of getting a firm enough result so that we could publish in a reputable journal, and try to end it that way" (emphasis added for the purpose of this exchange).

The last phrase in the above quotation is particularly significant. Garwin's group had circulated to other scientists and to Weber himself a paper by Irving Langmuir; it was quoted to me also. This paper deals with several cases of "pathological science"—"the science of things that aren't so." Garwin believed that Weber's work was typical of this genre; he tried to persuade Weber and others of the similarities. Most of the cases cited by Langmuir took many years to settle. Consequently, as a member of the Garwin group put it, "We just wanted to see if it was possible to stop it immediately without having it drag on for twenty years."

Garwin and Levine were worried because they knew that Weber's work was incorrect, but they could see that this was not widely understood. Indeed, the facts were quite the opposite. To quote a member of the group, "Furthermore, Weber was pushing very hard. He was giving an endless number of talks … We had some graduate students—I forget which university they were from—came around to look at the apparatus … They were of the very firm opinion that gravity waves had been detected and were an established fact of life, and we just felt something had to be done to stop it … It was just getting out of hand. If we had written an ordinary paper that just said we had a look and we didn't find, it would have just sunk without trace."

The second thing to note is that, as Allan correctly points out, there is no evidence in the published record that this is what Garwin and his team had in mind when they built and reported their experiment; this is not the kind of thing one puts in published papers or even in conference proceedings. Nor is it entered into the public record by those who were on the receiving end—and here we can repeat the quotations that Allan sets out above: "… as far as the scientific community in general is concerned, it's probably [Garwin's] publication that generally clinched the attitude. But in fact the experiment they did was trivial—it was a tiny thing … But the thing was, the way they wrote it up … Everybody else was awfully tentative about it … It was all a bit hesitant … And then [Garwin] comes along with this toy. But it's the way he writes it up you see" (Collins 1992, p. 92). "[Garwin's paper] was very clever because its analysis was actually very convincing to other people and that was the first time that anybody had worked out in a simple way just what the thermal noise from the bar should be … It was done in a very clear manner and they sort of convinced everybody" (Collins 1992, p. 92). In published works people don't mention that they were persuaded by a 'tiny thing' that was written up in a clever and convincing way.

Now, that Garwin had the decisive effect is not something I can prove any better than any other historical thesis, I can just make it more plausible rather than less by showing what Garwin and his team, and what many of those who were convinced, thought about it. But I want to insist that we need some kind of explanation of the demise of Weber's credibility along the lines of the Garwin intervention; the need for such an explanation is what we might call a theoretical necessity. A good analogy is Kuhn's discussion of paradigm shift: before the revolution there are lots of anomalies then people are persuaded to look at things a different way and the anomalies become supporting evidence. I am arguing that because experiments on their own can never be decisive where people are determined not to be persuaded, something has to done to make them look at the same matters in a different way and that this is what happened: my explanation for how it happened is Garwin's intervention but if one does not like that particular explanation one has to find some other tipping point. I say Garwin finally persuaded the community of gravity wave scientists that we're seeing Weber 'grasping and slipping' instead of 'clinging and climbing.' You need some kind of sociological perspective to see that such a tipping event is needed and you also need people to tell you, in ways they would never publish, that this is what they were trying to do and this is how they came to switch their perspective. Of course, when we switch from the theoretical necessity to the historical actuality it is going to be messy because social change is messy; there were probably some individuals who were more influenced by the slippage from 24 to 12 h as sidereal period and others changed their minds when the sidereal correlation disappeared and Weber certainly did not help himself with the computer error and the timing error. So the best I can do is to say that I think it was Garwin who crystallized all this but it is, at best, a thesis about a moment of historical change.

Franklin: For Collins the most significant evidence that he used in his accounts was provided by interviews,[21] whereas Franklin preferred the published record. These different pieces of evidence do not always give rise to the same picture. The more informal and strongly critical comments made by both Weber and his critics about the work of others do not appear in the published record.[22] Although such criticism

[21]In his more recent work, Collins has immersed himself in the gravity wave community and, as noted above, uses published sources (Collins 2011, 2013).

[22]In discussions with some of Weber's critics Franklin heard much more critical and far less polite comments about Weber and his work than appeared in the publications. For more discussion of differences between less formal comments and published work see Franklin (2013, pp. 234–236).

does appear in publications it is far more reserved. For Franklin the more considered presentation of the publications is of greater weight than comments made to an interviewer. The published record is the scientist's contribution to the permanent record of science. Franklin believes that it contains the best and strongest arguments that the scientists have available. The presentation of both views offers a more complete picture of this episode. It is unfortunate that there are not very many pairs of studies of the same episode from these different perspectives.[23] Although we are not suggesting a resumption of the "Science Wars" we do suggest that such conflicting studies can provide us with a better and more complete picture of the practice of science.[24]

Collins: Allan's moderate conclusion here is to be applauded but it seems to me that it implies we should do things my way if it is not impossible. It is not the case that I eschew the public record; in my work I use both where I can. Surely, Allan's earlier claim is the one we need to think about: he says the published record should be treated with greater weight since it is 'the scientist's contribution to the permanent record of science' and 'it contains the best and strongest arguments that the scientists have available.' To assess these claims we have to be sure about what we mean by 'the published record.' If we mean papers published in the journals, then we need to be very careful indeed. Medawar's (1964) 'Is the Scientific Paper a Fraud' tells us why: scientists do not report their experimental work in their papers, they report an idealized and bowdlerized version buttressed by a convincing 'literary technology.' For literary technology see Shapin (1984). It is, therefore, impossible to work out the detailed unfolding of the history of science from published professional papers—at least where experiment is at stake—and this is probably also true for the development of theories. Historians, of course, know this and try to use additional sources. They reach toward the informal by consulting comments published in conference proceedings—as Allan does. They also try to delve deeper by consulting laboratory notebooks if they can or correspondence. Allan has used both laboratory notebooks and correspondence, but only very infrequently and not in his study of the early gravity wave experiments. I, incidentally, did read correspondence as well as recording scientists' verbal descriptions of their work.

I realize, of course, that I am 'on a hiding to nothing' because most historians do not have an oral record to go on so my argument is not likely to be persuasive—that's a social influence on the history of science! But let me have another try at persuading even the likely readers of this volume. In 1996, Joe Weber published a paper that claimed a 3.6 sigma significance for correlations he had discovered between his 1970s gravitational waves and gamma-ray bursts. I went around my colleagues in

---

[23]Two such pairs are Pinch (1986) and Shapere (1982) on the solar neutrino problem and Galison (1987) and Pickering (1984) on the discovery of weak neutral currents.

[24]There is, of course, the possibility that a scholar might combine both approaches. The published record and interviews do not, of course, exhaust the possible sources of information. One might consult correspondence and laboratory notebooks, or attempt to reproduce the original calculations.

the interferometric gravitational wave detection community to find out what they made of it. The answer was that, literally, no-one had read it. That's what it means to have lost one's credibility so completely as to be excluded by the oral community and that is something else you won't find out from published sources.

References

Collins, H.M. 1975. The seven sexes: A study in the sociology of a phenomenon, or the replication of experiments in physics. *Sociology* 9(2): 205–224.

Collins, H.M. 1981a. Son of the seven sexes: The social destruction of a physical phenomenon. *Social Studies of Science* 11: 33–62.

Collins, H.M. 1981b. Stages in the empirical programme of relativism. *Social Studies of Science* 11: 3–10.

Collins, H.M. 1981c. What is TRASP: The radical programme as a methodological imperative. *Philosophy of the Social Sciences* 11: 215–224.

Collins, H.M. 1985. *Changing order: Replication and induction in scientific practice*. Beverley Hills and London: Sage Publications.

Collins, H.M. 1992. *Changing order: Replication and induction in scientific practice*, 2nd ed. Chicago: University of Chicago Press (1st ed. 1985).

Collins, H.M. 2004. *Gravity's shadow*. Chicago: University of Chicago Press.

Collins, H.M. 2009. Walking the talk: Doing gravity's shadow. In *Ethnographies revisited: Conceptual reflections from the field*, ed. A. Puddephatt, W. Shaffir, and S.W. Kleinknecht, 289–304. London: Routledge.

Collins, H.M. 2011. *Gravity's ghost*. Chicago: Chicago University Press. Paperback edition included in Collins 2013.

Collins, H.M. 2013. *Gravity's ghost and big dog*. Chicago: Chicago University Press.

Collins, H.M., and R. Evans. 2002. The third wave of science studies: Studies of expertise and experience. *Social Studies of Science* 32(2): 235–296.

Collins, H.M., and R. Evans. 2007. *Rethinking expertise*. Chicago: University of Chicago Press.

Collins, H.M., and T. Pinch. 1998. *The golem: What you should know about science*. Cambridge: Cambridge University Press.

Conan Doyle, A. 1967. *The annotated Sherlock Holmes*. New York: Clarkson N. Potter.

Franklin, A. 1990. *Experiment, right or wrong*. Cambridge: Cambridge University Press.

Franklin, A. 1994. How to avoid experimenters' regress. *Studies in the History and Philosophy of Modern Physics* 25(463–491).

Franklin, A. 1997. Calibration. *Perspective on Science* 5(31–80).

Franklin, A. 1998. Avoiding the experimenters' regress. In *A house built on sand: Exposing postmodernist myths about science*, ed. N. Koertge, 151–165. Oxford: Oxford University Press.

Franklin, A. 2002. *Selectivity and discord*. Pittsburgh: Pittsburgh University Press.

Franklin, A. 2007. The role of experiments in the natural sciences: Examples from physics and biology. In *Handbook of the philosophy of science: general philosophy of science: Focal issues*, ed. T. Kuipers, 219–274. Amsterdam: Elsevier.

Franklin, A. 2010. Gravity waves and neutrinos: The later work of Joseph Weber. *Perspectives on Science* 18: 119–151.

Franklin, A. 2013. *Shifting standards: Experiments in particle physics in the twentieth century*. Pittsburgh: University of Pittsburgh Press.

Galison, P. 1987. *How experiments end*. Chicago: University of Chicago Press.

Garwin, R. 1974. Detection of gravity waves challenged. *Physics Today* 27: 9–11.

Giles, J. 2006. Sociologist fools physics judges. *Nature* 442: 8.

Kennefick, D. 2007. *Travelling at the speed of thought: Einstein and the quest for gravitational waves*. Princeton: Princeton University Press.

Kennefick, D. 2014. *Relativistic lighthouses: The role of the binary pulsar in proving the existence of gravitational waves*. arXiv:1407.2164.

Levine, J. 2004. Early gravity-wave detection experiments, 1960–1975. *Physics in Perspective* 6(42–75).

Levine, J., and R. Garwin. 1974. New negative result for gravitational wave detection and comparison with reported detection. *Physical Review Letters* 33: 794–797.

Medawar, P.B. 1964. Is the scientific paper a fraud? In *Experiment: A series of scientific case histories*, ed. D. Edge, 7–13. London: BBC. First Broadcast in the BBC Third Programme.

Pickering, A. 1984. Against putting the phenomena first: The discovery of the weak neutral current. *Studies in the History and Philosophy of Science* 15(85–117).

Pinch, T. 1986. *Confronting nature*. Dordrecht: Reidel.

Shapere, D. 1982. The concept of observation in science and philosophy. *Philosophy of Science* 49: 485–525.

Shapin, S. 1984. Pump and circumstance: Robert Boyle's literary technology. *Social Studies of Science* 14(4): 481–520.

Shaviv, G., and J. Rosen (eds.). 1975. *General relativity and gravitation: Proceedings of the seventh international conference (GR7), Tel-Aviv University, June 23–28, 1974*. New York: Wiley.

Stoppard, T. 1967. *Rosencrantz and Guildenstern are dead*. New York: Grove Press.

Weber, J., M. Lee, D. Gretz, G. Rydbeck, V. Trimble, and S. Steppel. 1973. New gravitational radiation experiments. *Physical Review Letters* 31(12): 779–783.

Chapter 7
Pluralism in Historiography: A Case Study of Case Studies

Katherina Kinzel

Abstract In history the same historical episodes can be reconstructed from multiple perspectives, leading to different interpretations and evaluations of the same events, and sometimes even to different factual claims. In this paper, I analyze what I call "historiographical pluralism"—situations of conflict between different case studies of the same historical episodes. I address two interrelated questions: First, which features of historical reconstruction and representation give rise to such conflicts? Second, can we assess rival historical case studies and decide between them, thus restricting historiographical pluralism? As an answer to the first question, I highlight the selective and theory-laden character of historical representation and argue that the narrative dimension of historiography is central for the knowledge that a historical case study can convey. I then go on to analyze how—in practice—disagreement about historical facts emerges. I discuss four case studies paired around two historical episodes and show that conflicts arise from the selective, theory-laden, and narrative aspects of historical methodologies. The second question I answer by discussing different criteria for assessing historical accounts. I note a dilemma in the evaluation of historical reconstructions. On the one hand, there exist neutral and almost universally accepted evaluation criteria. But these criteria are weak and cannot always decide between conflicting accounts of the same episodes. On the other hand, there are stronger methodological criteria. Alas, they are often not neutral with respect to the substantial theoretical issues at stake in situations of conflict between different historical reconstructions. I conclude that because of this dilemma, we have to accept some degree of pluralism in historiography.

7.1 Introduction

The story of the Scientific Revolution in the 17th and 18th centuries has been told many times. It has been reconstructed in a discontinuous narrative by Koyré (1957), who described it as a fundamental intellectual transformation triumphing in the

K. Kinzel (✉)
Institut für Philosophie, Universität Wien, Vienna, Austria
e-mail: katherina.kinzel@univie.ac.at

T. Sauer and R. Scholl (eds.), *The Philosophy of Historical Case Studies*,
Boston Studies in the Philosophy and History of Science 319,
DOI 10.1007/978-3-319-30229-4_7

mathematization of nature. It has been told as an origin story by Butterfield (1949) for whom it marks the advent of modernity. Other authors have presented the story emphasizing continuities between modern scientific views and medieval and Renaissance knowledge practices. For instance, Crombie (1953) argued that experimental science had been practiced first by medieval natural philosophers. And Yates (1964) stressed continuities between the Hermetic-Cabalist traditions of natural magic and scientific empiricism. More recently, the prevalence of microhistory has led to a destabilization of big picture narratives, calling into question the very notion of the Scientific Revolution (Secord 1993). "There was no such thing as the Scientific Revolution, and this is a book about it" (Shapin 1998, p. 1), Steven Shapin claims in the introduction to his reconstruction of 17th century science.

Perhaps the existence of many and conflicting accounts of the Scientific Revolution is not surprising. After all, what is at stake here is nothing less than the origin of our modern world view, the identity of the scientific enterprise, and the status of science in Western society and culture (Cunningham and Williams 1993; Lindberg 1990; Porter 1986). Ideologically contested issues like these are bound to provoke disagreement. The existence of plural narratives of the Scientific Revolution may simply reflect the changing ideological evaluations of science and its place in society.

However, we encounter pluralism in the historiography of science not only when it comes to large scale historical transformations of great political and ideological significance. Local historical cases that, at least at first sight, appear to be ideologically innocuous have met with the same fate: they have been reconstructed several times, from a variety of different viewpoints, and they have come to support different philosophical conclusions. There exist rival case studies of events as local as a specific measurement procedure, an experimental derivation, or an episode of theory-choice. For example, Millikan's oil drop experiments which measured the charge of the electron have been reconstructed from competing sociological and rationalist perspectives (Barnes et al. 1996, pp. 18–45; Holton 1978; Franklin 1986, pp. 140–162). There exist diverging accounts of the historical fate of Mendel's experimentally derived laws of inheritance (for an overview, see Sapp 1990). And the victory of Lavoisier's oxygen-based over Priestley's phlogiston-based theory has been interpreted in light of different philosophical accounts of scientific change and theory choice—pluralist, structural-realist, rationalist, and sociological (Chang 2012; Ladyman 2011; Musgrave 1976; Kusch 2015). In some cases, the various reconstructions of the same historical events are compatible and complement each other. But in the examples mentioned above, the different case studies are in open conflict. They involve incompatible factual claims, give competing causal explanations, carry opposed epistemic evaluations, tell different narratives and reach rival philosophical conclusions.

This paper deals with the situations of conflict between different case studies of the same historical episodes. It addresses two questions: First, which features of historical reconstruction and representation give rise to such conflicts? Second, how can we assess rival historical case studies and restrict historiographical pluralism? Although I believe that my answers to these questions apply to historiography in general, I focus my discussion on case studies in the history of science and in history and philosophy of science (HPS).

In order to answer the first question, I discuss the interpretative and constructive dimension of historiographical methodology. My account highlights the selective and theory-laden character of historical representation and argues that the narrative dimension of historiography is central for the knowledge that a historical case study can convey. Based on this account, I analyze in detail four case studies paired around two historical episodes, and the methodological strategies employed in these case studies. I show that disagreement about historical facts emerges from the selective, theory-laden and narrative aspects of historical methodology.

The second question I answer by discussing different criteria for assessing historical accounts. I note a dilemma in the evaluation of historical reconstructions. On the one hand, there exist neutral and almost universally accepted evaluation criteria. But these criteria are weak. They cannot always decide between conflicting reconstructions of the same historical episodes. On the other hand, there are stronger methodological criteria that constrain historiographical pluralism more drastically. Alas, these strong criteria are often not neutral with respect to the substantial theoretical issues at stake in situations of conflict between historical accounts. Because of this dilemma, I argue, we have to accept some degree of pluralism in historiography.[1]

My paper has four parts. In the first part, I indicate on a general level which features of historical discourse give rise to pluralism. In the second part, I present a fine-grained account of disagreement in historiography by analyzing in detail four historical case studies. In the third part, I proceed to the problem of assessing rival case studies and discuss different historiographical evaluation criteria. In the fourth part, I apply these criteria to the case studies analyzed in the second part. The upshot of my discussion is that although historiographical pluralism is limited, it cannot be completely eradicated.

7.2 Sources of Historiographical Pluralism

What is it about the character of historical representation that enables substantially different retellings of the same events? How are competing historical reconstructions possible? Part of the answer is that historical discourse is a constructive and interpretative endeavor, and that historians can draw on a variety of different methodological strategies when reconstructing the past. In this section, I discuss, on a general and abstract level, three features of historical discourse that can help to understand why conflicting accounts of the same events are possible: (i) selectivity,

[1]Comparable issues arising within the natural sciences have been recently addressed in debates on scientific pluralism (Chang 2012; Kellert et al. 2006). Although I acknowledge that there may exist important parallels between pluralism in science and pluralism in historiography, in this paper I focus on the latter only. Note also that my aim in this paper is to provide a descriptive account of pluralism in historiography. I do not seek to answer the normative question whether historiographical pluralism is epistemically desirable or not.

(ii) theory-ladenness, and (iii) narrativity. I claim that these are not merely contingent features of historical discourse. In fact, they seem necessary. I shall discuss each in turn.

(i) Selectivity.

Like the Borgesian map of the Empire that has the size of the Empire, and coincides point-to-point with its territory, the complete historical account is an absurdity. All historical reconstructions are selective, and they are selective in three important ways. First, a historical account, unlike past reality, has a clearly marked beginning and an endpoint. From the infinite series of historical processes and events, the historical account selects an episode or case that is identified with a finite time-span. Second, once the time-span is determined, the historical account selects some events within that time-span and treats them as constitutive of the episode, while excluding others. Third, of the events included, some are highlighted, while others are relegated to a subordinate status.

Selectivity is a necessary feature of all representation. Recent discussions of representation in science have analyzed in great detail the selective choices that structure how scientific models represent their target domain. I will draw on these discussions, and in particular on "pragmatic accounts of scientific representation", in order to illuminate the ways in which historical accounts are selective.

In pragmatic accounts, representation is extended from a two-term relation that refers the representation to its target, to a three-term or four-term relation that also involves the user of the representation and the user's aims and purposes. For example, Ronald Giere states that representation always occurs in the context of use. "S uses X to represent W for purposes P" (Giere 2004, p. 743). Or, as Bas van Fraassen puts it, "[t]here are no representations except in the sense that some things are used, made, or taken, to represent some things as thus and so" (van Fraassen 2008, p. 23). Use encompasses a wide range of different factors: "the intention of the creator, the coding conventions extant in the community, the way in which an audience or viewer takes it, the ways in which the representing object is displayed, and so forth" (ibid.). The context of use also determines the selective choices made in the representation. The ways in which representations are selective are hence not arbitrary but systematically dependent on the context-specific relations between user, representation, and represented.

I want to suggest that historical accounts are similar to scientific models in that they select and highlight specific aspects of their target domain at the expense of others. And the selective choices of historiography are not arbitrary. They are structured according to the aims a historical account sets out to fulfill. Take Koyré's account of the Scientific Revolution as an example. One of Koyré's aims was to prove the relevance of non-testable metaphysical conceptions in theory creation, and to show that progress in science is driven by conceptual changes. As Rivka Feldhay points out, Koyré's historical narrative is perspectivally constrained. It includes and highlights some aspects of the historical process at the expense of others. Koyré represents history "as a series of texts involved in networks of dialogues" (Feldhay 1994, p. 37).

Focusing on the consistency of historical texts and the ideas embodied therein, he decides "to ignore the traces of production in the texts" (ibid.). His narrative brings ideas and texts into hermeneutic focus, but it excludes the social and cultural relations of their production. These inclusions and exclusions are clearly in line with his general aims. The actual selections made are aim-dependent.

Other examples can be generated with ease. If our aim is to explain historical change then the selection of discontinuous features of the historical episode under study, and the isolation of factors that may be seen to have caused or motivated these changes is the obvious strategy one should go by. If the aim is to create an experience of "historical otherness" then the features of the historical situation under study that stand out as absurd or unintelligible in light of current beliefs should be emphasized. If the aim is to study historical phenomena in their "longue dureé" then large-scale features of the historical period need to be prioritized. And so on.

The selective character of historical representation and the aim-dependence of selective choices constitute important sources of pluralism. If there are many different historiographical aims, then there can be multiple ways of reconstructing the same historical episode that satisfy different aims and make different selective choices about which events to include and which to highlight. Different selections lead to diverging accounts of the same historical events.

(ii) Theory-ladenness.

Historical accounts are not only selective in that they include certain historical events to the exclusion of others. On a more fundamental level, what it means to be a historical event and what it means to be a historical fact are outcomes of constructive processes. The facts of history are not simply found but have to be inferred from historical sources through complex inferential and interpretative processes. These processes, in turn, are structured by theoretical presuppositions. Hence, historical events and facts are theory-laden in the sense that they are partly constituted by theory (see also Kinzel 2015).[2]

Theoretical assumptions structure all stages of the inferential process from the sources to facts. To begin with, judgments regarding which types of sources are relevant for the reconstruction of a given episode, as well as judgments regarding which types of sources are reliable are made by reference to theoretical background knowledge. Then, after the sources have been chosen, they need to be interpreted in a consistent manner. In order to achieve a consistent interpretation the sources need to be related to each other; past events, actions, and meanings need to be inferred from them; relations (possibly causal) between the derived events need to be identified; and the events need to be assigned a certain significance with respect to

[2]I use the term "theory" broadly here, such that it includes all sorts of high-level conceptions. These conceptions need not be coherent systematic accounts; indeed, at times it may even be only single assumptions rather than elaborate theoretical constructions that are at issue. Also, I include methodological and basic philosophical commitments about the character of science and the nature of historical change among the sorts of theoretical assumptions that are relevant when talking about theory-ladenness in the historiography of science.

other past events or with respect to the present. Each of these interpretative maneuvers relies on theory. Moreover, there is not only an upwards inferential and interpretative process that leads from the sources to the facts, but also a downwards process in the concept-dependent identification of historical events. Each historical event is an event only under a description, and hence its identification is only possible if the identity conditions for the event are specified: "[E]vents may be sliced thick or thin, a glance may be identified as an isolated event or as an instance in an event. What the unit-event is depends on the telling of it" (Roth 1988, p. 9).

There are very likely many other forms of theory-ladenness in historiography. But these considerations suffice to indicate how theory-ladenness can be a source of pluralism: different theoretical assumptions and different methodological commitments will have consequences for the selection of relevant sources, for how historical events are reconstructed from the sources, for how they are interpreted, explained and evaluated, for the individuation of historical events, etc. Since historical facts are theory-laden, disagreement is likely to emerge between historical accounts that reconstruct the past on the basis of different theoretical assumptions.

(iii) Narrativity.

A third and final feature that leads to pluralism is the narrative character of historical discourse. The narratological tradition within the philosophy of history has long claimed that the peculiar form of historical representation—that which distinguishes historical reconstructions from other types of representational discourse—consists in the use of narrative. Rendering past events, states, and processes intelligible requires that a story be told about them. By becoming entrenched in a story historical events achieve significance and meaning and which story is told is relevant for the knowledge a historical account can convey.[3]

Hayden White observed that in order to build a historical account, the series of historical events (the chronicle) has to be molded into a story that characterizes these events in terms of beginning, transitional phase, and endpoint (White 1973, p. 5). And according to White, in this process, the choice of narrative form or mode of emplotment is crucial. This is because narratives "familiarize" historical events by relating them to already established plot structures. Historical accounts elicit understanding by referring the unknown back to already known themes and structures embodied in archetypal story types, such as comedy, tragedy, or satire. They integrate historical events into stories that are firmly entrenched in the cultural repertoire of known narratives and themes: "The effect of such encodation is to familiarize the unfamiliar, and in general this is the way of historiography, whose 'data' are always

[3]There is some disagreement as to whether the narrative structure of a historical account carries information, conveys knowledge and is properly representational, or whether narrative is a superfluous, merely "rhetorical" aspect of historical discourse. Different accounts of the role of narrative in historical representation have been developed by Carr (2008), Carroll (2001), White (1980). I cannot go into these debates here. However, I believe that my analysis in the next section will show that narratives do convey information about the past. In this sense, they should be seen as epistemic rather than "merely rhetorical."

immediately strange, not to say exotic" (White 1978, p. 49). Moreover, narratives convey information about the past by effecting closure. While the chronological series of historical processes and events is infinite, a narrative reaches an endpoint. Historical events become intelligible when the story reaches its resolution, when the questions raised at the beginning have been answered and the reader's expectations have been satisfied or disappointed (White 1980, pp. 24–27). By achieving closure, a narrative constitutes a more or less coherent, meaningful whole.

Historical narratives, according to White, do not provide the unique truth about the past. "[A]lternative, mutually exclusive and yet equally plausible" (White 1978, p. 55) narrative emplotments of the same events can be constructed. This is just the situation that we have been describing: the repertoire of culturally preexisting genres and story types is vast, and one and the same historical episode can be rendered intelligible in manifold ways by drawing on different story types and modes of emplotment.[4] About the same events, many different stories can be told.

The more general question that follows from the discussion of selectivity, theory-ladenness, and narrativity in historical discourse is how strong pluralism is or needs to be. White himself remains somewhat ambivalent about the strength of pluralism. In some passages he grants that there are epistemic constraints on what narratives can be plausibly told about a specific historical episode (White 1978, pp. 47–48, 59). In other passages, however, he suggests that there are no limits on narrativization at all; "we are free to conceive 'history' as we please, just as we are free to make of it what we will" (White 1973, p. 433). Here it appears that White is not only a pluralist, but also an anything goes relativist for whom there are no epistemic constraints whatsoever that would restrict what narratives we can meaningfully and plausibly tell.

My own pluralist thesis is not that radical. In later sections, I will show that there are important epistemic restrictions on the range of permissible alternative historical reconstructions. But before doing so, I want to examine historiographical pluralism in actu by studying in detail the ways in which rival historical methodologies lead to different accounts of past events.

7.3 The Structure of Disagreement: Four Case Studies

Pluralism is most interesting, or most controversial, when two conditions obtain; namely (a) when there exist conflicting accounts of the same historical episodes,

---

[4]Many aspects of White's narratological account are deeply problematic. On the one hand, his structuralist taxonomy of different modes of historical writing is static, artificial and irritatingly ahistorical. On the other hand, from insights into the constructive dimension of historiography, White draws radical conclusions about its subjective and fictional character. I share neither White's structuralist inclinations, nor his radical subjectivism, and wish to take from his reflections only the central theses that historical accounts have a narrative structure and that there can exist plural narrative emplotments of the same historical events.

and (b) when it is not obvious which of the different reconstructions is the correct, adequate or most plausible one.

In the previous section I discussed how historiography can fulfill condition (a) and discerned three abstract features of historical discourse that lead to pluralism. This section explores how these abstract features are realized in actual historiographical methodologies. It deals with the concrete structure of disagreement in four historical case studies: Harry Collins and Allan Franklin's rival accounts of the early gravitational radiation experiments and Alan Musgrave and Hasok Chang's different reconstructions of how phlogiston theory was abandoned. Exploring how these case studies fulfill condition (a), I try to remain as neutral as possible between the competing accounts. The problem of evaluation and the question to which extent the rival accounts satisfy condition (b) I address in later sections.

I structure my analysis of disagreement by introducing two levels on which differences between rival case studies can be observed; (I) the level of factual claims, and (II) the level of methodological strategies. I argue that differences arising on the level of factual claims can be traced to differences in methodological strategies. It is on the level of methodological strategies that the abstract features of historiography discussed above unfold their pluralist effects. As I will show in my analysis of concrete case studies, the reconstruction of past episodes of science involves (i) selective choices, (ii) theory-laden reconstruction procedures and (iii) techniques of narrative emplotment. Differences in selection, theory-ladenness and narrativization give rise to conflicts about what exactly happened in the historical episode under study.

This is, I believe, a general point about historiography: factual claims in history are always the result of complex methodological processes and these methodological processes always involve selection, theory-ladenness and narrativization. However, it needs to be noted that the case studies that I have chosen belong to the HPS context and that each of them comes with an explicit philosophical agenda. It may be argued that case studies in general history and in the history of science differ from those produced in HPS because they are not to the same degree committed to explicit philosophical doctrines. This is certainly correct. In general history and in professional historiography of science theory-ladenness does not typically take the form of explicit philosophical concepts being applied to the historical material. However, I believe this does not make the case studies produced in these fields less theory-laden. It only means that the theoretical and methodological assumptions that structure the reconstruction of the past, and the ways in which they do so, are more subtle in general history and in the historiography of science than they are in HPS.

7.3.1 The Gravity Waves Episode

I begin my analysis with the dispute over the interpretation of the high-flux gravity waves episode. Collins' reconstruction of the early attempts to measure gravitational radiation experimentally became central to the sociology of scientific knowledge canon. It presented a case for the prevalence of the "experimenters' regress" in cutting edge science and claimed that social factors play a vital role in the closure of scientific controversies (Collins 1985). A rival account of the episode was presented some years later by Franklin. Franklin claimed that the resolution of the debate was not brought about by social factors, but by the reasoned discussion of the available evidence (Franklin 1994). Not surprisingly, Collins rejected this alternative account vehemently (Collins 1994).

The two authors recapitulate their past dispute in a joint contribution to this volume. In it they explain that the nature of their disagreement has shifted since the original case studies were published. According to their own testimony, they now differ about some historical details and relatively mundane matters of methodology. But the "battle" over the social dimension of science, they claim, is over (see Franklin and Collins in this volume, pp. 6–6.5).

Note however, that even when that battle was still running, Collins and Franklin did agree on many of the technical details of the scientific controversy at issue, as well as on central historical events. The course of events that both authors agreed on can be summarized as follows: *In the late 1960s and early 1970s, Joseph Weber developed the first gravitational wave detectors and claimed to have acquired positive results. In the years to follow, other laboratories tried to replicate his results with slightly different experimental setups, but they did not manage to reproduce his observations. By the late 1970s, Weber's claim to have observed high fluxes of gravitational radiation had lost all credibility and was rejected by the scientific community.*

Despite their agreement on these points, Collins and Franklin's original historical accounts differed so radically that the two authors took the case of Weber to support two conflicting philosophical doctrines. How is this possible? In the following, I give a systematic analysis of the dispute between the two authors in terms of my account of selection, theory-ladenness, and narrative in historiography. I identify both obvious and subtle differences between the rival reconstructions.

(I) Factual claims.

Obviously, Collins and Franklin disagree on how and why Weber's claims were rejected. That is, they make different factual claims about the historical episode, different claims about what happened. Collins claims that the available evidence and rational arguments underdetermined the decision against Weber, and that, eventually, social processes led to the rejection of his results. According to him, the powerful rhetorical intervention of one of Weber's critics, scientist Q (Richard Garwin) was causally decisive in tipping the scales to Weber's disadvantage (Collins 1985, 93–95). Franklin claims the exact opposite. He thinks Q's rhetorical attack played only a minor role and argues that it was the sheer quantity of negative evidence against Weber that eventually led scientists to discard his results. According to him, rational deliberation was causally sufficient for a decision to be reached (Franklin 1994, 468–469). Both authors believe that the historical material in fact supports their respective views of the relevant causal relations.

(II) Methodological Strategies.

To understand how these conflicting factual claims come about, we have to take a look at the different methodological strategies that the two authors apply. In the following, I will argue that Collins and Franklin select different aspects of the past to be represented, reconstruct events on the basis of different philosophical assumptions and tell different narratives. This accounts for the factual disagreement between them.

(i) Selection.

Although agreeing on many of the technical details of the gravity waves controversy, Collins and Franklin differ on whether negative evidence was a decisive force in bringing the controversy to a close. Their different assessments of the role of the evidence correspond to differences regarding which aspects of the historical material they select and highlight.

When Collins reconstructs the scientific debates surrounding the experiments, he not only focuses on how the scientists who attempted to replicate Weber's initial results responded to his claims and arguments. He also presents in great detail how they responded to each other. And he reveals that scientists were in severe disagreement as to how to interpret and explain their failures to replicate Weber's findings. They found fault not only with Weber's experimental setup, but also with each other's experimental strategies (Collins 1985, pp. 84–88, 90–92). Representing scientists' mutual criticisms, Collins selects those aspects of the debate that indicate that there was profound disagreement in the scientific community.

In contrast, these aspects are almost completely absent from Franklin's reconstruction. Franklin presents the arguments in such a way that they fall into two opposed camps: Weber and his critics. The interrelations between Weber's critics and their mutual criticisms are not taken into account. Franklin is quite clear that, for him, the situation is one of agreement rather than disagreement: "The fact that Weber's critics might have disagreed about the force of particular arguments does not mean that they did not agree that Weber was wrong." (Franklin 1994, p. 472). So while Collins devotes much attention to the various points of disagreement between Weber's critics, Franklin selects and highlights points of agreement rather than disagreement.

Another salient difference regarding the selection of historical events concerns the role of scientist Q. Collins places Q at a central point in the narrative: Frustrated by the scientific community's hesitance to reject Weber's results, which he took to be mistaken from the beginning, Q sets out to destroy the credibility of Weber and his observation claims in a series of polemical attacks. In Collins' reconstruction, Q's rhetorical intervention constitutes the social cause that tips the scales against Weber and leads to the closure of the debate (ibid., 93–95).

Franklin, in contrast, emphasizes the continuity of negative results that existed before and after Q's intervention. According to him, it was the accumulation of negative results that led to the eventual rejection of Weber's claims, not the rhetorical intervention of one scientist.

Collins and Franklin's different factual claims can be seen to result from the selective choices they make. These choices structure which historical events and which aspects of the scientific debate under study are included and emphasized in their rival historical reconstructions.

(ii) Theory-ladenness.

Another basic difference between the two authors concerns their handling of the sources. Collins extends the realm of sources from the published record to also include interviews with the scientists involved, while Franklin puts the emphasis on the published material. These decisions are informed by theoretical assumptions which the authors themselves make explicit. Collins draws on interviews so as to avoid publication bias (Collins 1985, p. 498), while Franklin believes the published record to be more reliable than other sources (Franklin 1994, p. 465).

However, not only the selection but also the interpretation of the historical sources is theory-laden. It is obvious that Collins and Franklin rely on different theoretical resources when interpreting their historical material. Collin describes the disagreement among Weber's critics by referring to the concept of the experimenters' regress. The experimenters' regress occurs at the frontiers of enquiry when new phenomena are measured with new experimental apparatus. In these situations, ascriptions of when the apparatus is working properly hinge on whether it produces the wanted outcome, while, at the same time, what the correct outcome is becomes defined by reference to the quality of the experimental setup (Collins 1981, p. 34, 1985, p. 84).

Applying the theoretical concept of the experimenters' regress to the historical material, Collins can interpret the situation not only in terms of disagreement, but also in terms of contingency and open-endedness. He claims that the historical process could have taken a different trajectory than it actually did.

> Obviously the sheer weight of negative opinion was a factor, but given the tractability, as it were, of all the negative evidence, it did not have to add up so decisively. There was a way of assembling the evidence, noting the flaws in each grain, such that outright rejection of the high flux claim was not the necessary inference (Collins 1985, 91).

Franklin engages his reconstruction of the historical events with a philosophical agenda diametrically opposed to Collins'. He seeks to show that the resolution of the debate was not contingent, but the only rationally acceptable outcome. And he applies different theoretical resources to argue his point. Of particular importance is the concept of robustness. When Collins concluded that the plurality of interpretative options marked the situation as open-ended and contingent, Franklin believes that the different arguments reinforced one another. The fact that a series of slightly different experimental setups failed to replicate Weber's claims, according to Franklin, renders the negative results more robust and strengthens the argument against Weber (Franklin 1994, pp. 477–478).

Again, the differences in factual claims can be seen to rest on different theory-laden reconstruction procedures. Collins and Franklin select different sources and then interpret these sources by reference to different theoretical concepts. Most notably, Collins applies the concept of the experimenters' regress and interprets the historical situation in terms of contingency, while Franklin draws on a concept of robustness and describes the situation as one in which the negative evidence was decisive.

(iii) Narrative.

Finally, Collins and Franklin tell different stories of the historical events. The conflicting factual claims they make rest on different narrative emplotments of the episode. In order to establish this point, I propose a slightly unconventional practice of analysis: I will read the historical case studies as one would read a novel or a short story and apply some basic lessons from literary criticism. This will enable me to identify the narrative structure of the respective case studies and to show how different narrative structures carry different claims about the past.

Proceeding in this manner, Collins historical narrative is best described as an ironic tragedy. It resembles a tragedy because it tells the story of a social downfall and does so in discontinuous terms. According to the literary theorist Northrop Frye, a tragic plot is essentially a story of exclusion in which the hero is expelled or isolated from his society (Frye 1957, pp. 35–43). The story of Weber, as Collins presents it, is such a story: Weber is excluded by the society to which he tries to belong. Moreover, the tragic plot is usually discontinuous, characterized by a radical break: before and after Oedipus finds out that he killed his father and married his mother, before and after Lady Macbeth dies and Macbeth finally realizes he has been tricked by the witches. In Collins' narrative the discontinuity is Q's intervention. By structuring events in terms of discontinuity the tragic plot reflects the demand for identifying a causally decisive turning point.

What makes the tragedy an ironic one is that the hero is not causally responsible for his fate. As Frye explains, "the central idea of tragic irony is that whatever exceptional happens to the hero should be causally out of line with his character" (ibid., 41) The hero's demise is not brought about by a tragic hamartia: "Irony isolates from the tragic situation the sense of arbitrariness, of the victim's having been unlucky, selected at random or by lot, and no more deserving of what happens to him than anyone else would be" (ibid.). This sense of arbitrariness resonates with the interpretation of scientific closure that Collins defends: Weber's downfall is not a result of him being in error. What happened to him was at least partly due to circumstance and the story could well have ended differently. His was a contingent downfall. The historical claims Collins defends are thus embodied in the narrative structure of the account.

Not surprisingly then, Franklin's narrative is fundamentally different from that of Collins. It can be read as an adventure story. Adventure stories organize time in a strictly serial order. As Mikhail Bakhtin observes, the adventure story is constructed "as a series of tests of the main heroes, tests of their fidelity, valor, bravery, virtue, nobility, sanctity and so on" (Bakhtin 1986, p. 11). Typically, the hero emerges victorious from each test and the story closes with the hero's exaltation. Franklin's reconstruction resembles an adventure story in that it is strictly serial and accumulative. Moreover, the stages of the historical development are constructed as tests for Weber. Weber's claims are confronted with a series of problems to which he is forced to respond. In each test-situation the arguments for and against Weber are reconstructed in dialogical situations that juxtapose Weber's own considerations with those of his critics, so as to mimic an exchange of arguments. Weber, as the anti-hero of the story, emerges defeated from each test. Franklin's account turns the exaltation of the hero into his demise; his is an inverted adventure story.

As in the case of Collins, the plot structure carries historical claims. In Franklin's narrative, we witness the evidence against Weber—negative results, problems, doubts, criticism, counter arguments and errors—piling up as the story progresses. The "overwhelming negative evidence" (Franklin 1994, p. 472) against Weber makes it inevitable for a rational scientific community to reject his claims. The accumulative structure of Franklin's text thus carries his explanation of the events in terms of negative evidence rather than social processes.

7.3.2 The Chemical Revolution

I hope to have shown that differences in factual claims go back to differences in methodological strategies and that in these methodological strategies, the abstract features of historical representation discussed before unfold their pluralist effects. To further substantiate this claim, I turn to a second example, the Chemical Revolution. My discussion will be relatively brief and schematic. I discuss and compare two reconstructions of the episode, Musgrave's Lakatosian rational reconstruction, and Chang's attempt to mobilize the episode as a case in point for normative scientific pluralism. Here is what both authors can agree to have taken place in the Chemical Revolution: *The heyday of phlogiston-based explanations of combustion and calcination occurred between 1700 and 1790. In the early 1770s, Antoine Lavoisier began to develop an alternative explanatory framework that dispensed with phlogiston and instead postulated the existence of another substance, namely oxygen. Both the phlogiston and the oxygen theories enjoyed explanatory and predictive successes, as well as demonstrating appreciable problem-solving abilities. However, both frameworks also faced anomalies and failures. Precise weight measurements in later experiments favored the oxygen-based framework. Eventually, phlogiston theories were abandoned.* Beyond this basic agreement, Musgrave and Chang offer rival accounts of the processes through which phlogiston-based theories were replaced with oxygen-based ones, rival explanations of why this occurred and competing epistemic evaluations of the rationality and legitimacy of the victory of oxygen.

(I) Factual Claims.

Musgrave claims that phlogistonism constituted a degenerating research program and was rejected for that reason. According to him, the Chemical Revolution was a rational process (Musgrave 1976, pp. 205–206). Chang, by contrast, claims that the phlogiston theory was not clearly inferior to its competitor (Chang 2012, pp. 19–29), and that its potential had not been fully exhausted at the time of its abandonment (ibid., 43–48). For these reasons, the rejection of phlogiston was a non-rational and premature decision.

(II) Methodological Strategies.

As before, an analysis of methodological strategies will provide a better understanding of how such different factual claims can emerge.

(i) Selection.

Musgrave and Chang delineate the episode of the Chemical Revolution in different ways. Two aspects of their selective choices are particularly salient. First, Chang situates the rejection of phlogiston theory within the broader context of a long-term transformation of epistemic practices, the rise of what he calls the "compositionist" system of chemistry. According to Chang, the rejection of phlogiston theory was not rational in itself, but a mere epiphenomenon of this broader shift (Chang 2012, 36–42). Musgrave, in contrast, chooses to represent the concrete interactions between Priestley, Cavendish, and Lavoisier, but he excludes long-term transformations in chemists' practices from his account.

Second, Chang chooses to represent not only the actual historical events, but also what could have become of phlogiston theory had it not been abandoned. He presents a counterfactual history according to which phlogiston theory could have fostered scientific progress had it been retained. In this way, the counterfactual history becomes part of the episode under study.

It is by including the broader context of epistemic transformations in chemistry and the counterfactual benefits of phlogiston that Chang can claim the rejection of phlogiston theory to have been non-rational and premature: it was non-rational because it was brought about not by inherent deficiencies of the phlogiston theory, but by unrelated transformations in chemists' research practices. And it was premature because phlogiston had a potential that is revealed in the counterfactual scenario. Musgrave and Chang's different verdicts on the rationality of the Chemical Revolution are underpinned by the different selective choices they make.

(ii) Theory-ladenness.

A fundamental difference between Musgrave and Chang's reconstructions concerns temporality. Musgrave reconstructs the development of the phlogistonist and oxygenist rivals diachronically, distinguishing between different successor versions of the theories. He applies Lakatos' conception of competing research programs to the historical material. A Lakatosian research program consists in a diachronic series of successor versions of a theory.

Musgrave also uses the concept of progressive and degenerating research programs to evaluate the rivals. In his interpretation, both programs were successful before 1770. It was only between 1770 and 1785 that Lavoisier's oxygen theory began to outperform the phlogistonist program of Priestley and Cavendish. During this time span, the oxygen program developed in a coherent manner and each new version marked an increase in predictive power and theoretical growth. The program was progressive. The phlogiston program, in contrast, was confronted with increasing difficulties and degenerated (Musgrave 1976, p. 205). This evaluation then allows for the verdict that the rejection of the phlogistonist explanations was entirely rational: chemists at the time realized that the phlogiston-based system was degenerating and changed their allegiances.

Chang reaches a very different verdict. This is possible primarily because his reconstruction is systematic rather than temporal. Chang does not recount the successive steps in which the two theories developed. He reconstructs the rivals as holistically understood systems of practice. He analyses static and systematic features of the phlogistonist and oxygenist approaches, listing the questions the two systems addressed, the problems they found significant and the epistemic values they embodied (Chang 2012, pp. 19–28).

Comparing the two systems, Chang applies the concept of methodological incommensurability. According to his interpretation, both systems were able to solve the problems which they considered important in a manner that satisfied the epistemic values that they adhered to.

> It seems clear that each of the oxygenist and phlogistonist systems had its own merits and difficulties, and that there were different standards according to which one or the other was better supported by empirical evidence. [...] [B]oth systems were partially successful in their attempts to attain worthwhile goals and [...] there was no reason to clearly favor one over the other (Ibid., 29).

As in the above example, we can observe how differences in the theoretical assumptions that guide the historical interpretation translate into different factual claims: Musgrave engages in a diachronic reconstruction and interprets the situation in terms of Lakatosian confirmation theory, using the concept of research programs. On this basis he reaches the conclusion that the Chemical Revolution was a rational process. Chang, in contrast, interprets the situation in terms of the theoretical concept of methodological incommensurability and finds that the decision was not rational.

(iii) Narrative.

As in the example above, the historical claims are also brought across by the way the story is told. Musgrave's historical account is well described as a comedy. In a comedy, according to Frye, the complications, plot twists, and revelations that the hero has to live through before succeeding are more important in determining the course of events than the moral or intellectual qualities of the characters (Frye 1957, p. 170). Just like a comedy, Musgrave's narrative is driven not by the intentional action of the characters, but by unexpected plot twists. From the moment that Lavoisier's research program enters the stage, we are confronted with a series of sudden and unexpected changes of fate, some of which reveal a deep historical irony. Repeatedly, Lavoisier's opponents are also his helpers. They interpret their experiments in a way which enables Lavoisier to verify his prediction (of what would later be called oxygen). They correct Lavoisier's errors. They provide the insights and resources that help him to solve the anomalies that trouble his oxygen program (Musgrave 1976, pp. 194, 200–201). The comic emplotment in terms of unexpected plot twists accords with Musgrave's interpretation of the episode as rational. Scientific rationality exerts itself in the narrative as a Hegelian "List der Vernunft" (cunning of reason). Even when trying to disprove the oxygen theory, its opponents only contributed to its victory. And despite occasional contingencies and surprising plot twists, the more successful program wins eventually.

Chang's historical account draws on completely different narrative principles. The story he tells is not comic. To the contrary, the structure of his text is best captured if it is read as an elegy—an elegy to a promising theory that died an unjust death. Indications of such a reading can be found in Chang's own headlines, which read "The Premature Death of Phlogiston" (Chang 2012, p. 1) and "Why Phlogiston Should Have Lived" (ibid., 14). While life and death are no more than metaphors that facilitate easy comprehension of the main judgments that the historical argument is supposed to put forward, they are instructive. Like an elegy to a prematurely deceased hero, Chang's narrative mourns unrealized possibilities. It conjures up an image of what could have become of the deceased had they not passed away and seeks to establish that phlogiston, in line with Chang's normative pluralist agenda, should not have been abandoned. To prove this point, Chang engages in counterfactual history about what might have happened had phlogiston theory been maintained (ibid., 42–50). This counterfactual imagination enables the judgment that the death of phlogiston was not only rationally unwarranted (and in this sense unjust) but also premature if measured against the possibilities for innovation, development and discovery that it entailed. The elegiac practices of bemoaning and praising the deceased fit Chang's evaluation of the situation like a glove.

Having provided a detailed analysis of selected historical case studies, I hope to have made plausible two points. The first point is that there is not one unproblematic way of deriving historical facts from the sources. Rather, historians engage in complex methodological strategies in order to reconstruct, interpret, evaluate, and explain past events, and the methodological strategies of historiography involve selective choices, theory-laden interpretations and narrative emplotments.

Second, I hope to have shown that differences arising on the level of factual claims can be traced to differences in methodological strategies. Which selections are made when reconstructing historical happenings, which theoretical assumptions guide the interpretation and evaluation of past events, and which narratives structure the historical material has consequences for what factual claims a historical account can reach. Given the three features of historiography that I have highlighted, the existence of severe disagreement between case studies of the same episodes does not appear surprising.

7.4 Evaluating Historical Accounts

In my analysis above, I tried to stay as neutral as possible between the rival historical accounts, and did not present one side of the conflict as more plausible or better warranted than the other. In this section, I turn to the problem of whether and how we can assess historical case studies and settle conflicts such as the ones discussed.

As stated before, historiographical pluralism is most interesting if (a) it occurs between conflicting historical accounts, and (b) the alternatives on offer are plausible to roughly similar extents. The question as to whether and how we can assess competing historical case studies is of central importance to historiographical pluralism, because it bears on condition (b), the issue of comparative plausibility: the stronger our grounds for assessing case studies and for deciding between rival reconstructions, the weaker is our pluralist scenario. If we can always reach unequivocal decisions between competing accounts, then there is no room for pluralism in historiography. Or at least the more controversial forms of pluralism that occur between incompatible and conflicting reconstructions will be ruled out. On the other extreme, if we can never judge which of two or more alternative accounts is the most plausible, then we are confronted with a situation of extreme pluralism, or even anything-goes relativism.

My own approach takes a middle position between these two extremes. I argue that there indeed are epistemic considerations that allow for an evaluation of competing historical case studies. These considerations place restrictions on the space of permissible alternatives and hence restrict pluralism. However, they are not strong enough to always yield an unequivocal verdict as to which of two or more competing reconstructions to prefer. In some cases, a neutral decision between rival accounts may not be possible. In order to make my point, I begin by considering what types of epistemic considerations we can draw on in order to evaluate historical case studies. I distinguish between (α) basic and (β) complex evaluation criteria. Then I proceed to evaluating the above discussed case studies in terms of these considerations.

(α) Basic criteria.

When historians and philosophers of science discuss the merits of different historical reconstructions and case studies, their evaluation criteria often remain implicit.[5] However, there seem to exist a few rough and ready rules that one can draw on when assessing the quality of a historical reconstruction. Some of them are related to the practices of source criticism (how reliable are the sources used, how well are the known sources covered, and how varied is the evidence cited?). Others concern the composition of the historiographical text itself (is the historical argument consistent and intelligible?). Yet others relate to how well the historical reconstruction fits

[5]In the philosophy of science, epistemic criteria such as simplicity, variance of the evidence, surprising predictions, fruitfulness, and explanatory power are often thought to help scientists reach a verdict in situations of theory-choice. Since most of these criteria cannot be applied to historiography without difficulties, I develop my account of historiographical evaluation without substantially drawing on discussions of theory-choice in the philosophy of science.

within a broader system of knowledge (is the historical account consistent with accepted, incontrovertible background knowledge?). Standards regarding source-reliability and source-variance, internal consistency, and consistency with accepted background knowledge are relatively uncontroversial and they are commonly relied upon even when they are not made explicit. I refer to this type of evaluative standards as basic criteria.

The basic criteria have the advantage of being relatively neutral with regards to philosophical conflicts. That is, when using them to decide between conflicting historical accounts, we can usually be relatively certain that we are not already assuming a point at stake in the debate. Requirements for internal consistency merely raise demands concerning the logical or argumentative structure of the historical reconstruction. Such demands seem to be neutral in relation to the theoretical assumptions that might be at stake in a given conflict between rival historical reconstructions.

As described above, considerations regarding the reliability, variance, and completeness of the sources are not without theoretical presuppositions. Whether a historical account has covered the relevant sources to a sufficient degree and hence is "complete" depends upon existing background knowledge about existing sources, which in turn depends upon other available historical reconstructions, archival research, and so on. Moreover, what counts as a relevant source is contextually determined since it depends upon the aims and purposes of the historical account that draws on these sources. Selectivity and the restriction of covered sources is legitimate in principle, if it accords with and is conducive to the aims of the historical account. Completeness only refers to the completeness of the sources relevant to the satisfaction of a specific historiographical aim.

Nevertheless, a historical account that involves more varied sources can be considered more robust than an account that restricts its sources to a specific type. Also, there may be clear violations of the contextually understood completeness requirement. For example, if known sources that would be relevant to the historiographical aim but which are not in line with the argument that the historical account wishes to carry along are excluded, then the selective choices may be considered biased. This would strongly undermine the plausibility of the historical account in question. I believe that although they are context-dependent and theory-laden, evidential considerations such as the ones just presented can sometimes serve as neutral evaluation criteria. At a later point I provide an example for what an evaluation in terms of contextually determined source completeness can look like.

(β) Complex criteria.

But there are also more complex considerations that can be and often are used in the assessment of historical reconstructions. Complex criteria are evaluation standards drawn from debates about intricate historiographical issues such as contextualism, internalism, and externalism (are historical events adequately contextualized?); hermeneutics, understanding, and translation (are the historical actors' conceptions and understandings faithfully reconstructed and appropriately conveyed?); textual interpretation (is the original meaning of the text restored?); present-centeredness (are backwards-projections, anachronisms, and Whig-history avoided?); historical explanation (have the right causes been identified, has reductionism been avoided?); micro- and macrohistory (does the account address the correct level of description?) etc. Complex criteria, unlike basic ones, have the advantage that they are subject to explicit discussion. They are therefore relatively well understood and usually rendered explicit when they are used in the evaluation and critical assessment of a historical reconstruction.

On the downside, unlike basic ones, complex criteria are not generally agreed upon or uncontroversial. And disagreement regarding complex criteria can arise on at least two levels. First, it may not always be evident whether a complex criterion has been met. For example, it is not always evident whether illegitimate backwards projections and anachronisms have been avoided in a given reconstruction, or whether a historical account exhibits explanatory power. But second, and more importantly, the complex criteria themselves are contested.

For example, in discussions concerning Whig history and present-centeredness there is substantial disagreement regarding the identification of the exact vices that result from present-centered historiography of science. There is also disagreement concerning whether all or only some specific uses of present-day knowledge and categories in the interpretation and explanation of past science are to be avoided (Ashplant and Wilson 1988; Cunningham 1988; Cunningham and Williams 1993; Jardine 2000). Moreover, it has been suggested that the evaluation of past knowledge by present-day standards might not always be problematic (Tosh 2003), or at least that it is not as problematic as other practices that have come to be criticized under the heading of present-centeredness, for example, a triumphalist siding with the winners of past scientific debates (Chang 2009).

Regarding adequate levels of analysis and explanatory power, it is debated whether the capacity of the historiography of science to provide comprehensible explanations depends on micro-perspectives. Does historical explanation need to trace the local and particular causes that prompt specific historical events or can it be concerned with large-scale factors and processes as well? (For a useful discussion of the respective epistemic capacities of macro- and microhistorical perspectives, see Pomata (1998). More fundamentally, it is not even universally agreed upon that explanation, and in particular causal explanation, should be a central aim and method in the historiography of science. For example, the relevance of causal explanations in history has recently been denied by Daston and Galison (2007, pp. 34–37); for critical discussion, see Kinzel (2012).

Apart from not being generally agreed upon, the complex methodological criteria often are connected to theoretical assumptions about the character of science, or about the relations between past and present-day knowledge. They are connected to substantial philosophical issues and hence are seldom neutral with regards to philosophical debates.

The case is most obvious with methodological debates concerning adequate contextualization. Clearly, in the dispute between Collins and Franklin, questions regarding the method of adequate contextualization are intimately related to what is ultimately at stake in the conflict between them: the social nature of scientific decision-making. The other complex criteria are philosophically laden in a similar manner. Methodological debates over anachronism and Whig-history are related to philosophical questions regarding continuity and discontinuity in the history of science, scientific change, and progress. Considerations regarding explanation and understanding in history carry over into philosophical issues concerning the relations between reasons and causes, scientific rationality, and the driving forces of theory change. And to the degree that the complex criteria are not independent of philosophical positions and claims, they should also not be expected to be neutral with regards to the theoretical issues at stake in conflicts between rival historical reconstructions. When we are relying on a complex criterion in order to decide between conflicting historical accounts, we cannot always be sure that we are not already assuming a point at stake in the debate. In some cases, the failure of a historical account to satisfy a certain complex criterion may be more indicative of that same criterion being defined in a philosophically invested manner, rather than of a neutrally assessable flaw of the account in question.

7.5 Restrictions on Pluralism

Having distinguished between basic and complex criteria, I want to return to the four historical case studies analyzed earlier in order to evaluate how well they fare with respect to the neutral evaluation criteria we introduced. In this analysis, I seek to substantiate two claims. First, basic evaluation criteria reduce the space of permissible alternatives and hence restrict pluralism. But second, the verdicts that we can reach on their basis are relatively weak, and in order to reach a more definite decision, we would have to refer to complex criteria.

I begin with applying the basic evaluation criterion of internal consistency to the competing historical accounts of the Chemical Revolution. The criterion of internal consistency restricts the space of permissible alternatives, because, of the two case studies that I have discussed, only one meets its standard. Musgrave's account fails the consistency requirement. It involves a straightforward contradiction in its central factual claims. This contradiction emerges as follows: At the end of the historical narrative, and after having given his presentation of the historical development of the two competing research programs, Musgrave passes the following verdict on them:

> Between 1770 and 1785 the oxygen programme clearly demonstrated its superiority to phlogistonism: it developed coherently and each new version was theoretically and empirically progressive, whereas after 1770 the phlogiston programme did neither (Musgrave 1976, p. 205).

This verdict is indispensable for assessing the abandonment of phlogiston-based theories as rationally warranted. And yet, it does not accord with claims made earlier in the historical reconstruction. In particular, Musgrave had claimed that in 1775 experiments spoke as much against Lavoisier as they did in his favor (ibid., 196), and that in 1783 Priestley was having great predictive success with his phlogiston theory (ibid., 199). If these claims are correct, Musgrave's statement that after 1770 the phlogiston theory was starting to degenerate while the oxygen theory was progressing cannot be right. The historical facts he cites contradict his interpretation of the situation in terms of progressing and degenerating research programs. The claim that the choice for oxygen was rational by Lakatosian standards appears ill-grounded. Chang's account is less problematic in this respect. At least, it does not involve historical claims which directly contradict each other and hence it passes the basic criterion of internal consistency. Applying the basic criterion of internal consistency to the conflicting reconstructions leads to the exclusion of Musgrave's case study and hence restricts historiographical pluralism.

In the dispute between Collins and Franklin, evidential considerations become crucial. As mentioned above, the two authors draw on different types of sources. Collins goes beyond the published sources to also include extensive interviews with the scientists involved in the episode under consideration. In fact, in the presentation of his historical account interview excerpts are much more prominent than published material, since the interpretations and arguments surrounding the gravitational radiation experiments are primarily reconstructed on this basis. The consideration behind this strategy is that of circumventing publication bias and gaining insight into the reasoning processes of scientists before they are straightened to fit the format of a scientific journal. Collins explicitly criticizes Franklin for only drawing on the published record. According to Collins, the restriction to scientific publications is simply bad historiographical practice (Collins 1994, pp. 497–499). Is Franklin's decision to cover only the published sources indeed as problematic as Collins' suggests? Does Franklin fail the basic standards of completeness and variance of the evidence?

I think he does, although the situation is complex because of the theory-laden character of source selection. The main problem with Franklin's account is that it excludes a whole class of known sources which would in principle be relevant to the historical argument at stake. The restriction on published sources cannot be justified on the basis that they were the only ones available. On the contrary, the original account that Franklin wishes to disprove includes unpublished material. This means that Franklin draws on a subset of the types of sources used in the original account. How a less complete consideration of existing sources could be better suited for representing the actual process of scientific decision-making remains unclear and we should at least be skeptical whether Franklin's account passes the completeness requirement. However, Franklin justifies his restriction on source material by contending that the arguments that scientists themselves find to be the most convincing, the strongest reasons that they had for accepting or rejecting certain findings, are most likely to be found in the publications. According to Franklin, the publications display those reasons that scientists "wished to have made as part of the permanent record" (Franklin 1994, p. 465). Publications thus serve as a filter for scientists' actual beliefs. Franklin mobilizes the criterion of source reliability against the criterion of source completeness. He justifies his practice of excluding certain sources on the basis that they are less reliable than the ones he brings into focus.

The question is whether this stance is legitimate in light of Franklin's own representational aim, namely that of reconstructing the rational process through which scientists arrived at their verdicts about Weber's observation claim. To regard the restriction as legitimate in light of this aim, one would have to subscribe to the thesis that published results are the best indicators of the actual reasoning processes that brought them about, and that they are better equipped for this task than other types of sources, such as interviews, unpublished manuscripts, letters, and notebooks.

This assumption is almost universally rejected in the historiography of science. In particular, publications provide a retrospective account of scientific debates and most often, they are concerned with justifying the results of a debate rather than with tracing its development. Laboratory notebooks, correspondence, and other contemporaneous sources are not to the same degree tainted by retrospection. If one seeks to reconstruct the actual reasoning processes involved in theory creation, scientific development, and decision making, contemporaneous sources should be preferred over retrospective ones.

But perhaps even more worryingly for Franklin, he himself does not consistently uphold the methodological principle of primarily drawing on the published record. For example, in his case study of Millikan's oil drop experiments Franklin engages in great detail with Millikan's notebooks in order to interpret the former's experimental judgments as rationally justified (Franklin 1986, pp. 140–157). This makes the choice of sources for his reconstruction of the early gravitational waves episode seem arbitrary, rather than methodologically justified. Arguably, Collins' account fares better in this respect. One may point out though that, although Collins' sources are more varied than Franklin's, they are still not robust enough, since the strong emphasis on interviews is not sufficiently balanced with other types of source material. Moreover, if the trouble with citing from the published record is that publications offer only a retrospective view, it is not clear how interviews are superior. They too take a retrospective viewpoint. In the conflict between Franklin and Collins', neutral evaluation criteria favor Collins' reconstruction, but they do so only by a thin margin.

This is where I turn to my second claim. Neutral criteria can decide some historiographical conflicts, but the decisions they yield are relatively weak. A stronger decision could only be reached by drawing on some of the complex evaluation criteria.

There are two reasons why basic criteria are weak arbiters. First, they are only necessary but not sufficient for a historical account to be plausible. Therefore they act only as constraints on the space of alternatives, but do not single out one account as the correct or most plausible one. To illustrate this point, let us return to the dispute between Musgrave and Chang. The application of the internal consistency criterion excludes Musgrave's account from the range of legitimate historical reconstructions. But of course, this does not turn Chang's account into the one unequivocally correct representation of what happened in the Chemical Revolution. On the one hand, internal consistency is not the only evaluation criterion we can draw upon and there may be many reasons to be critical of Chang's reconstruction that have nothing to do with whether it is internally consistent or not (for two recent criticisms of Chang's account, see Blumenthal 2013 and Kusch 2015). On the other hand, I have only considered two of the many different and possibly conflicting reconstructions of the Chemical Revolution. There exist myriad alternative retellings of that episode (a comprehensive overview of the past 50 years of historical writing about the Chemical Revolution can be found in McEvoy 2010), and even without having analyzed them in detail, I contend that at least some of them will meet the basic criteria. These criteria restrict the space of possible alternatives, but they do not shrink it down so radically that it would only include one permissible account. Even after having applied them, there is still room for historiographical pluralism.

The second reason why the basic criteria are weak is that they serve to evaluate only specific case studies, but not more general principles of reconstruction, interpretation and narrative emplotment. While Musgrave's case study has been rejected on grounds of inconsistency, the Musgrave-type of historical analysis has not. Internal consistency is not endemic to a specific type of historical analysis. Could one not tell the story of the Chemical Revolution in a similar manner as Musgrave does, by reconstructing the diachronic development of phlogistonism and oxygenism as competing research programs and by emplotting historical events in a comic form, but without repeating his mistake? We may not be able to express with Musgrave's vigor the conclusion that after 1770 the oxygen-based program was clearly superior. But we could probably still tell the story of the success of the oxygen program as one in which scientific rationality prevailed through complex plot twists.

Or think about the debate between Collins and Franklin. We have seen that Franklin's reconstruction fails because the evidence he adduces is insufficient. However, when I analyzed the methodological differences between the two accounts, I argued that the most important differences do not merely concern the choice of sources, but rather, how historical facts are derived from these sources. I showed that Collins arranges and interprets his sources in such a way as to highlight disagreement and open-endedness, whereas Franklin reconstructs from his sources a historical situation of agreement and robust negative evidence. The fact that the set of covered sources is not exactly coextensive in Collins' and Franklin's reconstructions very likely facilitates them reaching diverging reconstructions and interpretations. However, is it not at least possible that such diverging reconstructions and interpretations could be reached even if the same sources were used? Maybe Franklin would have served his point better had he chosen the same sources as Collins, yet interpreted them according to his own methodological principles. The same seems to be true for narrative structure. Whether the story of the early searches for gravitational radiation is an inverted adventure story in which negative evidence piles up or whether it is a discontinuous tragedy does not seem to be completely determined by the available sources. We need to at least consider the possibility that Franklin could have told his adventure story on the basis of the same sources that were also used by Collins. Historians do enjoy some degree of freedom when it comes to choosing their methodological strategies and forms of narrative emplotment. The basic evaluation criteria help to identify the flaws in specific historical reconstructions. But they are not strong enough to rule out, on a more general level, specific methods of interpretation or forms of emplotment as clearly illegitimate.

In comparison, the complex criteria are significantly stronger as arbiters. First, considerations regarding methodological principles, such as contextualization, explanation and present-centeredness can restrict the space of permissible alternatives much more radically than basic criteria do. The set of historical accounts of the same episode that are internally consistent and that handle the known evidence in a satisfying manner will arguably be much larger than the set of historical accounts that, in addition to satisfying the basic criteria, are also appropriately contextualized (according to a specific understanding of relevant contexts), exhibit explanatory power (according to a specific criterion of explanatory power), avoid anachronisms (according to a specific definition of anachronism), etc.

Second, if we restrict the range of permissible methodological principles, we have ipso facto restricted the range of permissible types of historical reconstruction, not just the set of actually existing acceptable case studies. If we can show that the fault with Franklin's reconstruction goes beyond his handling of the sources, and that it lies in how he uses present-day knowledge in the interpretation and evaluation of Weber's arguments, then we have not only excluded Franklin's particular historical case study, but any historical reconstruction that draws on similar reconstructive and interpretative principles. If we can show that there is something wrong with the practice of rational reconstruction in Musgrave's account, then we have not only excluded this particular case study, but any account of the Chemical Revolution that draws on a Lakatosian conception of scientific rationality.

Hence, the complex methodological evaluation criteria are significantly stronger than the basic ones. But applying them brings us into the center of substantial philosophical conflicts about the nature of science, the relation between past and present knowledge, scientific rationality, theory change, and progress. The complex criteria are strong, but they are highly controversial and anything but neutral with regards to philosophical conflicts.

7.6 Conclusion

There appears to be a dilemma when it comes to evaluating historical reconstructions. On the one hand, there are basic evaluation criteria such as source-reliability, range of the evidence cited and internal consistency. These criteria are relatively neutral with regards to higher-level theoretical and philosophical conflicts. However, these neutral evaluation criteria are not very strong. They restrict pluralism but only to a relatively low degree. On the other hand, there are complex evaluation criteria that are stronger than the basic ones and restrict pluralism more radically. But the complex criteria are themselves contested and are seldom neutral with regards to the fundamental issues that are at stake in a conflict between different historical reconstructions. Put in a nutshell: neutral criteria are weak, strong criteria are not neutral.

This implies that we will have to live with some degree of pluralism in historiography, at least if we wish our decisions between competing historical accounts to be grounded in neutral criteria that are shared by everyone who participates in the debate. This pluralism will be limited because there are at least some neutral considerations that can serve to exclude unacceptable historical accounts. However, even after the clearly illegitimate historical reconstructions that do not meet the basic criteria have been excluded, there can still be plural historical reconstructions of the same historical episodes that support different philosophical doctrines.

Of course, we may not wish to remain neutral in our evaluations of historical reconstructions. A convinced social constructivist may well find dubious the methodological principles that Franklin employs, the interpretations he reaches, as well as his narrative strategies. The constructivist may wish to reject Franklin's historical account on the basis that it is internalist and present-centered and hence fails those complex methodological criteria that call for a more thorough contextualization and historicization of scientific debates. But in doing so, the constructivist has assumed some of the points at issue, namely that a reconstruction of the debate in terms of its technical contents only is deficient and that past beliefs should not be evaluated by present-day standards. The constructivist may have good reasons for holding these views, but a decision between conflicting case studies that is based on them is not a neutral decision. Mobilizing complex criteria in conflicts between historical accounts reinforces historiographical pluralism rather than eradicating it.

Acknowledgments For suggestions and comments on earlier drafts of this paper I am grateful to Martin Kusch, Veli Mitova, Martha Rössler, Katharina Bernhard, and Martin Strauss. I also want to thank the participants of the workshop "The Philosophy of Historical Case Studies" in Bern for their engaging discussions and criticism. Research leading up to this paper was made possible by grants from the Austrian Science Foundation (FWF) ("Contingency, Inevitability and Relativism in the History and Philosophy of Science", Project no.: P25069-G18) and the European Research Council (ERC) ("The Emergence of Relativism", Grant agreement no. 339382).

References

Ashplant, T., and A. Wilson. 1988. Present-centered history and the problem of historical knowledge. *The Historical Journal* 31(2): 253–274.

Bakhtin, M. 1986. The Bildungsroman and its significance in the history of realism (Toward a historical typology of the novel). In *Speech Genres and other Late Essays*, ed. C. Emerson, and M. Holquist, 10–59. Austin: University of Texas Press.

Barnes, B., D. Bloor, and J. Henry. 1996. *Scientific knowledge. A sociological analysis*. London: Athlone.

Blumenthal, G. 2013. On Lavoisier's achievement in chemistry. *Centaurus* 55(1): 20–47.

Butterfield, H. 1949. *The origin of modern science 1300–1800*. New York: The Free Press.

Carr, D. 2008. Narrative explanation and its malcontents. *History and Theory* 47(1): 19–30.

Carroll, N. 2001. Interpretation, history and narrative. In *Philosophical Essays*, ed. Beyond Aesthetics, 133–156. Cambridge: Cambridge University Press.

Chang, H. 2009. We have never been whiggish (about Phlogiston). *Centaurus* 51(4): 239–264.

Chang, H. 2012. *Is water H_2O? evidence, realism and pluralism*. Dordrechtht: Springer.

Collins, H.M. 1981. What is TRASP: The radical programme as a methodological imperative. *Philosophy of the Social Sciences* 11: 215–224.

Collins, H.M. 1985. *Changing order: Replication and induction in scientific practice*. Beverley Hills and London: Sage Publications.

Collins, H.M. 1994. A strong confirmation of the experimenters' regress. *Studies in the History and Philosophy of Science* 25(3): 493–503.

Crombie, A.C. 1953. *Robert Grosseteste and the origins of experimental science, 1100–1700*. Oxford: Clarendon Press.
Cunningham, A. 1988. Getting the game right: Some plain words on the identity and invention of science. *Studies in the History and Philosophy of Science* 19(3): 365–389.
Cunningham, A., and P. Williams. 1993. De-centring the 'Big Picture': The origins of modern science and the modern origins of science. *The British Journal for the History of Science* 26(4): 407–432.
Daston, L., and P. Galison. 2007. *Objectivity*. New York: Zone Books.
Feldhay, R. 1994. Narrative constraints on historical writing. The case of the scientific revolution. *Science in Context* 7(1): 7–24.
Franklin, A. 1986. *The neglect of experiment*. Cambridge: Cambridge University Press.
Franklin, A. 1994. How to avoid experimenters' regress. *Studies in the History and Philosophy of Modern Physics* 25: 463–491.
Frye, N. 1957. *The anatomy of criticism. Four essay*. Princeton: Princeton University Press.
Giere, R.N. 2004. How models are used to represent reality. *Philosophy of Science* 71(5): 742–752.
Holton, G. 1978. Subelectrons, presuppositions, and the Millikan-Ehrenhaft Dispute. *Historical Studies in the Physical Sciences* 9: 161–224.
Jardine, N. 2000. Uses and abuses of anachronism in the history of sciences. *History of Science* 38(3): 251–270.
Kellert, S.H., H.E. Longino, and C.K. Waters. 2006. *Scientific pluralism*. Minneapolis: University of Minnesota Press.
Kinzel, K. 2012. Geschichte ohne Kausalität. Abgrenzungsstrategien gegen die Wissenschaftssoziologie in zeitgenössischen Ansätzen historischer Epistemologie. Berichte zur. *Wissenschaftsgeschichte* 35(2): 147–162.
Kinzel, K. 2015. Narrative and evidence. How can case studies from the history of science support claims in the philosophy of science. *Studies in History and Philosophy of Science* 49: 48–57.
Koyré, A. 1957. *From the closed world to the infinite universe*. Baltimore: Johns Hopkins University Press.
Kusch, M. 2015. Scientific pluralism and the chemical revolution. *Studies in History and Philosophy of Science* 49: 69–79.
Ladyman, J. 2011. Structural realism versus standard scientific realism: The case of phlogiston and dephlogisticated air. *Synthese* 180(2): 87–101.
Lindberg, D.C. 1990. Conceptions of the scientific revolution from Bacon to Butterfield: A preliminary sketch. In *Reaprraisals of the Scientific Revolution*, ed. D.C. Lindberg, and R.S. Westman, 1–27. Cambridge: Cambridge University Press.
McEvoy, J. 2010. *The historiography of the chemical revolution: Patterns of interpretation in the history of science*. London: Pickering and Chatto.
Musgrave, A. 1976. Why did oxygen supplant phlogiston? Research programmes in the chemical revolution. In *Method and appraisal in the physical sciences. The critical background to modern science, 1800–1905*, ed. C. Howson, 181–210. Cambridge: Cambridge University Press.
Pomata, G. 1998. Close-Ups and long shots. Combining particular and general in writing the histories of women and men. In *Geschlechtergeschichte und allgemeine Geschichte. Herausforderungen und Perspektiven*, ed. H. Medick, and A.-C. Trepp, 99–124. Göttingen: Wallstein.
Porter, R. 1986. The scientific revolution: A spoke in the wheel? In *Revolution in history*, ed. R. Porter, and M. Teich, 290–316. Cambridge: Cambridge University Press.
Roth, P.A. 1988. Narrative explanations. The case of history. *History and Theory* 27(1): 1–13.
Sapp, J. 1990. The nine lives of Gregor Mendel. In *Experimental inquiries: Historical, philosophical and social studies of experimentation in science*, ed. H. Le Grand, 137–166. Dordrecht: Kluwer.
Secord, J.A. 1993. Introduction to 'The Big Picture'. *The British Journal for the History of Science* 26(4): 387–389.
Shapin, S. 1998. *The scientific revolution*. Chicago, London: University of Chicago Press.
Tosh, N. (2003). Anachronism and retrospective explanation. In Defence of a present-centered history of science. *Studies in the History and Philosophy of Science* 34(3): 647–659.

van Fraassen, B.C. 2008. *Scientific representation. Paradoxes of perspective*. Oxford: Clarendon Press.
White, H. 1973. *Metahistory. The historical imagination in nineteenth-century Europe*. Baltimore, London: Johns Hopkins Press.
White, H. 1978. The historical text as a literary artifact. In *The writing of history. literary form and historical understanding*, ed. R.H. Canary, and H. Kozicki, Madison: University of Wisconsin Press.
White, H. 1980. The value of narrativity in the representation of reality. In *The content of the form. Narrative discourse and historical representation*, 1–25. Baltimore: Johns Hopkins University Press.
Yates, F.A. 1964. *Giordano Bruno and the Hermetic tradition*. Chicago: University of Chicago Press.

Chapter 8
Contrasting Cases: The Lotka-Volterra Model Times Three

Tarja Knuuttila and Andrea Loettgers

Abstract How do philosophers of science make use of historical case studies? Are their accounts of historical cases purpose-built and lacking in evidential strength as a result of putting forth and discussing philosophical positions? We will study these questions through the examination of three different philosophical case studies. All of them focus on modeling and on Vito Volterra, contrasting his work to that of other theoreticians. We argue that the worries concerning the evidential role of historical case studies in philosophy are partially unfounded, and the evidential and hermeneutical roles of case studies need not be played against each other. In philosophy of science, case studies are often tied to conceptual and theoretical analysis and development, rendering their evidential and theoretic/hermeneutic roles intertwined. Moreover, the problems of resituating or generalizing local knowledge are not specific to philosophy of science but commonplace in many scientific practices—which show similarities to the actual use of historical case studies by philosophers of science.

8.1 Introduction

Philosophers of science make frequent use of case studies, and the use of case studies has become even more prevalent in recent years with the more marked practice–orientation of even mainstream philosophy of science. Yet the case study methodology is rarely discussed by philosophers of science, and even more rarely is their own use of case studies reflected on. Some philosophical reflections on case studies

T. Knuuttila (✉)
Department of Philosophy, University of South Carolina, Columbia, USA
e-mail: tarja.knuuttila@helsinki.fi

A. Loettgers
Center for Space and Habitability, University of Bern, Bern, Switzerland
e-mail: andrea.loettgers@unige.ch

A. Loettgers
Department of Philosophy, University of Geneva, Geneva, Switzerland

T. Sauer and R. Scholl (eds.), *The Philosophy of Historical Case Studies*,
Boston Studies in the Philosophy and History of Science 319,
DOI 10.1007/978-3-319-30229-4_8

(e.g. by Shrader-Frechette and McCoy 1994; Morgan 2012, 2014; Ankeny 2012), discuss the use of cases in scientific discourses but not as vehicles of philosophical theorizing. The lack of reflection among philosophers concerning their own use of case studies seems curious, since, on the face of it, the philosophical use of cases might seem problematical—unless they were understood as a mere means of illustration. The issue, of course, concerns generalization. Philosophical reflection often moves on a general conceptual level, and the question is how a single case, or any limited number of cases, for that matter, is going to give us general insights. In other words, is there a gap between specific historical cases and general philosophical theorizing?

Pitt (2001) thinks that this is indeed the case: according to his analysis, the use of historical cases as evidence for philosophical theorizing is highly problematical. In particular, their use is subject to the following dilemma. On the one hand, if the cases were used to back up a general philosophical claim, the question is how one, or a few, episodes of science can really establish it, and, furthermore, to what extent they have been fabricated in order to fit the philosophical thesis in question. On the other hand, if one starts from the case study and attempts to work one's way up to interesting philosophical conclusions, it is far from clear how this is supposed to be happening. Pitt's rather skeptical conclusions concerning the possible role of historical case studies in philosophy has generated a debate with some of counterarguments and qualifications. Pitt concludes: "even very good case studies do no philosophical work. They are at best heuristics" (Pitt 2001, p. 373). In response to Pitt, Burian (2001, p. 387) argued that Pitt's "dilemma is a false dilemma." He demonstrates with reference to some examples drawn from the history of molecular biology that case studies help to shed light on such styles of scientific work and modes of argumentation that had not, so far, received due recognition in standard philosophical analyses. According to Burian, Pitt's dilemma only applies if we take the doubtful view of philosophy of science as aiming at the discovery of a universal or objective scientific method. He sees more fruitful philosophical work done in particular contexts, offering only limited, fallible generalizations–yet having also transformative power in regard to our view of science.

Writing a decade later, in reviewing the history of history and philosophy of science (HPS) and the confrontations between the two disciplines, Schickore (2011) takes up the argument between Burian and Pitt. She agrees with Pitt's conclusion that it is unjustified to generalize on the basis of one or a few episodes to wholesale claims concerning science—adding that the same applies for attempts to discard any such claims drawing only on a case or two (a genre that has been particularly vigorous in science and technology studies). What she does not believe in are attempts at rescuing the case study methodology by using sets of longitudinal or comparative case studies—an approach she attributes to Burian. Instead, Schickore thinks that there is something fundamentally wrong in approaching philosophical analysis in terms of the practice of natural science, which seeks to confront the theory with evidence. According to such "confrontation model", philosophy of science is developing a theory of science and the role of history of science is to provide data for its confirmation or falsification. In contrast to this confrontation model, Schickore advances a hermeneutical account that generates understanding of how scientific concepts, norms, and practices have developed. The hermeneutic circle of Schickore's proposal contains both historical and philosophical insights and allows the modification of concepts through the work on historical cases.

Recently, Kinzel (2015) has portrayed this discussion by distinguishing between the evidential and hermeneutic approaches.[1] She points out, rightly we think, that the confrontation model does not work in the natural sciences either, and so argues for the evidential role of historical case studies. They are, according to her, capable of providing some evidential support for philosophical theorizing, provided that their theory-laden narrative and constructed nature are properly understood. The above short description of the discussion spawned by Pitt's skepticism concerning the philosophical use of historical case studies shows that the discussion has tended to revolve around the issues concerning generality versus locality of the claims supported by case studies, and the hermeneutic versus evidential functions of historical cases. In particular, it seems that all the participants of the debate (except perhaps Burian) take it that the hermeneutic/narrative/theory-laden nature of historical case studies limits their evidential role.[2] But does this need to be the case? Is it possible to conceive of historical (and empirical) case studies in a way that does not see the interpretative, theory-laden nature of historical case studies as a limitation for their evidential role for philosophical reasoning? In short, is it possible to conceive of a particular philosophical way of using historical case studies that combines their evidential-cum-theoretical nature? And how would such an account stand in terms of the question concerning the general versus local nature of case-based reasoning?

In what follows, we will approach these questions through the examination of three different philosophical case studies, one of which was authored by us. As philosophers of science, having studied and presented several case studies (e.g. Knuuttila 2006, 2013; Loettgers 2007; Knuuttila and Loettgers 2013, 2014), we will engage in this article in self-reflection of our own method presenting one of our case studies. Our endeavor is motivated by the realization that the discussion on the use of historical case studies within philosophy of science so far has largely proceeded on a rather general level. Consequently, a more case-based investigation of the philosophical use of historical cases may seem in order.

The three different case studies we will consider focus on modeling and on the work of the Italian physicist and mathematician Vito Volterra—they all take his version of the Lotka-Volterra model as a point of departure. The model is a staple of philosophical discussion about mathematical modeling. What binds the three case studies together is the fact that both our own discussion (Knuuttila and Loettgers 2016) and that of Scholl and Räz (2013) were written in part as responses to Weisberg's (2007) influential "Who is a Modeler?". In this chapter, Weisberg contrasts Volterra's version of the Lotka-Volterra to Mendeleev's periodic table claiming that of the two scientists only Volterra was a modeler. Weisberg then uses the contrast

[1]Kinzel gives also Chang (2012) as an example of a proponent of a hermeneutic approach.

[2]Some vestiges of the confrontation model seem to be at play here, even in the rejection of it.

between Volterra's and Mendeleev's styles of theorizing to support his claim that modeling is a distinct form of theorizing.

What is especially interesting about the three studies on Volterra's modeling is that all of them contrasted his work to that of another theoretician to deliver their philosophical points. The contrasts used were different, however. Scholl and Räz (2013) contrasted Volterra's modeling to Darwin's theory of the formation and distribution of coral atolls in the Pacific Ocean (also discussed by Weisberg 2007). In our work, we contrasted Volterra's version of the Lotka-Volterra model to Lotka's version of it (Knuuttila and Loettgers 2013). We suggested that the contrast between Volterra and Lotka provides interesting material for evaluating Weisberg's claims concerning modeling since Weisberg only considered Volterra's construction of the Lotka-Volterra model. Our case studies showed that although Volterra and Lotka presented models that, from a formal point of view, looked identical, they nevertheless followed different kinds of modeling strategies (Knuuttila and Loettgers 2016).

In what follows, we will first discuss Michael Weisberg's contrasting cases, turning then to Scholl and Räz's (2013) article that makes use of a different contrast in arguing against Weisberg's position. Our own case study follows. In the discussion and concluding chapter, we will contrast and compare the three different case studies with each other, asking what kind of philosophical insight they offer, and how they do it. In particular, we will suggest that the worries concerning the evidential role of case studies, and their contextualization, are partially based on an inadequate understanding of how historical case studies are used in actual philosophical practice. As philosophical discourses are often tied to conceptual analysis and development, the evidential and hermeneutic roles are intertwined in case-based reasoning. Cases are often studied in order to advance a philosophical argument, and in such use case studies provide both evidence and interpretative resources for exploring and developing philosophical concepts and theories. In this work, various ways of expanding the depth and coverage of the case-based argumentation may be used, one of which is making use of the vehicle of contrasting different scientific cases.

8.2 Modeling as Indirect Representation

In his "Who is a Modeler?" Weisberg (2007) argues that many standard philosophical accounts fail to distinguish between different forms of theorizing. What he is interested in is articulating modeling as a distinct theoretical practice. According to him, the goals, procedures, and representations employed by modelers and other kinds of theorists differ. In particular, he distinguishes between two types of theorizing: modeling and abstract direct representation. Modelers, according to Weisberg, study real-world phenomena through the detour of creating hypothetical simplified entities, models. That is, they practice the art of *indirect representation*. In contrast, the theorists practicing *abstract direct representation* strive to represent the data or real-world phenomena directly.

The central philosophical thesis of "Who is a Modeler?" revolves, then, around the notion of indirect representation and the peculiar way models relate to real-world phenomena. Models form a class of theoretical representations that are not constructed by representing the real target systems. The consideration of the latter first enters the process of modeling at a later stage that runs counter to the traditional representational idea of models as inherently *models* of some real target systems.[3] The claim that model construction happens *before* the possible real target systems are being considered challenges the traditional understanding of models as representations of some real-world target systems. Much of the scientific discussion has (explicitly or implicitly) taken models as prototypical scientific representations and proceeded to analyze the notion of representation through modeling (see, e.g., Bailer-Jones 2009; Contessa 2007; da Costa and French 2000; French and Ladyman 1999; Frigg 2010; Giere 2004; Morgan and Morrison 1999; Maeki 2009; Suárez 2008). Weisberg departs from such assumptions in pointing out the diversity of ways in which theoretical representations may be built, and related to real world systems (see also Weisberg 2013).

In making the distinction between modeling and abstract direct representation, Weisberg redirects the focus from models to the activity of *modeling*. He proceeds in three stages. Firstly, a model is being constructed, after which, secondly, the modeler refines, analyses, and articulates its properties and dynamics. Only at the third stage, the relationship between the model and any real-world target system is assessed, if such an assessment is deemed necessary. Often modelers are predominantly interested in studying model systems themselves, and so the relationship between model and any real-world target may be left implicit at best. This characteristic feature of contemporary modeling practice tends to go unnoticed if models are understood in the traditional representational fashion.

Interestingly, after characterizing abstract direct representation and modeling in a preliminary fashion, Weisberg proceeds to analyze this distinction in a more fine-grained manner making use of two cases. He contrasts Vito Volterra's modeling style of theorizing from abstract direct representation as exhibited by Dimitri Mendeleev's periodic table. In his argument, Weisberg makes only use of Volterra's articles published in 1926 in Italian and English (Volterra 1926a, b). According to Weisberg, Volterra studied post-World War I fish populations in the Adriatic Sea by "imagining a simple biological system composed of one population of predators and one population of prey" (Weisberg 2007, p. 208). He attributed to this hypothetical system only a few properties writing down a couple of differential equations to describe their mutual dynamics. The word "imagining" used by Weisberg captures the difference between the procedures of direct and indirect representation. He underlines that Volterra did not arrive at these model populations by abstracting away from properties of real fish, but rather constructed them by stipulating certain of their properties (Weisberg 2007, p. 210). In contrast to Volterra, Weisberg claims, Mendeleev built his Periodic Table through abstractions from data in order to identify some central

[3]In recent discussion on modeling, the idea of models as fictions has been entertained by several authors, e.g. Suárez (2008).

factors of chemical behavior. Thus, in contrast to modelers constructing hypothetical systems, he was trying to "represent trends in real chemical reactivity, and not trends in a model system" (Weisberg 2007, p. 215, footnote 3).[4]

Both abstract direct representation and indirect representations abstract, approximate, select, and idealize, so the difference between the two does not hang on these procedures of scientific representation. What distinguishes modeling from abstract direct representation is that it proceeds by describing another simpler, hypothetical system. Consequently, models should be considered independent objects in the sense of being independent from some determinable real target systems.[5] This implies also an important difference on how one gains knowledge via modeling and abstract direct representation. Namely, Weisberg claims that in abstract direct representation "anything the theorist discovers in her analysis of the representation is a discovery about the phenomenon itself, assuming that it was represented properly. There is no extra stage where the theorist must coordinate the model to a real phenomenon'—as is the case with modeling (Weisberg 2007, pp. 226–227). Although it seems that Weisberg exaggerates the extent to which abstract direct representation can allow the direct study of the phenomenon, yet such difference may explain why the discussion on scientific representation has focused on modeling. Models seem to provide the hard case that the various accounts of scientific representation are designed to address.

As already mentioned, Weisberg's account pays attention to the fact that models are often studied quite apart from any representational relationships that they might have to real-world systems. Moreover, many scientific models are far too simple to be considered as models of some actual target systems, although they may bear similarities to them. Secondly, scientists also study models of phenomena that are not known to exist. If it is better understood why these phenomena do not exist, one has also gained some understanding of the phenomena that do exist. In both of the aforementioned situations, Weisberg claims that "it is clear that the model and only the model is the object of study" (Weisberg 2007, p. 223).

While Weisberg's thesis of modeling as indirect representation thus brings to the fore some aspects of modeling largely neglected by the philosophical discussion so far, the account seems lacking in some crucial respects. This becomes visible if we examine how Weisberg thinks a modeler can be recognized: "To judge whether or not a particular theorist is a modeler," argues Weisberg, "[w]e will actually need to know something about how the theory was developed and how the modeler set

---

[4]Godfrey-Smith (2006) likewise distinguishes between indirect representation and abstract direct representation—and invokes examples in trying to account for the difference between the two strategies of theorizing. Godfrey-Smith studies two influential books on evolutionary theory: Leo Buss's *The Evolution of Individuality* (1987) and Maynard Smith and Szathmáry's *The Major Transitions in Evolution* (1995). Buss examines the "actual relations between cellular reproduction and whole-organism reproduction in known organisms" (Godfrey-Smith 2006, p. 731), while Maynard Smith and Szathmáry put forth "idealized, schematic causal mechanisms."

[5]On models as independent or autonomous entities, see also Morgan and Morrison (1999) and Knuuttila (2005).

about trying to represent the world" (Weisberg 2007, p. 222). Let us consider, then, what Weisberg says about how Volterra went about developing his version of the Lotka-Volterra model:

> Volterra began his investigation of Adriatic fish not by looking directly at these fish or even the statistics gathered from the fish markets, but by constructing a model. This is characteristic of the first stage of modeling. *He imagined a population of predators and a population of prey, each with only two properties. Setting this idea to paper, he wrote down equations specifying the model that he had imagined.* (Weisberg 2007, p. 222, emphasis added).

We find this a gross simplification of actual modeling practice and its reliance on the already established computational methods and representational tools. In our case study, we will study how Volterra and Lotka constructed their respective models. As we will argue, Volterra does not qualify as a prime example of a modeler, since he pursued the essential or sufficient components of the real predator-prey system. Although what he eventually accomplished suits Weisberg's claims about modeling, his original intentions were nonetheless different. Lotka provides a better example of a modeler, but for reasons that are not discussed by Weisberg. Lotka started from a systems theoretical perspective, developing a general model template, which he applied to the analysis of biological and chemical systems. This kind of approach is becoming prevalent in modeling complex systems. It does not start from imagining simplified hypothetical systems (still somehow connected to some particular real-world systems) but from applying cross-disciplinary computational templates to various subject matters (cf. Humphreys 2002, 2004). Before going into our case studies, we will discuss the critique of the thesis of indirect representation by Scholl and Räz (2013). They approach modeling in terms of "insufficient epistemic access" and use as contrast cases Volterra and Darwin, a comparison already made by Weisberg (2007). However, Scholl and Räz are mainly referring to Volterra's much later work, co–authored with D'Ancona (Volterra and D'Ancona 1935), whereas Weisberg relies exclusively on Volterra's original articles from 1926 (Volterra 1926a, b). On our view, this may partially explain their different interpretations of Volterra.

8.3 Modeling Is Not About Indirect Representation

Although Weisberg launches the contrast between abstract direct representation and indirect representation mainly in terms of the contrast between Volterra and Mendeleev, later in his article he also discusses Darwin's theory of the origin and distribution of the coral reefs as an example of abstract direct representation (Weisberg 2007, pp. 227–228). Scholl and Räz (2013) focus on the contrast between Volterra's and Darwin's causal reasoning in their critique of Weisberg's thesis of indirect representation. Despite their critique, Scholl and Räz's interpretation of Volterra is also very much in line with what Weisberg claims. They point out that according to Volterra and D'Ancona their "deductive approach" does not attempt to "extract ecological laws directly from experimental data or observation." Instead, it proceeds on a "constructive path" through hypotheses about basic causal relationships, integrating them "into a system of interactions" (Scholl and Räz 2013, p. 120). Moreover, they adopt from Weisberg the distinction between dynamical and representational fidelity (see below). Where the two interpretations differ most, is how Weisberg, on the one hand, and Scholl and Räz, on the other hand, define modeling.

While Weisberg uses the contrast between Volterra's and Mendeleev's (and Darwin's) approaches to elucidate further his notion of modeling as indirect representation, Scholl and Räz extract their notion of modeling from Volterra and D'Ancona (1935), arguing that Volterra and D'Ancona chose a modeling approach because direct methods were not available for the problem they were studying. What would have been the direct methods?[6] Volterra and D'Ancona (1935) distinguish and discuss three different "direct" methods. Each of the methods has their own limitations when it comes to the studying of predator and prey dynamics. Firstly, there is the experimental method of studying individual causes in isolation in controlled conditions. Experiments on individual animals under laboratory conditions would allow for causal inference, but ecologists study interactions of entire *populations* of animals, making this approach deemed unsuitable by Volterra and D'Ancona. Consequently, the second alternative would be to overcome this limitation by transferring the method of causal inference to ecology by performing breeding experiments on entire populations.

In order to perform such controlled breeding experiments several requirements have to be fulfilled: A space whose dimension has to be proportional to the size of the animal is needed to run the experiment, and the length of the experiment has to comply with the life expectations of the animals and their breeding cycles. Finally, the environmental conditions would need to be controlled. Such experiments were actually performed by the Russian biologist Gause (1935) who used micro-organisms instead of fish. Micro-organisms have two important advantages: they do not need a lot of space and have short generation times. Under controlled laboratory conditions, Gause was able to explore how the prey micro–organisms developed in isolation, and how the situation changed when the predator micro–organisms were added. As promising as these experiments seemed to be to Volterra and D'Ancona, they turned out not to be able to replicate the situation modeled by the Lotka-Volterra model.

Finally, the third possibility considered by Volterra and D'Ancona, field experiments, seemed infeasible because of the large number of uncontrollable factors interfering with the population dynamics and would require intensive effort.

Based on the problems, or outright impossibility, of direct methods, Volterra and D'Ancona then argue in favor for modeling, i.e. for their "deductive method." This leads Scholl and Räz to conclude that it was precisely the "insufficient epistemic access" that forced the two scientists to resign to a modeling approach. Such an approach is described by the two authors as follows: "We begin with hypotheses about basic causal relationships and integrate them into a system of interactions. Then we check whether the constructed system, the model, is applicable to the target system" (Scholl and Räz 2013, p. 120). This procedure of constructing and apply-

[6]We are here largely following the discussion by Scholl and Räz (2013).

ing a hypothetical model is very similar to Weisberg's indirect modeling approach. However, they proceed to claim, in contrast to Weisberg, that Darwin faced the same problem of insufficient epistemic access in his attempt to causally explain the formation, origin, and distribution of coral reefs and atolls in the Pacific Ocean. And so he was, by their criteria, also a modeler: "Darwin's investigation has all the hallmarks of model-based science … The subsidence of islands and the growth of corals occur over hundreds of thousands of years, distributed over the entire Pacific Ocean, and so we can have no hope of directly investigating the process" (ibid., 127).

The starting point of Darwin's modeling process was an island surrounded by a fringing reef. According to his observations, corals prefer warm and shallow waters, surrounding volcano islands. In imagining the hypothetical model, Darwin proceeded from fringing reefs to barrier reefs. The formation of barrier reefs starts with an island that sinks down. As a natural consequence also the corals will submerge under the water and die. These dead corals provide the basis for new growing corals, which will be now further away from the island because the island got smaller by sinking down. If the island keeps subsiding the coral reef keeps growing on its own foundation up the point where the island is completely under the water and forms therefore a coral atoll. In such step-by-step reasoning, Darwin was able to "render all the known forms of coral islands" (ibid., 128).

The main difference between Volterra and D'Ancona and Darwin is not due to the method of modeling, according to Scholl and Räz, but rather the fact that Darwin was able to provide an "how actually" model of the formation of coral reefs and islands in contrast to the "how possibly" models of Volterra and D'Ancona. Darwin's account was able to "mirror"—a not so fortunate choice of word by Scholl and Räz—the causal structures of the target system, due to his careful observation of the coral reef formation at its different stages. Scholl and Räz depict Darwin's quest from a "how possibly" model, via empirical demonstration of the hypothesized mechanism to adducing empirical evidence "in support of the claim that the model faithfully represents the actual causal processes responsible for the growth of coral atolls" (ibid., 130). As a result, his model was representationally faithful instead of succeeding only to reproduce empirical phenomena. Such dynamical fidelity was what Volterra and D'Ancona were eventually only able to accomplish.

Although thus Weisberg as well as Scholl and Räz are largely in agreement on their analysis of Volterra, they end up presenting his modeling exercise in different terms. In what follows, we will argue, presenting yet another case study on Volterra, that their disagreement can be partially settled by paying attention to the fact that they make use of different writings of Volterra. Whereas Weisberg focuses on the papers in which Volterra published his version of the Lotka-Volterra model for the first time (Volterra 1926a, b), Scholl and Räz discuss at length the work that appeared nearly a decade later representing a mature state of a research program inspired by the early papers of Volterra and D'Ancona (1935). Furthermore, using Alfred Lotka's very different design of the Lotka-Volterra model as a contrasting case, we draw attention to the actual tools of modeling that both Weisberg as well as Scholl and Räz have glossed over in their analysis. The differences between Volterra and

D'Ancona, and Darwin are to a large degree due to the fact that the former were engaged in mathematical modeling. Such theoretical activity is largely dependent on mathematical tools and methods that are often interdisciplinary by their nature.[7]

8.4 The Design of the Lotka-Volterra Model by Volterra

Weisberg begins his story of the origin of the Lotka-Volterra model with the problem presented by Umberto D'Ancona (1896–1964) to the world-renowned mathematical physicist Vito Volterra (1860–1940) in 1925. D'Ancona, a marine biologist and Volterra's son-in-law, had made a statistical study of the Adriatic fisheries over the period 1905–1923. The data showed an unusual increase in predators during the final period of the First World War and immediately after, when fishing was hindered by the war. D'Ancona's aim was to get mathematical support for the thesis that cessation of fishing was favorable for predator fish. Thus Volterra set out to "mathematically explain" D'Ancona's data on "temporal variations in the composition of species" (Volterra 1927, p. 68). He had no prior experience of fisheries, yet this problem sparked his longer-term research program on the inter-species dynamics that culminated in *Leçons sur la théorie mathématique de la lutte pour la vie* (Volterra 1931) and *Les associations biologiques au point de vue mathématique* (Volterra and D'Ancona 1935). However, the way Volterra went about modeling the predator-prey system can be traced further back in time.

8.4.1 Volterra and the Mathematization of Biology and Social Sciences

Already decades before the formulation of the Lotka-Volterra model, Volterra was interested in the mathematization of biology and social sciences. At the opening of the academic year at the University of Rome in 1901, he delivered an Inaugural Address entitled *On the Attempts to Apply Mathematics to the Biological and Social Sciences* (Volterra 1901). In this talk, Volterra spoke in favor of translating "natural phenomena into arithmetical or geometrical language" and by doing so opening "a new avenue for mathematics" within biology and social sciences. An important ingredient in this transformation process was provided by mechanics: biology and social sciences should be mathematized according to the example provided by mechanics. For Volterra it constituted "together with geometry, if not the most brilliant then surely the most dependable and secure body of knowledge" (ibid., p. 251).

On the other hand, physics at the beginning of the 20th century was ridden by the apparent failure of the mechanistic world-view. This led Volterra to remark that instead of the "illusions about giving a mechanical explanation of the universe"

[7]Our case studies are based on our earlier work (Knuuttila and Loettgers 2016).

one should "more modestly, [be] satisfied by analogy, and especially mathematical analogy" (Volterra 1901, p. 255). He thought that a large part of mathematical physics would still be usable, especially differential equations.[8]

The application of mathematics to social science and biology involved, for Volterra, transforming qualitative elements into quantitative measurable elements, measuring the variations, idealizing and abstracting the systems and processes under investigation, representing them with differential calculus, and forming hypotheses in the same fashion as in mechanics. The goal of idealization and abstraction was to identify the "fundamental parameters" governing the "change in the corresponding variable elements of the phenomena" (ibid., p. 255). Volterra saw in economics a good example of a science modeled on mechanics:

> The concept of *Homo economicus*, which has prompted so much discussion and provoked such enormous difficulty that there are still those who refuse to accept it, comes so naturally to our mechanist that he is surprised at the suspicions aroused by this abstract, schematic being. He sees in *Homo economicus* a concept similar to those that, by long habit, have become familiar to him. He is used to idealizing surfaces as frictionless, wires as inextensible, solids as undeformable, and to substituting perfect liquids and gases for the natural kind. Not only has he made a habit of all this, but he also knows the advantages of doing so. (Volterra 1901, p. 252).

Yet it appears contradictory to transfer the modeling methods and concepts of mechanics to another entirely different areas of study and strive, simultaneously, to capture the fundamental factors behind the phenomena in question. Is there a reason to suppose that the mechanical approach works in such fields as biology, or social sciences, taking into account the complexity of the phenomena they study? As we will show in the next section, this was precisely the reason why Volterra had to resort to "the method of hypothesis" in modeling biological associations.

8.4.2 Volterra's Method of Hypothesis

According to his methodological ideals, Volterra embarked on accounting for D'Ancona's statistical data by "isolating those factors one wishes to examine, assuming they act alone, and by neglecting others" (Volterra 1927, p. 67). First, he distinguished between "external" and "internal" causes. External causes were "periodic circumstances relating to the environment, as would be those, for example, which depend upon the changing of the seasons, which produce oscillations of an external character in the number of the individuals of the various species" (Volterra 1928, p. 5). Volterra was focusing on internal causes that had "periods of their own which add their action

[8] Volterra had started his scientific career as a mathematician and had made important contributions to the theory of calculus. This work is summarized in Volterra's book *Theory of Functionals and of Integral and Integro-Differential Equations* (Volterra 1930).

to these external causes and would exist even if these were withdrawn" (ibid.). However, this was just a starting point for him since already at the beginning he had a larger picture in mind. He went on to model more complicated cases, adding also some effects of the environment. The Lotka-Volterra model was merely one of the basic models of *biological associations* with which Volterra referred to stable associations that "are established by many species which live in the same environment" (Volterra 1928, p. 4). In the paper in which he presents the Lotka-Volterra model for the first time (Volterra 1926b, 1928),[9] he begins from a consideration of one species alone and then adds other species. The first association he models is that between two species which contend for the same supply of food. After that he formulated the Lotka-Volterra model on two species, one of which feeds upon the other.

Although Volterra strove to separate external and internal causes, he admitted that they could be interrelated in a myriad of ways. Interacting species in a changing environment constitutes a problem of a much higher degree of complexity than the systems studied in classical mechanics. The mathematical methods and techniques of mechanics could not be applied off-hand to the study of the predator-prey dynamics. Even if the variations observed in populations living in the same environment showed some well-known characteristics observed in many mechanical systems, such as oscillatory behavior, it was unclear, which were the components of the system and in which ways they interacted. Consequently, Volterra faced the following dilemma: On the one hand, the complexity of the system had to be rendered manageable, enabling the use of certain mathematical tools. On the other hand, the available mathematical tools and methods exhibited a serious constraint on the kinds of structures and processes that could be studied. Volterra reflected on this situation in the following way:

> But on the first appearance it would seem as though on account of its extreme complexity the question might not lend itself to a mathematical treatment, and that on the contrary mathematical methods, being too delicate, might emphasize some peculiarities and obscure some essentials of the question. To guard against this danger *we must start from the hypotheses*, even though they be rough and simple, and give some scheme for the phenomenon. (Volterra 1928, p. 5, emphasis added)

Since he could not isolate internal causes from external causes, due to the complexity of the interactions between the components of the system of interest, he constructed a hypothetical system with the help of certain assumptions concerning them. Some of these assumptions were directly due to the application of differential calculus to the problem of predation (see also our discussion on the method of isolation in Sect. 5.1). The central assumptions made in the construction process of the model were:

---

[9]Volterra (1928) is a partial English translation of the Italian original (Volterra 1926b); in the following, references are made to the 1928 translation.

- The numbers of species increase or decrease in a continuous way, which makes them describable by means of differential equations.
- Birth takes place continuously and is not restricted to seasons. The birth-rate is proportional to the number of living individuals of the species. The same assumption is made for the death rate.
- Homogeneity of the individuals of each species, which neglects the variations of age and size.

Thus Volterra concentrated exclusively on the dynamics between predators and preys by formulating a simplified hypothetical system consisting solely of "*the intrinsic phenomena*" due to the voracity and fertility of the co-existing species (Volterra 1927, p. 68). This strategy of "starting from the hypotheses" allowed Volterra to apply well-known mathematical tools and methods to the study of biological associations.

8.4.3 The Construction of the Lotka-Volterra Model by Volterra

In deriving the Lotka-Volterra equations, Volterra started out from a situation in which each of the species is alone. In this situation, he assumed, the prey would grow exponentially and the predator in turn would decrease exponentially, because of missing food resources. Translated into the language of mathematics, the development of prey and predator populations is described by the following two differential equations describing the change in time t of the prey (N_1) and predator (N_2) populations:

$$\frac{dN_1}{dt} = \varepsilon_1 N_1, \qquad \frac{dN_2}{dt} = -\varepsilon_2 N_2, \tag{8.1}$$

where $\varepsilon_1, \varepsilon_2 > 0$ are constants. Integration of the two differential equations leads to an exponential increase of the prey and an exponential decrease of the predator population, with $N_{i,0}$ referring to the numbers of individuals at time 0.

$$N_1(t) = N_{1,0}e^{\varepsilon_1 t}, \quad N_2(t) = N_{2,0}e^{-\varepsilon_2 t}. \tag{8.2}$$

Exponential growth or decrease is the simplest way of describing the development of a population in time. It does not take into account any environmental influences or the obvious fact that there is must be an upper limit of the population sustainable by the resources provided by the environment. To allow for the interaction between prey and predator populations, Volterra introduced a coupling term in each equation. The combined predator and prey system is described by the following set of differential equations:

$$\frac{dN_1}{dt} = (\varepsilon_1 - \gamma_1 N_2)N_1, \tag{8.3}$$

$$\frac{dN_2}{dt} = (-\varepsilon_2 + \gamma_2 N_1)N_2. \tag{8.4}$$

The interaction between predator and prey is now described by the terms involving the product $N_1 N_2$, which introduces a non-linearity into the system in addition to the coupling of the two differential equations. The proportionality constant γ_1 links the prey mortality to the number of prey and predators, and the constant γ_2 links the increase in predators to the number of prey and predators. One of the possible solutions to these coupled non-linear differential equations is an oscillation in the numbers of predator and prey. Concerning those oscillating solutions Volterra wrote:

> From the analytical viewpoint, it is to be noted that the study of fluctuations or oscillations of the number of individuals of species living together, [...] falls outside the ordinary study of oscillations, because in these researches we had to deal generally with non-linear equations, whereas the classical study of the theory of oscillations involves linear equations. (Volterra 1928, p. 23)

The mathematical analysis of the resulting equations gave Volterra some important results—including a solution to D'Ancona's problem concerning the relative abundance of predatory fish during the war years. Volterra summarized his results in what he called the "three fundamental laws of the fluctuations of the two species living together" (Volterra 1928, p. 20). The third law states that if an attempt were made to destroy the individuals of the predator and prey species uniformly and in proportion to their number, the average number of the prey would increase and the average number of the predator would decrease.[10] As regards fisheries, this "law" was anticipated already by Lancaster (1884).[11] Volterra himself quoted Darwin: "If not one head of game were shot during the next 20 years in England, and at the same time no vermin were destroyed, there would in all probability be less game than at present, although hundreds of thousands of game animals are now annually shot" (Volterra 1926a, p. 559; Darwin 1882, pp. 53–54). For Volterra his long-term research on "biological associations" was a contribution to the Darwinian theory of struggle for existence (see Volterra 1931; Volterra and D'Ancona 1935).[12]

To appreciate the importance of mechanical analogies in the construction of Volterra's model one can, firstly, consider the way he treated predation. He assumed that the increase and decrease of predator and prey populations (Eqs. 8.3 and 8.4) are linear with respect to the product of N_1 and N_2, i.e. γ_1 and γ_2 are constants. To justify this assumption Volterra drew an analogy to mechanics by using the so-called "method of encounters" according to which the number of collisions between the particles of two gases is proportional to the product of their densities.[13] Thus Volterra

---

[10]For this so-called Volterra principle, see Weisberg and Reisman (2008).

[11]Lankester suggested that to protect edible prey-fish their enemies should be destroyed in the same proportion as the adult prey fish were "removed" (Lancaster 1884, p. 416).

[12]On Volterra's Darwinism, see Scudo (1992).

[13]Volterra made use of the method of encounters also in his study of the demographic evolution of a single species: There he applied the method of encounters to mating.

assumed that the rate of predation upon the prey is proportional to the product of the numbers of the two species.

Secondly, in generalizing his account to take into consideration the different kinds of interactions and multiple species, Volterra utilized mechanical analogies in various ways (e.g. Volterra 1926a, 1927, 1931). For instance, making use of the concept of friction in mechanics he made a distinction between two types of biological associations, conservative and dissipative ones (Volterra 1926a, 1927). Conservative systems are analogous to frictionless systems in mechanics. In conservative associations, the oscillations produced by the interactions of the species remain constant like in the Lotka-Volterra model. In dissipative associations, the fluctuations of the species are damped due to the friction caused by the interaction between individuals of the same species (which takes into account the effects of a population's size on its own growth). These cases display a parallel to the cases of damped and undamped harmonic oscillator in mechanics. Although conservative associations have very appealing mathematical properties, Volterra thought that dissipative associations are more realistic approximations of the natural situation than the conservative ones (Volterra 1928, p. 47).

The tension between applying the concepts and mathematical techniques suggested by classical mechanics and the aim to construct more realistic models marked Volterra's long research program on biological associations. He spent the rest of his life, more than a decade, formulating more elaborate models, taking into account different kinds of associations and situations, and making extensive use of modeling methods borrowed from mechanics. Already in his original 1926 article (Volterra 1926b), he also considered the cases of any number of species which either contended for the same food or some of which fed upon the others. One year after the publication of the original Italian article, Volterra (1927) also introduced integro-differential equations in an attempt to take into account the delayed effects of feeding on reproduction.[14] Finally, in a group of papers published in 1936 and 1937, Volterra made use of the calculus of variations in an attempt to provide a synthesis of his theory of biological associations along the lines of analytical mechanics. It is worth citing at length his explanation of this agenda:

> The second part [of Principes de biologie mathématique] begins with a conservation of demographic energy, according to which there are two sorts of energy, one actual and one potential, which transform mutually the one into the other. The principle is the analogue of the principle of conservation of mechanical energy. It is followed by the enunciation of the three laws relating to the biological fluctuations, the experimental verification of which has been investigated by several naturalists. The success, which has attended their efforts is well known.
>
> Everybody knows the importance of Hamilton's principle in mechanics and in all the domains of physical science. An analogous variation principle can be found in biology, and from it one can deduce the fluctuation equations in the canonical Hamiltonian form and also in the form of a Jacobian partial differential equation. [...]

[14]Today Volterra is mostly known for the Lotka-Volterra equation. For a discussion on how Volterra's various models anticipated several theoretical advances in theoretical ecology, see Scudo (1971).

> Hamilton's principle leads to the principle of least action (Maupertuis). There exists also in biology a closely related principle, which may be called the principle of least vital action. Its analytical form is such that it requires the existence of a true minimum, a state of affairs which does not always hold good in the analogous case in mechanics. (Volterra 1937b, p. 35)[15]

The reference to the experimental verification in the quotation above is important. Being faithful to his earlier methodological pronouncements, Volterra was also interested in testing his theories on empirical data, although he typically kept the mathematical, technical treatments and the empirical accounts separate from each other.[16] Apart from his collaboration with D'Ancona, he was also engaged in active correspondence with other biologists and scientists that served as an attempt for him to verify his theoretical findings empirically (see Israel and Gasca 2002).[17] Volterra rejected the idea of formulating mathematical models that could not be tested empirically and he insisted that all the variables introduced in the mathematical formalizations should be measurable. This eventually led him into a disagreement with D'Ancona, who was skeptical of Volterra's quest for empirical validation. He argued that Volterra's models were rather interesting theoretical working hypotheses able to stand on their own (see Israel 1991, 1993). Volterra's preference of grounding hypotheses in empirical research is displayed also by his reply to Alfred Lotka (Israel and Gasca 2002). Lotka had claimed priority for the Lotka-Volterra model on the basis of his *Elements of Physical Biology* (1925) published in 1925. Volterra (1927) acknowledged Lotka's priority, but stressed that he had formulated principles concerning "sea-fisheries," implying that this was not what Lotka had focused on. Indeed, Lotka derived his version of the Lotka-Volterra model in an entirely different way than Volterra. His approach is in a sense more contemporary than that of Volterra's, pointing towards complex systems theory and its use across the disciplines.

8.5 The Design of the Lotka-Volterra Model by Lotka

In contrast to Vito Volterra, the other author of the Lotka-Volterra model, Alfred Lotka (1880–1949), struggled to gain recognition for his work from the scientific community throughout his life. In addition to being a mathematician and statistician he had a background in physics, physical chemistry, and biology. In his work, Lotka integrated concepts, methods, and techniques from those various fields, developing a

---

[15] A partial English translation of this paper can be found in Scudo and Ziegler (1978).

[16] For example in Volterra (1934a, 1936) he discussed the connection between his theories and biological data.

[17] The biologists with whom Volterra corresponded included Georgii F. Gause, R.N. Chapman, Jean Régnier, Raymond Pearl, Karl Pearson, D'Arcy W. Thompson, William R. Thompson, Alfred J. Lotka, and Vladimir A. Kostitzin. The correspondence of Vito Volterra on mathematical biology also provides interesting material as regards modeling methods. Among other things, it provides material on the choices between deterministic and probabilistic approaches; between continuous and discrete models; between closed-form solutions and numerical solutions, and between qualitative and quantitative models.

modeling approach that could be characterized as a precursor for a systems approach. The developers of general systems theory and cybernetics, von Bertalanffy (1968) and Wiener (1948), were inspired by Lotka's work, especially his book *Elements of Physical Biology* (1925). Herbert Simon characterized Lotka "as a *forerunner* whose imagination creates plans of exploration that he can only partly execute, but who exerts great influence on the work of his successors" (Simon 1981, p. 493). For Simon, Lotka's book provided insight into how mathematics could be fruitfully applied in the social sciences. Lotka is also regarded as the founder of mathematical demography, and exerted a great influence on ecologist Eugene P. Odum, who counts as the founder of systems ecology. As the above discussion on Lotka's influence on the development of various systems theoretic approaches already hints at, his design of the Lotka-Volterra model was opposite to that of Volterra. Instead of starting from different simple cases and generalizing from them, he developed a highly abstract and general *model template* that could be applied to modeling various kinds of systems.

8.5.1 Physical Biology According to Lotka

Lotka aimed at developing a "physical biology," employing "physical principles and methods in the contemplation of biological systems" (Lotka 1925, p. viii). He was, however, skeptical of applying the most idealized cases of mechanics to biological systems. His main focus was on the evolution of biological systems, which he defined as follows: "Evolution is the history of a system undergoing irreversible changes" (Lotka 1925, p. 49). This definition does not exclude reversible processes, although Lotka argued that all real processes are irreversible. Reversible processes were for him idealizations. The evolution of a system in time is characterized, according to Lotka, by an increase in entropy. Physical biology was in turn "a branch of the greater discipline of the *General Mechanics of Evolution*" (ibid.).

Another important impulse for Lotka's program of physical biology came from the success of physical chemistry by the end of the 19th century (Servos 1990). Physical chemistry functioned as a model science for Lotka in much the same way as mechanics did for Volterra. Based on the conviction "that the principles of thermodynamics or of statistical mechanics do actually control the processes occurring in systems in the course of organic evolution" (Lotka 1925, p. 39), Lotka set out to apply the methods, techniques, and concepts from thermodynamics and statistical physics to the study of the evolution of biological systems. He realized, however, that biological systems are too complex to allow any straightforward application of thermodynamics. Lotka attempted to overcome this problem by introducing a generalized approach, which can be best understood as a kind of systems approach. The model later dubbed as the Lotka-Volterra model was just one application of Lotka's systems approach.

Apart from mechanics and physical chemistry, also the field of energetics had an impact on Lotka's theorizing. Energetics as a specific theoretical field originated in the 19th century in the works of Helm (1898) and Ostwald (1893) and others. It aimed at the development of a generalized theory based on the concept of energy. In a broader context, the movement can be understood as a reaction against the mechanistic world-view. In addition to being one of the main spokesmen of energetics, Ostwald (1893) was also one of the founding fathers of physical chemistry. From energetics Lotka took the idea of conceptualizing the components of systems as energy transformers in an analogy to heat engines (energy transformers could be organisms, chemical elements, etc.).

Energy transformers and the processes linked to them constituted what Lotka called the *Micro-Mechanics* of a system. *Macro-Mechanics* on the other hand encompassed the redistribution of mass between the components of the system. This distinction is similar to thermodynamics and statistical mechanics where, according to Lotka, the *Macro-Mechanics* examines the "phenomena displayed by the component aggregates in bulk", and the *Micro-Mechanics* "is centered primarily upon the phenomena displayed by the individuals of which the aggregates are composed" (Lotka 1925, p. 50). Thus Lotka attempted, at the same time, to apply thermodynamics and statistical mechanics to biology and to formulate a general approach that could overcome the problems inherent in drawing direct analogies between different disciplines—as Volterra had done.

8.5.2 Lotka's Systems Approach and the Lotka-Volterra Model

In his version of the Lotka-Volterra model, the model concerned macro-level phenomena. In order to describe the general dynamics on the macro level, Lotka started out from the law of mass action used in chemistry to describe the behavior of solutions. Lotka introduced the law by using the example of a system consisting of 4 g-molecules of hydrogen, 2 g-molecules of oxygen, and 100 g-molecules of steam, at one atmosphere pressure, and a tempreature of 1800 °C. The equation describing the evolution of this system is of the following form:

$$\frac{1}{v}\frac{dm_1}{dt} = k_1\frac{m_2^2 m_3}{v^3} - k_2\frac{m_1^2}{v^2}, \tag{8.5}$$

where v is the volume, m_1 is the mass of steam, m_2 the mass of hydrogen, and m_3 the mass of oxygen. The coefficients k_1 and k_2 are characteristic parameters of the reaction such as temperature and pressure. Lotka was not interested in this particular equation but in the more general statement implied by the equation according to which "the rate of increase in mass, the velocity of growth of one component, steam (mass m_1), is a function of the masses m_2 and m_3, as well as of the mass m_1 itself, and of the parameters v (volume) and T (temperature)" (Lotka 1925, p. 42). He then went on to write the equation in a more general form:

$$\frac{dX_i}{dt} = F_i(X_1, X_2, \ldots, X_n; P, Q). \tag{8.6}$$

This equation describes evolution as a process of redistribution of matter among the several components X_i of the system. Lotka called this equation the "Fundamental Equation of Kinetics" where the function F describes the physical interdependence of the several components. P and Q are parameters of the system. Q defines, in the case of biological systems, the characters of the species variable in time and P the geometrical constraints of the system such as volume, area, and extension in space.

Lotka had introduced this general approach in two articles published already 5 years before *Elements of Physical Biology* appeared in print. Interestingly, in both of these articles there appears a pair of equations that has the same form as what Volterra independently arrived at some years later. The first of these was entitled "Analytical note on certain rhythmic relations in organic systems" (Lotka 1920a) and the second paper "Undamped oscillations derived from the law of mass actions" (Lotka 1920b). In the first of the papers, the equations are applied to the analysis of a biological system, and in the second paper they are applied to a chemical system.[18] The title of the second paper refers explicitly to the law of mass action. In contrast to Volterra, who first considered the simplest models of interaction and then generalized the results to any number of species, Lotka started out from very general considerations and only after he had formulated his general equation did he turn to specific cases, such as the Lotka-Volterra model.

A further, important element in Lotka's design of the Lotka-Volterra model were the methods he introduced to analyze and calculate the dynamic behavior of the systems he had described. Having formulated the fundamental equation of kinetics, Lotka showed that without knowing the precise form of the function describing the interaction between the components, the properties related to the steady states of the system can already be studied. Lotka assumed that both the environment and the genetic constitutions are constant, after which, by means of a Taylor series expansion, he calculated the possible stationary states of the system. He was able to show that, in general, the system will exhibit one of the following three behaviors with increasing time: First, the system asymptotically approaches an equilibrium; second, it performs irregular oscillations around an equilibrium; or third, it performs regular oscillations around the equilibrium. He then applied the fundamental equation to the case of two species, one of which feeds upon the other, arriving at the following equations:

$$\frac{dN_1}{dt} = (\varepsilon_1 - \gamma_1 N_2)N_1, \tag{8.7}$$

$$\frac{dN_2}{dt} = (-\varepsilon_2 + \gamma_2 N_1)N_2, \tag{8.8}$$

[18]Lotka dealt with the rhythmic effects of chemical reactions already in his earlier writings, see e.g. Lotka (1910).

which are the same as Volterra's equations. They constitute a set of non-linear coupled first-order differential equations, which cannot be solved analytically. Therefore, Lotka's general method of calculating the stationary states became a valuable tool for dealing with such sets of coupled differential equations. As already mentioned, he claimed priority for the model on the basis of his *Elements of Physical Biology* (1925). The reason for this might be that in (Lotka 1920a) he draws the Lotka-Volterra equations from his general equation inspired by chemical dynamics without any discussion of *empirical* biological systems. In *Elements of Physical Biology* he applies the equations to the study of a host-parasite system, citing also Thomson (1922) and Howard (1897) on this topic. In the third part of the book, the fundamental kinetic equation is also used to study various other cases, such as the spreading of malaria.

As was the case with Volterra, also Lotka's program went much further than the development of what became known as the Lotka-Volterra model. In fact, his visions went far beyond ecology. His book *Elements of Physical Biology* is a unique conceptualization of the manifold biological, physical, and chemical processes and their complex interactions in the world surrounding us. The organic world becomes a giant energy transformer in which the general kinetic equations provide the mathematical tool for describing the distribution and the transfer of energy between the components of the world. Biological systems, according to Lotka's vision, were to be treated identically with physical systems: it all boiled down to the study of transformations of matter and energy.

8.6 Discussion

Lotka and Volterra worked along the same lines, taking inspiration of physical sciences in modeling biological systems, and eventually they presented the same model. Yet they arrived there following different kinds of modeling strategies. While Volterra was making repeatedly use of analogies taken from physical sciences, Lotka was more wary of this kind of procedure and adopted instead a more general, template-based approach. He did not set out to explain some specific dynamics associated with, for example, the spreading of malaria. He focused on more general characteristics of evolving systems, which he defined as follows:

> [...] an evolving system is an aggregation of numbered or measured components of several specified kinds, and which observes and registers the history of that system as a record of progressive changes taking place in the distribution, among those components, of the material of which the system is built up. (Lotka 1925, p. 41)

The template Lotka constructed is a mathematized form of this description of evolving systems. All systems, which show some kind of dynamics, should in principle be describable by means of the template. Such analysis would not be restricted to certain species or populations, but concerned various kinds of transformer types, organic and inorganic.

This difference in the respective modeling strategies of Lotka and Volterra has profound implications concerning the interpretation of their versions of the Lotka-Volterra model. Volterra approached modeling from the perspective of the causal explanation of real mechanisms, presenting his model in terms of fully specified equations governing the dynamics of the system in question. This approach enables the ecological interpretation of the coefficients, but simultaneously makes it a gross simplification of the biological reality. Lotka's formulation recognizes the implausibility of completely specifying the full functional forms of the equations governing an ecological system, or any other complex system, for that matter. Within a local neighborhood of an equilibrium, the full equations are approximated by the Taylor series expansion (see Haydon and Lloyd 1999, pp. 205–206).

Having elucidated the historical roots of the Lotka-Volterra model, we will now turn to the philosophical discussion on Volterra and modeling. Let us recall that both Weisberg (2007) as well as Scholl and Räz (2013) considered Volterra as a modeler, but on different grounds. Much of what they say on Volterra's theoretical approach is congruent, however, even though they make use of different parts of his work. Weisberg relies exclusively on Volterra's (1926a, 1926b) original publications in Italian and English. Although Scholl and Räz refer to Volterra's early (1926a, 1928) publications, they draw most heavily on the arguments presented in his later work co-authored by D'Ancona (1935). As we have shown, as Volterra's research program on the biological associations progressed, he started to pay more and more attention to the empirical verification of his models. Thus the somewhat different takes on Volterra's work by Weisberg and Scholl and Räz can be partly explained by their focus on the different phases of his work.

Where Weisberg's and Scholl and Räz's studies part concerns not so much Volterra's actual claims, but rather the more philosophical question of whether Volterra's method of hypothesis can be considered to represent a unique style of theorizing (that Weisberg calls modeling). While Weisberg seems to be working with an already established intuition of modeling that he then chooses to exemplify by contrasting Volterra with Mendeleev and Darwin, the focus of Scholl and Räz is on causal inference. They derive their working notion of modeling as arising from the insufficient epistemic access from the later writings of Volterra (1935) without any explicit attention to the philosophical discussion on modeling.[19] This differs from Weisberg's treatment. He does, after discussing the different case studies, explicate at length what kind of features "an account of models adequate for characterizing the practice of modeling must have" (Weisberg 2007, p. 221).

It is noteworthy that apart from agreeing with Weisberg about the indirect nature of Volterra's theoretical endeavour, Scholl and Räz adopt, furthermore, his distinction between representational and dynamical fidelity. What their criticism eventually boils

[19]Our reading of Scholl and Räz (2013) differs somewhat from their later reading of their own article (this volume). In their original article, they do not clearly offer causal inference as a contrast to modeling (that would provide an alternative for Weisberg's contrast between modeling and abstract direct representation). According to them "much of our discussion will focus on *models of causal structures*" (Scholl and Räz 2013, p. 117, emphasis added). Their earlier focus was on causal inference in general and modeling a strategy to deal with insufficient epistemic success.

down to, is to showing how Darwin succeeded in what Volterra did not: establishing a "trajectory from "how possibly" (dynamical fidelity without representational fidelity) to "how actually" (dynamical fidelity and representational fidelity)" (Scholl and Räz 2013, p. 131).[20] For Scholl and Räz, then, modeling does not need to stay at the level of indirect reasoning and production of only dynamically accurate models, and so indirect representation needs not to be the mark of modeling. Yet Weisberg would not necessarily disagree with them. His thesis concerns the way the model is *developed*, that is, the indirect strategy of representing a hypothetical system. In that stage, the dynamical fidelity may function as the most important guide. But it does not mean that models could not be developed into representationally more accurate descriptions of actual target systems. That is what Volterra (and D'Ancona) attempted, to some degree. So there seems not to be too big a difference between Weisberg's and Scholl and Räz's claims concerning Volterra's work and modeling.

Scholl and Räz (this volume) seem to be willing to render the contrast between their and Weisberg's account clearer by claiming that Volterra and D'Ancona "write that for their own investigation they would have *preferred* an experimental approach, which would have allowed for direct causal inferences in the system under scrutiny." We find it doubtful that a world-renowned mathematician and theoretical physicist, whose outspoken goal was to mathematize social science and biology would have preferred the experimental approach.

In our view it is more likely that Volterra and D'Ancona's discussion of the rationale of their mathematical approach is due to Volterra's methodological views according to which the empirical verification of theories was important—indicating that Volterra cannot be conceived as a modeller making use of a purposeful strategy of indirect representation. Yet, at retrospect, what Volterra eventually accomplished can be approached as an instance of modeling despite his methodological pronouncements. As we discussed in our case study, Volterra's primary aim, already expressed in his Inaugural Address (Volterra 1901) was to isolate the "fundamental parameters" of the predator-prey system. In actual practice he started right away from certain assumed factors and from the hypothesis that the oscillations in the fishery data might be accounted for solely by these factors and the resulting interaction of the two species.

From our point of view, the crucial question concerns the reasons as to why Volterra (and D'Ancona) did not achieve more representationally accurate models? Was there something about their method that sets it apart from Darwin's and Mendeleev's achievements? We think that there is a more profound reason for why Volterra adopted an indirect modeling strategy that neither Weisberg, nor Scholl and Räz discuss. Namely, as we showed through our case study, Volterra was interested in the mathematization of biology and social sciences, and in utilizing the tools and methods of mechanics in this task. The way he proceeded to model biological

[20] Scholl and Räz adopt the distinction between "how possibly" and "how actually" from the discussion on mechanistic explanation (Machamer et al. 2000), a discussion that has been up until recently relatively disinterested in modeling and that considered models as explanation sketches only (see Knuuttila and Loettgers 2013).

associations made heavy use of analogies to mechanics that enabled him to transfer mathematical tools and methods of physics to biology. The strategy of constructing a simpler hypothetical system to which only some properties are assigned is due to the goal of mathematizing the problem at hand. Such method had proven successful within physics, but Volterra was only too aware of the complexity of the problem posed by biological associations in a natural environment with various kinds of perturbations. The dynamical fidelity was important to Volterra precisely since his was an attempt of explaining the oscillations only by internal causes, i.e. solely by the interaction between the two species. A major part of contemporary mathematical modeling across different disciplines is precisely providing these kinds of hypothetical explanations.[21] In Knuuttila and Loettgers (2016) we argue that methods-drivenness and outcome-orientedness are characteristic features of such modeling exercises.

Yet, as our study concerning the different designs of the Lotka-Volterra model shows, modeling is nevertheless not any unitary theoretical strategy, as Weisberg seems to be claiming. Lotka's strategy for using the model templates[22] drawn from statistical mechanics and physical chemistry was different from that of Volterra's. For Lotka the general systems approach he developed provided the justification for his modeling approach. As we have discussed, his approach anticipates many contemporary modeling practices in which the model templates developed in complex systems theory are applied across the variety of disciplines, studying both natural and social systems. Interestingly, in the context of the study of complex systems, the Lotka-Volterra model was "rediscovered" as one of the basic simple models that afforded the study of complex systems (e.g. May 1974).

8.7 Conclusion: Theoretical Case-Based Philosophical Practice

How do philosophers of science make use of historical case studies? In particular, are their accounts of historical cases necessarily purpose-built and lacking in evidential strength as a result of putting forth and discussing philosophical positions? In order to find this out, we have examined three different philosophical case studies on Vito Volterra's work, one of which is our own. The interesting outcome of this exercise is that while the philosophical conclusions of the three case studies are different, they largely agree on their interpretation of Volterra's work! The comparison of these case studies discussed (by Weisberg 2007; Scholl and Räz 2013; Knuuttila and Loettgers

---

[21] The "how-possible" may be a bit misleading expression in this context, since what Volterra accomplished was an alternative explanation for the prevailing explanations that attributed the fluctuations to external causes.

[22] A model template is an abstract conceptual idea concerning usually a certain kind of interaction and associated with particular mathematical forms and computational methods, see for further discussion Knuuttila and Loettgers (2014).

2016) does not lend credence to the idea that they argue for their philosophical claims by simply confronting them with historical cases—or constructing the cases according to their preferred theory (cf. Pitt 2001). Does this mean that historical case studies in philosophy of science should be understood as interpretative activity investigating scientific concepts, norms and practices, as Schickore has suggested? Our answer to this question is positive, but we do not see that it would need to imply either rejecting the evidential role of case studies (cf. Schickore 2011), or compromising, or lessening, their evidential value (cf. Kinzel 2015). In our view, philosophers of science usually use case studies as vehicles of theoretical reflection, as resources in examining, questioning, and developing philosophical concepts and accounts. In that use evidential and hermeneutic roles go hand in hand, informing each other. Thus the three case studies presented in this paper serve as examples of case-based theoretical philosophical practice that is underlined by the way each of them uses the strategy of contrasting partially similar, and partially different scientific examples. The use of the contrasting example highlights the conceptual distinctions made.

How should one, then, understand the philosophical case-based theoretical practice? The first thing to notice is that it is difficult to recognize it, if one approaches philosophy as an activity that aims only at a general/rational reconstruction of scientific activity (although that would also need historical and empirical knowledge, if only to recognize what constitutes, in fact, successful science). It seems that this kind of conception of the philosophy of science lies behind the various iterations of the claims that scientific case studies cannot give evidential support for philosophical positions. But clearly, philosophical theorizing also contains an important descriptive component as well as being often more local and tentative in nature—as the practice-oriented philosophy of science has recently shown.

If one looks at the use of historical case studies as vehicles for philosophical theorizing, nothing very special seems to be going on there. Also in scientific research using case studies as a resource for investigation one has to negotiate the relationship between the generalizable insights and the context-specific details. Even in the physical sciences single cases and experiments are sufficient for theorizing, as Shrader-Frechette and McCoy (1994) point out in their study of case-based reasoning within ecological sciences. According to them, ecologists often prefer case-specific knowledge coupled with conceptual and methodological analysis to "ecological theorizing based on untestable principles and deductive inferences drawn from mathematical models" (p. 244). The case-specific local knowledge allows various kinds of inferences: they can be local to local, or local to many. In local to many reasoning, the local knowledge is desituated to a more generic level, or used to construct typical representatives or exemplars (see Morgan 2014).

What needs to be recognized is that most scientific knowledge is local, or constructed from the local knowledge, being subject to various initial conditions and environmental contexts. Moreover, while case studies provide a springboard for theorizing and generalization, they are often also used to question earlier held theoretical views, or their generality. The case study methodology has also advantages that spring forth from the way the evidential is woven together with theoretical. A historical case study typically presents "a complex, often narrated, account that ... contains some of the raw evidence as well as its analysis and that ties together many different bits and pieces in the study" (Morgan 2012, p. 668). Thus narrative becomes a way to deal with what Morgan calls "evidential density," which contributes to theoretical development by offering rich resources for a critical study of different theoretical perspectives. This evidential richness is clearly one of the benefits of case studies: many relevant factors do not need to be abstracted away or shielded, as with laboratory studies and mathematical modeling. Consequently, it seems a mistake to try to tease the evidential dimension of case studies apart from their conceptual and interpretative content. Both are woven together in the theoretical narrative that aims to integrate different kinds and bits of evidence by showing their interdependence (Morgan 2012, p. 675). This theoretical-cum-conceptual modality of case studies is so strong that even when case studies succeed to identify a novel interesting phenomenon, like the "street corner society" (Whyte 1943), "the community's response was to understand the phenomena revealed as potentially generic" (Morgan 2012, p. 673).

It is our claim, then, that the key to the epistemic value of case studies, in philosophy of science, like in both natural and social sciences, lies in the way they weave together different kinds of evidence with the conceptual analysis and theoretical development. The observation that the same cases may be interpreted differently does not seem to us so grave an objection, since, as pointed out by numerous scholars, case studies often breed new interpretations of the same cases, as well as attempts to confirm the results by new case studies and independent data (e.g. Shrader-Frechette and McCoy 1994). The three cases provide a good example of this practice of presenting related case studies. As we have argued, they are largely in agreement concerning Volterra's work,[23] and the theoretical and interpretative element can most clearly be located to how Volterra's work is contrasted with the work of other scholars: Mendeleev, Darwin, and Lotka.

Finally, we remain skeptical of the idea that to justify case study methodology, the cases should be typical of their kind—how do we know the typical without any cases?—or somehow important or critical. The scientific record does not seem to lend support to these kinds of claims either. Ankeny (2012) argues concerning developmental biology that many model organisms used as kind of "cases" are now regarded as presenting typical patterns of phenomena, while they were often originally selected for study for other reasons, such as convenience or ease of experimental manipulation. And even when they turned out atypical, they still continued to provide a focal point in the field, permitting investigation of variations in phenomena or processes. We see this kind phenomenon taking shape also with the three case studies on Volterra, each of which presents variations on the theme of modeling, delineated with the help of contrasting Volterra's work with that of other theorists.

---

[23] Even though the differences between the three case studies with respect to Volterra's work were more substantial, such underdetermination of theories by data would be a common feature of other scientific practices, too.

References

Ankeny, R.A. 2012. Detecting themes and variations: The use of cases in developmental biology. *Philosophy of Science* 79: 644–654.

Bailer-Jones, D. 2009. *Scientific models in philosophy of science*. Pittsburgh: University Press.

Burian, R. 2001. The dilemma of case studies resolved: The virtues of using case studies in the history and philosophy of science. *Perspectives on Science* 9: 383–404.

Buss, L. 1987. *The evolution of individuality*. Princeton University Press.

Chang, H. 2012. *Is water H_2O? evidence, pluralism and realism*. Boston studies in the philosophy and history of science. Dordrecht: Springer.

Contessa, G. 2007. Scientific representation, interpretation, and surrogative reasoning. *Philosophy of Science* 74: 48–68.

da Costa, N.C.A., and S. French. 2000. Models, theories and structures: Thirty years on. *Philosophy of Science* 67: 116–127.

Darwin, C. 1882. The origin of the species by means of natural selection. Murray.

French, S., and J. Ladyman. 1999. Reinflating the semantic approach. *International Studies in the Philosophy of Science* 13: 103–21.

Frigg, R. 2010. Models and fiction. *Synthese* 172: 251–268.

Gause, G. 1935. Studies in the ecology of the orthoptera. *Ecology* 11: 307–325.

Giere, R.N. 2004. How models are used to represent reality. *Philosophy of Science (Symposia)* 71: 742–752.

Godfrey-Smith, P. 2006. The strategy of model-based science. *Biology and Philosophy* 21: 725–740.

Haydon, D., and A. Lloyd. 1999. On the origins of the Lotka-Volterra equations. *Bulletin of the Ecological Society of America* 80: 205–206.

Helm, G. 1898. Die Energetik. Verlag von Veit.

Howard, L.O. 1897. A study in insect parasitism: A consideration of the parasites of the white-marked tussok moth with an account of their habits and interrelations, and with descriptions of new species. U.S. *Department of Agriculture Technical Bulletin* 5: 1–57.

Humphreys, P. 2002. Computational models. *Proceedings of the Philosophy of Science Association* 3: S1–S11.

Humphreys, P. 2004. *Extending ourselves: Computational science, empiricism, and scientific method*. Oxford University Press.

Israel, G. 1991. Volterra's analytical mechanics of biological associations (second part). U.S. *Department of Agriculture Technical Bulletin* 5: 1–57.

Israel, G. 1993. The emergence of biomathematics and the case of population dynamics: A revival of mechanical reductionism and darwinism. *Science in Context* 6: 469–509.

Israel, G., and Gasca, A.M. 2002. *The biology of numbers: The correspondence of Vito Volterra on mathematical biology*. Birkhäuser.

Kinzel, K. 2015. Narrative and evidence: How can case-studies from the history of science support claims in the philosophy of science? *Studies in History and Philosophy of Science Part A* 49: 48–57.

Knuuttila, T. 2005. Models, representation, and mediator. *Philosophy of Science* 72: 1260–1271.

Knuuttila, T. 2006. From representation to production: Parsers and parsing in language technology. *Sociology of Science* 25: 41–55.

Knuuttila, T. 2013. Science in a new mode: Good old (theoretical) science vs. brave new (commodified) knowledge production. *Science and Education* 22: 2443–2461.

Knuuttila, T., and Loettgers, A. 2013. The productive tension: Mechanisms vs. templates in modeling the phenomena. In *Representations, models, and simulations*, ed. P. Humphreys, and C. Imbert, 3–24. Routledge.

Knuuttila, T., and A. Loettgers. 2014. Varieties of noise: Analogical reasoning in synthetic biology. *Studies in History and Philosophy of Science Part A* 48: 76–88.

Knuuttila, T., and Loettgers, A. 2016. Modelling as indirect representation? the Lotka-Volterra model revisited. *The British Journal for the Philosophy of Science*.

Lancaster, E.R. 1884. The scientific results of the exhibition. *Fisheries Exhibition Literature* 4: 405–442.

Loettgers, A. 2007. Model organisms, mathematical models, and synthetic models in exploring gene regulatory mechanisms. *Biological Theory* 2: 134–142.

Lotka, A. 1910. Contribution to the theory of periodic reactions. *The Journal of Physical Chemistry* 14: 271–274.

Lotka, A. 1920a. Analytical note on certain rhythmic relations in organic systems. *Proceedings of the National Academy of Art and Science* 42: 410–415.

Lotka, A. 1920b. Undamped oscillations derived from the law of mass action. *Journal of the American Chemical Society* 42: 1595–1598.

Lotka, A. (1925). *Elements of physical biolgy*. Williams and Wilkins.

Machamer, P.K., L. Darden, and C. Craver. 2000. Thinking about mechanisms. Philosophie of. *Science* 67: 1–25.

Maeki, U. 2009. Missing the world: Models as isolations and credible surrogate systems. *Erkenntnis* 70: 29–43.

May, R. 1974. Biological populations with nonoverlapping genertations: Stable cycles, and chaos. *Science* 186: 645–647.

Maynard-Smith, J., and E. Smathmáry. 1995. *The major transitions in evolution*. Oxford: Oxford University Press.

Morgan, M. 2012. Case studies: One observation or many? justification or discovery? *Philosophy of Science* 79: 667–677.

Morgan, M. 2014. Resulting knowledge: Generic strategies and case studies. *Philosophy of Science* 81: 1012–1024.

Morgan, M., and Morrison, M. 1999. Models as mediating instruments. In *Models as mediators. Perspectives on natural and social science*, ed. M. Morgan, and M. Morrison, 97–146. Cambridge University Press.

Ostwald, W. 1893. Lehrbuch der allgemeinen Chemie: Chemische Energie. Wilhelm Engelman.

Pitt, J.C. 2001. The dilemma of case studies: Toward a heraclitian philosophy of science. *Perspectives on Science* 9: 373–382.

Schickore, J. 2011. More thoughts on HPS: Another 20 years later. *Perspectives on Science* 19: 453–481.

Scholl, R., and T. Räz. 2013. Modeling causal structures: Volterra's struggle and Darwin's success. *European Journal for Philosophy of Science* 3: 115–132.

Scudo, F. 1971. Vito Volterra and theoretical ecology. *Theoretical Population Biology* 2: 1–23.

Scudo, F. 1992. *Vito Volterra, ecology and the quantification of Darwinism*. Accademia Nazionale dei Lincei: Atti del Convegno internazionale in memorie di Vito Volterra. Rome.

Scudo, F.M., and J.R. Ziegler. 1978. *The golden age of theoretical ecology: 1923–1940*. Berlin: Springer.

Servos, J.W. 1990. *Physical chemistry from ostwald to pauling: the making of a science in America*. Princeton University Press.

Shrader-Frechette, K.S., and E.D. McCoy. 1994. How the tail wags the dog: How value judgments determine ecological science. *Environmental Values* 3: 107–120.

Simon, H. 1981. *The sciences of the artifical*. MIT Press.

Suárez, M. (2008). Scientific fictions as rules of inference. In *Fictions in science: philosophical essays on modeling and idealisation*, ed. M. Suárez, 258–270. Routledge.

Thomson, W.R. 1922. Théorie de l'action des parasites entomophages. les formules mathématiques du parasitisme cyclique. *Comptes Rendus Academie des Sciences Paris* 174: 1202–1204.

Volterra, V. 1901. On the attempts to apply mathematics to the biological and social sciences. In *The Volterra chronicles: Life and times of an extraordinary mathematician 1860–1940*, ed. J.D. Goodstein, 247–260. AMS.

Volterra, V. 1926a. Fluctuations in the abundance of a species considered mathematically. *Nature*, CXVIII:558–560.

Volterra, V. 1926b. Variazioni e fluttuazioni del numero d'indivui in specie animali conviventi. *Memorie della R. Accademia Lincei* 2: 31–113.
Volterra, V. 1927. Letter to nature. *Nature* 119: 12.
Volterra, V. 1928. Variations and fluctuations of the number of individuals in animal species living together. *Journal du Conseil International Pour l'Exploration de la Mer* 3: 3–51.
Volterra, V. 1930. *Theory of functionals and of integral and integro-differential equations*. Blackie and Son.
Volterra, V. (1931). Leçons sur la Théorie Mathématique de la Lutte Pour la Vie. Gauthier-Villars.
Volterra, V. 1936. La théorie mathématique de la lutte pour la vie et l'expérience (a propos de deux ouvrages de C.F. Gause). *Scientia* LX: 169–174.
Volterra, V. 1937a. Applications des mathématiques à la biologie. *L'Enseignement Mathématique* XXXVI: 297–330.
Volterra, V. 1937b. Principes de biologie mathématique. *Acta Biotheoretica* III: 1–36.
Volterra, V., and D'Ancona, U. 1935. *Les associations biologiques au point de vue mathématique*. Hermann.
von Bertalanffy, L. 1968. *General system theory: Foundations, development, applications*. George Braziller.
Weisberg, M. 2007. Who is a modeler? *British Journal for the Philosophy of Science* 58(2): 207–233.
Weisberg, M. 2013. *Simulation and similarity: Using models to understand the World*. Oxford University Press.
Weisberg, M., and K. Reisman. 2008. The robust Volterra principle. *Philosophy of Science* 75: 106–131.
Whyte, F. 1943. *The street corner society*. The social structure of an Italian slum: The University of Chicago Press.
Wiener, N. 1948. *Cybernetics or control and communication in the animal and the machine*. MIT Press.

Chapter 9
Gone Till November: A Disagreement in Einstein Scholarship

Tim Räz

Abstract The present paper examines an episode from the historiography of the genesis of general relativity. Einstein rejected a certain theory in the so-called "Zurich notebook" in 1912–13, but he reinstated the same theory for a short period of time in the November of 1915. Why did Einstein reject the theory at first, and why did he change his mind later? The group of Einstein scholars who reconstructed Einstein's reasoning in the Zurich notebook disagree on how to answer these questions. According to the "majority view", Einstein was unaware of so-called "coordinate conditions", and he relied on so-called "coordinate restrictions". John Norton, on the other hand, claims that Einstein must have had coordinate conditions all along, but that he committed a different mistake, which he would repeat in the context of the famous "hole argument". After an account of the two views, and of the reactions by the respective opponents, I will probe the two views for weaknesses, and try to determine how we might settle the disagreement. Finally, I will discuss emerging methodological issues.

9.1 Introduction

The present paper examines an episode from the historiography of general relativity (GR) that exhibits methodological problems of history and philosophy of science. These problems emerge because there are two competing accounts of an important episode from Einstein's long path to GR.[1] The fact that two competing accounts exist is not particularly exciting in itself—many episodes from the history of science have been retold in many, often incompatible ways, and a plurality of accounts need not be a sign of fundamental methodological problems. Plurality can be due to different sources; sometimes new sources are discovered; the same episode can be presented from different vantage points, and approached with different questions; new scientific

[1]Most articles relevant to the present paper can be found in Janssen et al. (2007a, b).

T. Räz (✉)
Universität Konstanz, FB Philosophie, 78457 Konstanz, Germany
e-mail: tim.raez@gmail.com

T. Sauer and R. Scholl (eds.), *The Philosophy of Historical Case Studies*,
Boston Studies in the Philosophy and History of Science 319,
DOI 10.1007/978-3-319-30229-4_9

knowledge can deepen our understanding of an episode, making previous accounts obsolete. This will lead to different accounts of the same historical episode in a natural and unsurprising manner.

The present case is different. The two competing views of the episode exhibit a considerable unity in perspective, methods, and sources. The Einstein scholars defending the diverging views worked over a period of ten years on a reconstruction and interpretation of one of the most important sources of Einstein scholarship, the "Zurich notebook".[2] This long and close collaboration—I will call it the "genesis collaboration"—resulted in a jointly authored book, the four-volume *opus magnum* on the genesis of GR, Renn (2007). Despite the close collaboration, diverging views of crucial turning points of the genesis of GR have emerged. A contribution by John Norton defends a "minority view" of the episode in question, while the rest of the genesis collaboration, notably Jürgen Renn, Tilman Sauer and Michel Janssen defend what I will call the "majority view".

What is the disagreement? Einstein rejected a certain theory in the Zurich notebook in 1912–13, while the very same theory was reinstated for a short period of time in the November of 1915. Why did Einstein reject the theory in the notebook, and why did he change his mind in November 1915? The majority claims that there is one major difference between the first and the second period, Einstein's awareness of so-called "coordinate conditions". This is a now-standard mathematical procedure for bringing the field equations of GR into correspondence with Newtonian gravitational theory. The majority argues that Einstein was unaware of coordinate conditions at the time of the notebook, and that he relied on so-called "coordinate restrictions", which severely limited the generality of the field equations and made the theory unacceptable. Only when he became aware of coordinate conditions did the theory become acceptable again. Norton finds this story implausible. He claims that Einstein must have had the modern notion of coordinate condition all along, but that he committed a different, more elaborate mistake. Importantly, this is a mistake that Einstein would repeat in the context of the famous "hole argument", a major roadblock on the path to the final theory of GR.

I will discuss methodological issues that arise from this non-trivial disagreement. Is this a dispute that cannot be settled despite a unity of evidence and methods? There is hope, I will argue, that we can dissolve the dispute. We can do better in the interpretation of the available evidence, in the reconstruction of the scientific and mathematical context of Einstein's struggle, and we can challenge the internal consistency of the two views. However, there are also fundamental methodological problems that we have to navigate. There are boundary conditions of rationality that enter into the reconstruction of historical episodes, for which there is no clear-cut justification.

I provide a short introduction to the history of GR and to the most important concepts in the upcoming section. I have tried to make the technical subject-matter of the episode accessible to non-specialists as much as possible. I then give an account

[2]See the introduction in Janssen et al. (2007a) for remarks on the collaboration between Jürgen Renn, Tilman Sauer, Michel Janssen, John D. Norton, and John Stachel.

of the two views, and of the reactions to the views by the respective opponents. After reviewing the arguments, I will probe the two views for weaknesses, and try to determine how we might settle the disagreement based on the available evidence, and on other considerations. Finally, I will discuss emerging methodological issues based on the previous discussion, and on methodological remarks by the parties involved.

9.2 A Bird's Eye View of the Episode

The story of the genesis of GR can be told in the form of a drama in three acts.[3] The main character is Einstein, with appearances by other famous physicists and mathematicians, notably his friend Marcel Grossmann, as well as various mathematical and physical theories and concepts. The premise of the drama is that not all is well in the house of gravitational physics. There is tension between the new theory of special relativity (SR), which constrains all physical theories and has a built-in finite speed of light, and the old Newtonian gravitational theory (NGT), which works by instantaneous action-at-a-distance. NGT will have to change, but how?

Enter Einstein, who sets out to reconcile the two theories by formulating a relativistic theory of gravitation. The core piece of the new theory will be field equations that generalize the gravitational Poisson equation. Gravitational field equations tell us how gravitation and matter, energy, and momentum hang together. The first two acts of the drama go relatively smoothly. In the first act, Einstein formulates the equivalence principle, which establishes a connection between accelerated reference frames and gravitational fields. In the second act, he finds that the best way to represent the gravitational potential in the field equations is by the metric tensor, which generalizes the notion of distance in Euclidean space to distance in space-times with variable curvature.

At the beginning of act three, Einstein has learned how to represent distance in space-time by the metric tensor, and he knows that energy-momentum can be represented using the energy-momentum tensor.[4] All that is left to do is to find the gravitational field equations, which tell us how space-time is influenced by the distribution of matter, energy, and momentum. Mathematically, the missing element of the field equations is a differential operator, which acts on the metric and thereby tells us how the metric and the energy-momentum tensor hang together. An appropriate differential operator generalizes the Laplace operator of the Poisson equation. This is where the reversal of fortune sets in. Einstein tests various candidates, straightforward generalizations of the Laplace operator, and also other candidates. However, none of them fits the bill. In a state of desperation, Einstein turns to Marcel Grossmann, his mathematician friend, for help.

[3] Stachel (2007) has given an account of the genesis of GR in this form. The present section serves as an introduction; technical details are mostly relegated to footnotes.

[4] A detailed account of how the third act unfolded can be found in Renn and Sauer (2007).

Grossmann is indeed able to help.[5] He finds a mathematical theory, the "absolute differential calculus", proposed by the Italian mathematicians Gregorio Ricci-Curbastro and Tullio Levi-Civita; this calculus is a framework that provides candidate differential operators for the field equations. The candidates are generally covariant, i.e., they keep their form under arbitrary coordinate transformations. The single most important object is the Riemann tensor; every possible generally covariant differential operator can be constructed from it. At the end of the drama, in November 1915, it will turn out that the absolute differential calculus and gravitational theory were right for each other all along. However, in 1912, Einstein and Grossmann do not realize this and the final, correct field equations have to wait behind the scenes.

Einstein and Grossmann use the Riemann tensor to derive the so-called Ricci tensor, a promising candidate.[6] However, Einstein soon rejects the Ricci tensor as unsuitable for the field equations. This is a mistake, because the reasons for the rejection are ill conceived. The root of the problem has to do with the correspondence principle: The new field equations have to be put in correspondence with the old, classical gravitational field equation—the classical case should be recovered as a limiting case of the new, general theory. In order to do this, one has to consider other, intermediate cases, such as weak gravitational fields. At this point, Einstein is already experienced in handling such intermediate cases, but this experience does not serve him well: It generates wrong expectations about the form that special cases should take. The rejection of the Ricci tensor is a consequence of these wrong expectations.

Einstein then turns to the so-called "November tensor",[7] a little brother of the Ricci tensor. The November tensor can be found by decomposing the Ricci tensor into two summands—one of these is the November tensor. It is not a generally covariant object, but its covariance group still includes some accelerated reference frames. At this point, the drama gets confusing. Einstein is able to show that the November tensor does not run into the same difficulties that had led him to eliminate the Ricci tensor. Despite this apparent progress, Einstein eliminates the November tensor as well, and it is not considered any further in the Zurich notebook. What prevented Einstein from investigating the November tensor further? What is more, years later, in November 1915, he returned to the November tensor and used it to formulate a version of general relativity. What made the November tensor acceptable again in November 1915? In later recollections, Einstein stated reasons for rejecting

[5]The collaboration between Einstein and Grossmann resulted in several publications, most importantly the so-called "Entwurf" ("outline") theory (Einstein and Grossmann 1995), which contains the first detailed exposition of tensor calculus in the context of GR. The *Entwurf* theory does not yet formulate the final, correct field equations of GR; see Sauer (2014) for an account of Grossmann's contribution to GR.

[6]From here on, the story can only be reconstructed on the basis of the Zurich notebook. This part of the drama is now well understood thanks to the genesis collaboration; see Janssen et al. (2007a, b). The following account of Einstein's struggle is based on Norton (2007, Sect. 1).

[7]The name was coined by the genesis collaboration; the November tensor became prominent in November 1915. It first appears on p. 22R of the Zurich notebook; see Fig. 9.1. I use the standard pagination; see Janssen et al. (2007a), Klein et al. (1995) for a facsimile of the notebook and Janssen et al. (2007b) for the commentary. Note that a facsimile of the Zurich notebook is also available online at Einstein Archive Online.

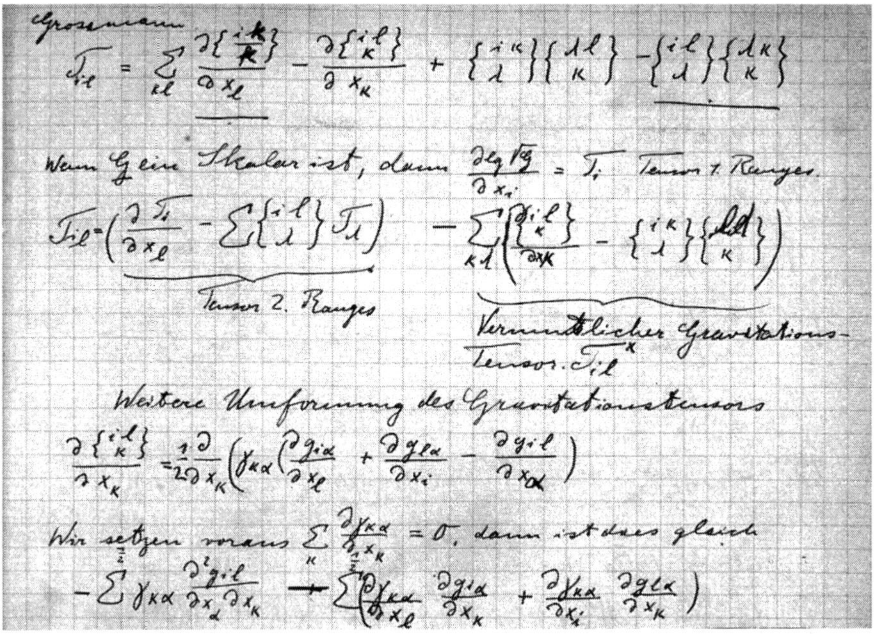

Fig. 9.1 Top portion of page 22R of Einstein's Zurich notebook (Einstein Archives Call No. 3-006). The *first line* shows the Ricci tensor $\mathscr{T}_{il}$, next to Grossmann's name. The Ricci tensor is then split up into two parts; the second part, labelled "Vermutlicher Gravitationstensor", is the so-called "November tensor." ©The Hebrew University of Jerusalem, Albert Einstein Archives; reproduced with permission

the November tensor. However, Einstein's explanations are not entirely satisfactory; his later recollections cannot fully resolve the puzzle.

Up to this point, the genesis collaboration agrees on how the drama unfolded, but now, the views start to diverge. The disagreement concerns the reason why the November tensor was rejected in the Zurich notebook. If we compare Einstein's calculations involving the November tensor in the Zurich notebook in 1912–13, and in November 1915, there is one important difference. In the notebook, Einstein uses coordinate conditions in a way that differs from the modern usage. Coordinate conditions are a standard operation to recover the old gravitational theory from the new field equations. However, some of Einstein's calculations do not make sense from a modern point of view. Did Einstein deliberately apply coordinate conditions in a way that deviates from modern usage, or was he not aware of the modern usage? The two views under discussion disagree about how this question should be answered. The majority claims that Einstein did not have the modern notion of coordinate conditions at the time of the Zurich notebook. Norton, on the other hand, claims that Einstein was aware of coordinate conditions. Consequently, we get two different accounts of what led Einstein to reject the November tensor in the notebook.

9.3 The Two Views

We now turn to the two competing explanations of what went on in the November episode. The account given here is based on the detailed exposition in Norton (2007).[8] Norton first presents the majority view, and then his own account. The reason why I use Norton's account is that it explicitly discusses, and accentuates, the contrast between the two views, while the other contributions do not focus on this disagreement. I will later turn to reactions of the majority view at the end of this section; this would serve as a sufficient corrective if Norton's account of the majority view were biased.

9.3.1 The Majority View

The majority view is that Einstein took the field equations to have a special, weak field form not just in some particular coordinate system, but in general. This implies that Einstein had to reject the November tensor, as it does not have the required form. In order to understand this explanation, two different ways of using coordinate systems have to be distinguished.

9.3.1.1 Coordinate Conditions Versus Coordinate Restrictions

A generally covariant theory holds for arbitrary coordinate systems. However, in order to apply the equations to concrete situations, one has to introduce special coordinate systems. Coordinates can be introduced in different ways. One possibility is to specify differential equations that the coordinates have to satisfy. This will fix a coordinate system only up to coordinate transformations that leave the differential equations invariant. Einstein may have used coordinates in two different ways in the notebook.

Coordinate conditions in the modern sense are a standard procedure if one wants to recover, say, the Newtonian limit. Newtonian gravitation in its standard formulation is not generally covariant, only covariant under Galilean transformations. Therefore, one may impose conditions on field equations of broader covariance if one wants to recover the Newtonian limit. Coordinate conditions do not restrict the covariance of the field equations; they are only used in the context of obtaining the Newtonian limit, e.g., when the imposition of a weak field assumption is not sufficient to recover Galilean covariance.

The examination of the Zurich notebook reveals that Einstein used coordinates in a second, non-standard way. The genesis collaboration has called this non-standard use *coordinate restrictions*. The introduction of coordinates into field equations always

[8]There is an accessible presentation of Norton's view in Norton (2005). See Janssen et al. (2007a, p.11), for a brief overview of the evolution of Einstein scholarship concerning this episode.

yields a new expression with restricted covariance. If coordinates are introduced as coordinate restrictions, the resulting expression is interpreted as the new field equations, which are not only valid under particular circumstances, but taken to be the gravitational field equations as such. The field equations before the application of coordinate restrictions are just an intermediate step in the derivation of the real field equations.

9.3.1.2 Coordinate Restrictions for the November Tensor

Einstein shows on p. 22R that if the so-called "Hertz condition"[9] is applied to the November tensor, it reduces to the expected weak field form, an important intermediate step to the Newtonian limit. However, Einstein did not use the Hertz condition on p. 22R as a coordinate condition, but as a coordinate restriction. This can be seen by examining a calculation on p. 22L. There, Einstein writes down two conditions, the Hertz condition, and the condition for "unimodular transformations",[10] the covariance group of the November tensor. He then calculates the covariance of the Hertz condition under unimodular transformations, i.e., he determines the covariance group of the November tensor combined with the Hertz condition. This calculation does not make sense if he wants to use the Hertz condition as a coordinate condition.[11] Subsequently, Einstein discards the Hertz condition. There is a second instance in the Zurich notebook where Einstein used coordinate restrictions, not coordinate conditions.

9.3.1.3 The Majority's Explanation

According to the majority view, the concept of coordinate restriction explains the difference between the situation in the Zurich notebook and the situation in November 1915 as follows. The November tensor does not have the form required for the Newtonian limit. If Einstein was unaware of the possibility of using coordinate conditions, this means that the November tensor was unacceptable as a candidate for the field equations, but it could still be used as an intermediary for a candidate with restricted covariance. Einstein therefore used the Hertz condition as a coordinate restriction; this produced a new candidate gravitation tensor with restricted covariance. It is documented that in November 1915, Einstein had acquired the notion of coordinate condition, and the November tensor became acceptable.

[9]The name was coined by the genesis collaboration. It figures prominently in correspondence between Einstein and Paul Hertz; see, e.g., Renn and Sauer (2007, p.184).

[10]Unimodular coordinates require that the determinant of the Jacobian of the coordinate differentials are equal to one.

[11]Only Galilean covariance is needed for the Newtonian limit, but it is easy to get Galilean covariance, because the condition is invariant under linear transformations, which implies Galilean covariance. However, Einstein does not eliminate terms from his calculation that would vanish under linear transformations. Therefore, he is not after linear transformations.

The majority and Norton agree that in the Zurich notebook, Einstein used coordinates in a way that can only be interpreted in terms of coordinate restrictions. Both views presuppose that Einstein did not use the Hertz condition as a coordinate condition. However, the majority view also assumes that Einstein was unaware of the possibility of interpreting the application of the Hertz condition to the November tensor as a coordinate condition. It is on this point that the majority and Norton disagree.

9.3.2 *Norton on the Majority View*

The majority assumes that Einstein was unaware of coordinate conditions. How plausible is this assumption? According to Norton, this question cannot be settled on the basis of the available evidence. There are instances where Einstein used coordinate restrictions in the notebook. However, in other cases, it is unclear how he interpreted the use of coordinates; the "harmonic coordinates"[12] are an example for the latter. Einstein did not check the covariance group of the harmonic coordinates. Norton concludes that nothing in the notebook precluded Einstein from being aware of coordinate conditions.

9.3.2.1 "Vermutlicher Gravitationstensor"

One central piece of evidence speaking against the assumption that Einstein was unaware of coordinate conditions can be found on p. 22R of the notebook; see Fig. 9.1. Einstein splits the Ricci tensor into two parts, one of which is the November tensor. He marks the November tensor with the label "presumed gravitation tensor" ("Vermutlicher Gravitationstensor"). Norton writes that, at this point, it must have been clear to Einstein that the November does not have the necessary form to reduce to the Newtonian limit without a further condition; he was sufficiently experienced to see this immediately. But if this were the case, the label would be inappropriate. The November tensor would not be the "presumed gravitation tensor", but just another intermediate step on the way to the right candidate; it would play the same role as the Ricci tensor.

Now, this could be an oversight on Einstein's part; he could have assigned the label in haste. However, according to Norton, this is implausible. For one, the page is neatly written, more like a summary than a hasty calculation. Also, Einstein had probably discussed the November tensor with Grossmann at this point—Grossmann's name appears on the page. Furthermore, the Hertz condition also appears on this page. Einstein's hope may have been that the November tensor reduces to the right

[12]Einstein used harmonic coordinates to recover the weak field form of the metric in the context of the Ricci tensor. Harmonic coordinates were known in the mathematical literature as "isothermal coordinates" at the time of the notebook.

Newtonian limit with the help of the Hertz condition. However, if he interpreted the Hertz condition as a coordinate restriction, then the November tensor would not be the "presumed gravitation tensor", but just an intermediary.

9.3.2.2 Evidence Against Coordinate Conditions?

Is there any clear evidence that Einstein was unaware of coordinate conditions at the time of writing the notebook? Norton argues that this is not the case. First, Einstein did not mention problems with coordinate conditions later on. This is relevant because he frequently commented on mistakes he committed during the genesis of GR, from the hole argument to the wrong assumption that the static metric has to be spatially flat. However, he never mentioned problems with coordinate conditions later on. Second, coordinate conditions are not needed in order to recover the Newtonian limit of the *Entwurf*; this explains why he never mentioned problems with coordinate conditions in this context. However, he also failed to mention them in other contexts where they may have been relevant. Third, Einstein had shown that he was aware of different ways of using coordinates. A lack of awareness of how to use coordinate conditions is implausible, because coordinates and their use was one of Einstein's motivations for the construction of a generalized theory of relativity in the first place.

According to Norton, all of this suggests, or at least leaves open the possibility, that Einstein was aware of the possibility of using coordinate conditions in the notebook. Could Einstein have considered both coordinate conditions and coordinate restrictions at the same time? If Einstein was aware of coordinate conditions, new puzzles have to be solved. It is the purpose of Norton's account to spell out how the events surrounding the November tensor unfolded if Einstein was aware of coordinate conditions.

9.3.3 Norton's View

Two problems have to be solved if we assume that Einstein was aware of coordinate conditions at the time of the notebook. First, we have to explain why Einstein was unable to recover the Newtonian limit of the November tensor using the Hertz condition, because this calculation features on p. 22R. Second, we have to explain why he stopped using coordinate conditions in combination with the November tensor in the notebook. Of course, not any explanation will do. If coordinate restrictions are not the explanation, Einstein must have made some other mistake. It would be easy to just invent an additional error—but this is not sufficient, as Norton points out: "The real difficulty is to establish that the error was really committed" (Norton 2007, p. 748).

There is one misconception that satisfies these requirements. Einstein explicitly defended this misconception, and admitted later on that it had been a mistake. It is the mistake of "attributing an independent reality to coordinate systems",[13] which was later central in the infamous hole argument. This mistake might also explain the puzzles of the November episode.

Norton conjectures that this misconception already shows up in the notebook. This would explain why Einstein gave up on the use of coordinate conditions on p. 22R of the notebook, and why he accepted the restricted covariance of the *Entwurf* equations. He only reversed his mistake in November 1915. If Einstein made this mistake in the context of the November tensor, the result would be that if a coordinate condition is applied, the covariance of the theory is reduced, and coordinate conditions are not only valid in the context of the Newtonian limit, but in general. Einstein may have realized this on p. 22R, and therefore abandoned the use of the Hertz condition as a coordinate condition.

Norton's conjecture presupposes that the mistake of attributing an independent reality to coordinates was tacit. The mistake was only made fully explicit when Einstein withdrew the hole argument in late 1915. If he had realized that he attributed an independent reality to coordinates in the notebook, he would not have endorsed the position. It is plausible that Einstein was not clear on this point, as he later had problems to spell out what exactly had gone wrong in the hole argument.

9.3.3.1 The Hole Argument

Norton's conjecture is based on an analogy to the hole argument. Einstein came up with the hole argument in late 1913. It served as an argument against the requirement that the field equations of general relativity should be generally covariant.

The argument runs as follows.[14] Take a region of space-time that is free of matter. In this region, the only field that can make a physical difference is the metric. The metric g is a function of the coordinate system x, written $g(x)$. Now, we can choose different coordinates x', which agree with x everywhere except in the region free of matter, where they deviate smoothly from x. The transformation from x to x' will yield a different representation of the metric, written $g'(x')$. Einstein interpreted $g'(x')$ to be the same field as $g(x)$, written in different coordinates—one and the same metric has different components in different coordinate systems; this does not constitute a physical difference. If we accept general covariance, we can also write the metric g' in terms of the *original* coordinate system x. However, this gives rise to the problem that the new metric in terms of the original coordinates, $g'(x)$, deviates from the $g(x)$ inside the designated space-time region: it yields a different field, despite being expressed in the same coordinates. This constitutes a physical indeterminacy, which is unacceptable; the culprit is general covariance, which, therefore, has to be rejected.

---

[13]This formulation is used in Norton (2007) for this particular misconception. I will use it as a technical notion in the present paper. It does not apply to the mistake of, say, using coordinate restrictions instead of coordinate conditions.

[14]See Norton (2011) for a discussion of the hole argument.

The hole argument is defective, as is well known. The mistake is to attribute a different physical meaning to the two solutions $g(x)$ and $g'(x)$. They are just mathematically different expressions of the same physical field. The source of the mistake is to (implicitly) attribute an independent reality to the coordinate system x. If the coordinate system would pick out space-time points uniquely and independently of g and g', then a disagreement between g and g' would be due to physical properties of the space-time point. However, space-time points are only individuated in virtue of the metric field. The difference is mathematical, not physical.

9.3.3.2 Norton's Conjecture

Norton's conjecture is that Einstein made the same mistake in the case of the November tensor: he attributed an independent reality to coordinates. As a consequence, the covariance of the theory was restricted to the coordinate conditions used to recover the Newtonian limit.

Einstein's reasoning might have proceeded along the following lines. The November tensor, while not generally covariant, is covariant under unimodular transformations. In the notebook, Einstein examined the transformation from Minkowski coordinates, x_{SR}, to uniformly rotating coordinates, x_{ROT}. This transformation is nothing but a change of coordinates. It yields a different expression for the metric: Starting from the Minkowski metric $g^{\text{SR}}(x_{\text{SR}})$, one arrives at a different metric $g^{\text{ROT}}(x_{\text{ROT}})$ in rotating coordinates. Given that the Minkowski metric is a solution of the November theory, and because uniformly rotating coordinates are unimodular, $g^{\text{ROT}}(x_{\text{ROT}})$ is also a solution of the November theory. This is not yet problematic.

However, it is only unproblematic insofar as g^{SR} and g^{ROT} are solutions in *different* coordinate system. This is where Einstein might have made a mistake by interpreting the transformation to g^{ROT} differently: He may have (implicitly) presupposed that g^{SR} and g^{ROT} both have to be solutions in the *same* coordinate system. He used the Hertz condition to bring the November tensor into the form required for the Newtonian limit. By checking whether both g^{SR} and g^{ROT} are compatible with the Hertz condition, he implicitly assumed that these two expressions needed to be compatible with the same coordinate system, namely the coordinates x_{HERTZ} compatible with the Hertz condition. He thus expected that $g^{\text{SR}}(x_{\text{HERTZ}})$ and $g^{\text{ROT}}(x_{\text{HERTZ}})$ both have to be solutions to obtain the Newtonian limit if one uses the Hertz condition. This, however, is impossible, because the Hertz condition is not compatible with g^{ROT}. This may have led to the rejection of the November tensor.

The problem is that Einstein used the Hertz condition to check compatibility of different expressions of the metric. This effectively limited the covariance of the theory to the covariance of the coordinate conditions of the Newtonian limit—the covariance of the Hertz condition in the present case—and there is no longer a difference between coordinate conditions and coordinate restrictions. Consequently, at this point, Einstein turned to using coordinate restrictions, as they had the advantage of yielding simplified field equations. This also explains why Einstein checked the covariance of the Hertz condition—he simply attributed a physical meaning to this condition.

9.3.3.3 Evidence for Norton's Conjecture

There is no direct evidence for Norton's conjecture in the Zurich notebook or in the *Entwurf*, i.e., it is unclear whether Einstein attributed an independent reality to coordinate systems at this time. We only have evidence that he did so in the context of the hole argument. Norton finds that the conjecture is compatible with Einstein's pronouncements on coordinate systems between 1912 and 1915 and with his attitude towards general covariance. One piece of evidence speaking in favor of Norton's conjecture is a letter to de Sitter, in which Einstein ties the lack of rotational covariance of the *Entwurf* field equations to the rejection of the hole argument. The conjecture establishes a direct connection between the hole argument and rotational covariance.

Norton locates the strength of the conjecture in its explanatory power, under the assumption that Einstein was aware of coordinate conditions—we have seen Norton's reasons for assuming that Einstein was aware of coordinate conditions in Sect. 9.3.2 above. The conjecture explains why Einstein gave up on using coordinate conditions in the notebook, it explains why he later thought that he did not succeed in deriving the Newtonian limit from the November tensor, despite a calculation that appears to show the contrary, and it explains his indifference towards general covariance in the *Entwurf* phase prior to the hole argument.

9.3.4 The Majority on Norton's View

The majority fraction of the genesis collaboration has not reacted to Norton's conjecture in detail. The majority maintains that the distinction between coordinate conditions and coordinate restrictions was sufficient for Einstein's rejection of the November tensor in the notebook, and they attribute the revival of the November tensor in 1915 to Einstein's realization that he could use coordinate conditions. Here are two reactions of the majority to Norton's conjecture.

9.3.4.1 Jürgen Renn

Renn (2004, Sect. 2) discusses the question as to why Einstein abandoned promising differential operators, such as the Ricci and the November tensor, in the notebook. Renn asks whether Einstein may have been unaware of coordinate conditions in the Zurich notebook—is it possible that "Einstein could have been guilty of such a trivial error?" (Ibid., p. 12). He points out that Einstein used harmonic coordinates in the notebook in the context of the Ricci tensor, which suggests that Einstein was aware of coordinate conditions. The calculations surrounding the November tensor, however, tell a different story; they show that Einstein's understanding of coordinate conditions differs from the modern view. In particular, Einstein checked the transformation group of coordinate conditions, which does not make sense according to the modern view; see Sect. 9.3.1 above. What could have induced Einstein to think that coordinates impose real restrictions on the field equations?

Renn briefly discusses Norton's answer to this question. He characterizes Norton's account as attributing a deep and "conceptual, if not metaphysical" (Ibid., p. 14) error to Einstein. Renn is not convinced by Norton's account:

> The evidence available makes it, in my view, implausible that this was indeed Einstein's pitfall in early 1913. If he committed an error conceptually close to the hole argument then it becomes incomprehensible why, as the historical documents indicate, Einstein only formulated this argument as late as summer 1913, and from then on regarded it as the life belt of the 'Entwurf' theory, while, before that, he considered its lack of being generally covariant as a shameful dark spot (Ibid., p. 14).

Renn mentions the Besso memo in support of this claim. The Besso memo, probably written on the 28th of August 1913, contains a preliminary version of the hole argument.[15] Renn concludes that the hole argument, or the reasoning underlying the hole argument, is not the "original sin" leading to the abandonment of the November tensor, and that the only viable explanation is that Einstein really did not have the modern notion of coordinate conditions.

9.3.4.2 Michel Janssen

Janssen (2007) comments on Norton's argument in the context of the Besso memo, which contains an early version of the hole argument, as well as reasons for rejecting it. Janssen writes: "It is my belief that Einstein used coordinate restrictions in the Zurich Notebook simply because he did not yet have the modern understanding of coordinate conditions. No further explanation is needed. Consequently, I am skeptical about Norton's conjecture" (Ibid., p. 828). Janssen does not elaborate on why he believes that Einstein did not yet have the modern notion of coordinate conditions, and he does not give further arguments against Norton's conjecture.

In sum, the majority is skeptical of Norton's solution, but there is no sustained engagement with Norton's arguments. Both Renn and Janssen point out that the hole argument, which is in the background of Norton's conjecture, has a philosophical, conceptual, or even metaphysical ring to it. This might indicate that Renn and Janssen consider Norton's proposal to be somewhat speculative.

[15]The argument for dating the Besso memo is given in Janssen (2007).

9.4 How to Resolve the Disagreement

Can the dispute between the majority and Norton be resolved? I agree with Norton that there is, at present, no evidence that could definitively settle the issue. However, I am optimistic that progress can be made. In this section, I will suggest several ways in which the debate can be brought forward. This will prepare the ground for the discussion of the more fundamental methodological issues in the next section.

Here is a sketch of the disagreement. The main point of contention is whether Einstein was aware of coordinate conditions at the time of the notebook. The distinction between coordinate conditions and coordinate restrictions, emphasized by the majority view, only explains Einstein's rejection of the November tensor if he was not aware of coordinate conditions. If it were possible to decide whether or not Einstein was aware of coordinate conditions, then the disagreement would simply disappear. I will reexamine this point in Sect. 9.4.1. Norton claims that Einstein must have been aware of coordinate conditions. The central piece of evidence for this claim is the "presumed gravitation tensor"; this is the topic of Sect. 9.4.2. If Einstein was aware of coordinate conditions, the change from coordinate conditions to coordinate restrictions is not accounted for by the distinction between coordinate conditions and restrictions, and a different explanation for the use of coordinate restrictions is needed. Now, Norton's conjecture comes into play. The occurrence of coordinate restrictions is explained by Einstein's mistake of attributing an independent reality to coordinate systems, a mistake he made in the context of the hole argument later on. By implicitly assigning independent reality to coordinates, Einstein collapsed the distinction between coordinate conditions and coordinate restrictions in the notebook. Norton comments on the (theoretical) virtues of his and the majority's explanation in various passages. This will be the topic of Sect. 9.4.3.

9.4.1 Einstein's Knowledge of Coordinate Conditions

Did Einstein have the notion of coordinate conditions at the time of the notebook? Both views seem to agree that, prior to November 1915, there is no instance where Einstein clearly used coordinate conditions in the modern sense.[16] However, direct evidence is not all that matters. We also have to take into consideration in how far coordinate conditions were available in the literature at the time of the notebook. If the notion was available, and if only in part, the claim that Einstein was aware of coordinate conditions gains plausibility: he simply had to check the relevant literature.

Both the majority fraction and Norton mention this point only in passing. There is hardly any discussion of the relevant physical and mathematical literature. A point mentioned by both parties is that Einstein took harmonic coordinates from the mathematical literature. Harmonic coordinates were known as "isothermal coordinates" in

[16]See Janssen and Renn (2007, Sect. 1.5) for an argument to this effect. This argument is neutral with respect to the disagreement discussed here.

the literature on differential geometry such as Bianchi (1910) and Wright (1908). We know that these works were familiar to Einstein. But how relevant is this particular kind of coordinates to the modern notion of coordinate conditions?

To answer this question, we have to examine the two sources just mentioned. Here is a brief recapitulation. In Bianchi (1910, Chap. 3), a textbook on differential geometry, it is shown that we can find a coordinatization of a surface such that the line element takes a particularly simple form. Such a parametrization exists if the second Beltrami parameter vanishes; this is mentioned by Einstein on p. 19L, as Norton points out. Bianchi also discusses the geometrical significance of these systems of curves. Wright (1908) is a monograph on quadratic differential forms; "isothermal systems of curves" are discussed in the context of applications of invariant theory. Wright also states that isothermal systems are tied to the vanishing of the second Beltrami parameter.[17] This shows that harmonic, or isothermal, coordinates were well understood mathematically.

However, the mathematical notion of harmonic coordinates, and the notion of coordinate condition, are quite far apart. Most importantly, the mathematical literature considered above is completely silent on the issue of using coordinates in a physical context.[18] There is no discussion of using coordinates to obtain, say, the Newtonian limit—this is not surprising; after all, these are works on differential geometry and invariant theory, not on physics. However, adapting coordinates to particular situations is the key ingredient of the modern notion of coordinate conditions. When considering the Newtonian limit, we can use a particular set of coordinates for our field equations, which does not affect the generality of the equations. This idea does not feature in the mathematical literature. It would be more fruitful to search the physical literature for seeds of the notion that Einstein needed.

What does this mean for the two diverging views? If the key ingredient to the modern notion of coordinate condition was not available in the literature, Einstein had to find the notion on his own. However, there are not many traces of this search. Thus, the story in Janssen and Renn (2007, Sect. 1.5) that the modern notion was forced upon Einstein only in 1915 gains plausibility as an account of how Einstein did arrive at the modern notion. This speaks in favor of the majority view.

9.4.2 *Evidence Against the Majority View: "Vermutlicher Gravitationstensor"*

Norton adduces one central piece of evidence against the majority view: the labelling of the November tensor as the "presumed gravitation tensor" ("Vermutlicher Gravitationstensor") on p. 22R of the notebook; see Sect. 9.3.2 above. Norton argues that if

[17]Note that Ricci and Levi-Civita (1901) discuss isothermal *surfaces*.

[18]Relevant parts of the modern notion may be discussed elsewhere in the mathematical literature. There are useful remarks on the history of "Euclidean geometry by means of general coordinates" in Veblen (1927, p.66).

we adopt a literal reading of this label, then the November tensor itself is the gravitation tensor, and not an intermediate step on the way to a different gravitation tensor. This implies that Einstein would not apply a coordinate restriction to the November tensor, but a coordinate condition.

How convincing is this argument? Unfortunately, there is no response by the majority, and the commentary on the notebook in Janssen et al. (2007b) does not further elaborate on the label. What are the possible explanations for labelling this part of the expression as the "presumed gravitation tensor"? We cannot reject Norton's explanation and still claim that the label is accurate. But other explanations for the label are possible.

Norton writes that the page is neatly written, indicating that p. 22R may have served as a summary of a calculation or a discussion. One alternative explanation is that the label served as a mnemonic device: Einstein simply wanted to mark this part of the expression as relevant, as opposed to its other parts, and to be used as a *basis* for a "presumed gravitation tensor". On this account, the label would have a contrastive role instead of a descriptive one. Maybe it would have been tedious to write down that the expression should be the *basis* for the gravitation tensor, because it may have been clear at this point that this is out of the question. This interpretation does not refute Norton's explanation, but it show that other explanations of the label are possible.

A different way of deciding whether or not Norton's explanation is plausible is to consider other instances of labelling in the notebook. Here is an example. Einstein labels a different object as "vermutlicher Gravitationstensor" on p. 9L.[19] The context of this second occurrence of a candidate gravitation tensor is different—the issue on p. 9L is the gravitational stress-energy tensor. Also, the handwriting on p. 9L is less tidy than on p. 22R. It is not clear whether the use of the label on p. 9L speaks in favor or against Norton's interpretation of the label on p. 22R, but discussing other instances of labelling in the notebook might convey a feeling for Einstein's usual practice.

9.4.3 Theoretical Virtues: Simplicity and Explanatory Power

An issue that is repeatedly discussed by Norton is the simplicity of the two views. He thinks that the majority view is the simpler account of why Einstein considered the November tensor to be untenable in the notebook. The majority view can account for Einstein's actions in virtue of just one distinction, that between coordinate conditions and coordinate restrictions, and there is no need to explain why Einstein used coordinate conditions in some contexts, and not in others. Norton, on the other hand, tells a more complex story, involving a mistake that was made twice, but only appeared in writing once, in the context of the hole argument.

[19] Note that the two labels are identical in German (up to the capital letter), but translated differently in the commentary; see Janssen et al. (2007b, p.555andp.647).

Of course, Norton's view is more complex for a reason—it assumes that Einstein was aware of coordinate conditions, and the view is therefore able to account for the available evidence under this assumption. Much hinges on this starting point: Norton's view rejects the premise of the majority view; this necessitates the introduction of a more complicated explanation. If the assumption of Einstein's awareness of coordinate conditions holds water, then the complexity of Norton's view does not speak against it.

Norton also discusses a second notion of simplicity, a quantitative parsimony of mistakes. On this notion of simplicity, Norton's view is simpler than the majority view, because it does not "multiply mistakes beyond necessity". The majority view has to attribute an "elementary blunder" to Einstein, that of not being aware of coordinate conditions. On Norton's view, Einstein committed the elaborate mistake of attributing an independent reality to coordinate systems in the context of the November tensor, and on top of this, Einstein repeated this very mistake in the context of the hole argument. Norton thus proposes a kind of "common cause explanation": Both the rejection of the November tensor, and the hole argument, are due to the same kind of blunder.

How convincing is this "common cause explanation"? One way to criticize it is along the lines of Renn (2004); see Sect. 9.3.4: Under the assumption that the mistake was already at work in the notebook, Renn contends, it should have had other observable effects, which, however, we do not find. This is a problem of (unobserved) consequences of the mistake conjectured by Norton.

A different line of criticism is to call into question that we are in fact dealing with just one mistake. One problem might be that the mistake is not exactly the same in the case of the November tensor and in the case of the hole argument; the two situations are only analogous. However, if it is not exactly the same mistake in both cases, should we still count it as *one* mistake? Norton emphasizes that the analogy is quite strong. There is not only a qualitative, but a formal parallel between the two mistakes, as we have seen in Sect. 9.3.3. If the parallel between the two situations really is that strong, the claim of quantitative parsimony seems legitimate.

This observation brings a different problem of Norton's common cause explanation to the fore. His view is based on a sophisticated Einstein, who does not commit elementary blunders, especially when it comes to coordinate systems. Norton therefore has to presuppose that the mistake of attributing an independent reality to coordinates was implicit—if Einstein had been completely aware of the ramifications of the mistake, he would not have made it. However, if the erroneous reasoning is only implicit, the tie to Einstein's later mistake in the hole argument gets weaker. We appreciate the link between the two mistakes because the parallel is made explicit. If Einstein would have seen the analogy between the mistakes as presented by Norton, and given Einstein's competence when it comes to coordinates systems—would he still have embraced it? If the parallel between the two occasions of the mistake was clear to Einstein, it is less plausible that Einstein would have made the mistake. If, on the other hand, the parallel is unclear, the common cause explanation gets weaker, and it is dubious that just one mistake was committed. There is a tension between

Einstein's committing the mistake consciously, and the force of the common cause explanation.

9.5 Methodological Issues

In the previous section, I suggested how we might settle the disagreement in substance. Now I will take a step back and reflect on the nature of the disagreement. Can we settle the dispute on the basis of the available evidence? Is the disagreement a matter of personal taste, or is there a fundamental methodological difference between the two views?

The nice thing about the present case is that the disagreement arose within a group that has worked in close collaboration on the same evidence, shares a large part of historical methodology, and has tried to come up with a common interpretation of the evidence. The dispute is not rooted in the fact that the two views take different sources into account; rather, the disagreement concerns the interpretation of evidence. This makes the problem of deciding between the two views hard, and interesting.

Norton, and the majority fraction of the genesis collaboration, have repeatedly commented on their methodology. These methodological remarks will be my starting point. I will focus on two lessons we can learn from the present case. The first lesson is positive: The dispute is not a mere matter of taste—progress is possible. The second lesson is more guarded; it points out a fundamental methodological problem, the question as to how much rationality we should ascribe to a historical actor.

9.5.1 Reconstructing Einstein

In the introduction of his paper, Norton characterizes the problem of deciding between the two views as a puzzle that lies on the boundary between the clear and the obscure: It is possible to formulate candidate solutions, but there is not enough evidence to reach a final verdict. What is at stake is not the evidence, but the interpretation of the evidence, and theoretical considerations. Norton suggests that we should evaluate the plausibility of the different views. He writes that when we invoke plausibility, "our personal Einsteins speak as much as evidence" (Norton 2007, p. 745).

Is a "personal Einstein" really that important in Norton's analysis? On closer inspection, it is not. Norton certainly assigns weights to evidence in a different way than the majority, but he is always careful to defend these weights. Arguments decide the outcome of the dispute, not subjective factors. Instead of different "personal Einsteins", I prefer to think of different, argued reconstructions of Einstein that are subject to critical evaluation. These arguments can be probed further in order to advance the discussion—we can push the boundaries of the unknown. Above, we saw three ways of probing the reconstructions of Einstein.

First, if direct evidence cannot settle an issue, we may take further sources and background knowledge into account. Did Einstein know about coordinate conditions at the time of the notebook, or did he not? I have suggested in Sect. 9.4.1 that we may look at the context if we cannot decide this question on the basis of the notebook. We can try to determine how likely it would have been to know about coordinate conditions, given the mathematical and physical background. If the notion of coordinate condition was not available at the time, and if there is no evidence of how Einstein acquired the notion, it gets more implausible that he had the modern notion.

Second, once we have proposed a certain reconstruction, it can be scrutinized anew on the basis of the available evidence. One example is the labelling of the "presumed gravitation tensor"; see Sect. 9.4.2 above. Norton's argument is based on the assumption that Einstein was accurate when it comes to labelling. We can now revisit the notebook and compare this instance of labelling with other instances, and thereby decide whether Einstein really is that accurate, or whether we can come up with an alternative explanation that trumps Einstein's accuracy. A second example of reevaluating a given reconstruction is Renn's point that the mistake of attributing an independent reality to coordinates should have had observable consequences before the hole argument.

Third, we can check the internal consistency, or plausibility, of the reconstructions. I argued in Sect. 9.4.3 that there is a tension in Norton's account between the quantitative parsimony of mistakes on the one hand, and how carefully the mistaken argument proceeds on the other.

9.5.2 *Historical Errors: A Dilemma*

The episode under dispute depends on the attribution of mistakes to the historical actor, Einstein. This generates methodological problems. Norton formulates one of these problems as follows: "Of course it is always possible to invent hidden errors varying from the trivial slip to the profound confusion, tailor made to fit this or that aberration" (Norton 2007, p. 748). If we are interested in explaining the actual course of events in a historical episode, the indiscriminate introduction of errors threatens to trivialize the account.

Norton argues that this problem can be overcome by a quantitative parsimony of mistakes. His account of the episode satisfies this constraint: "What is appealing about the conjecture is that it requires us to posit no new errors" (Ibid., p. 781). According to Norton's conjecture, we do not have to multiply mistakes beyond necessity, because Einstein repeated the mistake of attributing an independent reality to coordinate systems in the context of the hole argument. I have already pointed out a material problem of Norton's conjecture in Sect. 9.4.3: the mistake might not be exactly the same in both situations, but only analogous.

However, there is an even more fundamental problem lurking in the background. Why should we attribute as few errors as possible to Einstein *in principle*? Isn't it natural that scientists commit mistakes? Isn't it problematic to presuppose that historical actors proceed in a quasi-rational manner? On the one hand, we can explain any historical episode if we presuppose the right kind of error at the right moment in history. On the other hand, do we have good reasons to minimize the amount of errors we conjecture in our historical accounts? It appears that we face a dilemma. The second horn of the dilemma has been forcefully formulated by Michel Janssen (2007, p.832):

> So, to put it somewhat bluntly, whenever one encounters a passage containing what on the face of it looks like an error on Einstein's part, the strategy is to look for an interpretation in which the apparent error is the manifestation of some deep conceptual difficulty that had to be overcome before general relativity as we know it could be formulated.

The worry might be that we end up with a reconstruction of Einstein that is too rational in that it only deviates from the perfect path of discovery if Einstein encounters "deep" difficulties. Janssen discusses this issue in Sect. 5 of his paper against the background of Einstein scholarship since the 1980s. Before Norton's and Stachel's groundbreaking interpretation of the notebook, historical reconstructions of the genesis of general relativity attributed trivial errors to Einstein. For example, these early accounts assumed that Einstein did not know that if one transforms the components of the metric using coordinate transformations, the resulting expression is not physically different from the untransformed metric. Norton and Stachel ruled out this possibility, thus avoiding the first horn of the dilemma. Janssen finds that while this was an improvement, Norton got too close to the second horn of the dilemma with his "excessively acute Einstein". Janssen prefers a different, "opportunistic" Einstein, who did not follow up on inconsistencies if they threatened his pet principles.

There is probably no silver bullet for this dilemma. We should avoid the implausible attribution of trivial errors to Einstein, but also steer clear of an overly charitable interpretation, or of "overly complex errors".[20] We can only avoid these pitfalls by scrupulously reconstructing Einstein from the available evidence.

9.6 Conclusion

We have seen two accounts of the same historical episode. Both are based on the same evidence, and still they disagree. The reason for the disagreement lies, first, in the weight assigned to the evidence. For example, Norton emphasizes the case of the "presumed gravitation tensor"; the majority view does not discuss this point. A second source of disagreement is a different view on the (background) knowledge we may attribute to Einstein, or when and how Einstein acquired this knowledge. Did Einstein know about the freedom to apply coordinate conditions? Here, both camps

[20] It would be desirable to get a better systematic understanding of the role of errors in this episode. Such an understanding might be gained on the basis of the so-called "dynamical inferential conception" of the application of mathematics, proposed in Räz and Sauer (2015). This framework systematizes different kinds of mistakes that can be made in the context of applying mathematics to empirical problems.

have merely sketched the context. Third, the two views have, implicitly and explicitly, emphasized different theoretical virtues. The majority emphasizes one distinction as crucial, while Norton has a more intricate story that connects the notebook to one other important episode in the genesis of GR. Norton's position might be more speculative and, therefore, also more susceptible to criticism.

The methodological discussion has shown that fundamental methodological issues play a role in the disagreement as well. On the one hand, the majority view attributes a mistake to Einstein that may seem elementary from a modern perspective. On the other hand, Norton's view constructs an elaborate mistake, which would persist for some time and resurface later. While the first view may run the risk of telling too simple a story and trivialize the episode, the other may be conceived as painting a picture of Einstein that is too rational.

Despite these difficulties, there is reason for hope; progress is possible at all points. We can try to decide on the relevance of evidence by comparing similar cases; we can make an effort to reconstruct the background knowledge; we can adduce philosophical and psychological theories in order to understand how the transfer of knowledge from one field to another works, and to understand what kind of mistake we may attribute to historical actors. All of this will lead to an improved reconstruction of Einstein.

Acknowledgments I thank John Norton and Raphael Scholl for comments on previous drafts of the paper, and Tilman Sauer for comments and fruitful discussions concerning the genesis of GR.

References

Bianchi, L. 1910. *Vorlesungen über Differentialgeometrie*, 2 ed. Teubner.

Einstein, A., and M. Grossmann. 1995. *Entwurf einer verallgemeinerten relativitätstheorie und einer theorie der gravitation*, 302–343. In Klein et al. (1995).

Janssen, M. 2007. *What did Einstein know and when did he know it? A besso memo dated August 1913*, 785–838. In Janssen et al. (2007b).

Janssen, M., Norton, J.D., Renn, J., Sauer, T., and J. Stachel. 2007a. *The genesis of general relativity. Vol. 1. Einstein's Zurich notebook: Introduction and source*. Dordrecht: Springer.

Janssen, M., Norton, J.D., Renn, J., Sauer, T., and J. Stachel. 2007b. *The genesis of general relativity. Vol. 2. Einstein's Zurich notebook: Commentary and essays*. Dordrecht: Springer.

Janssen, M., and J. Renn. 2007. *Untying the knot: How Einstein found his way back to field equations discarded in the Zurich notebook*, 839–925. In Janssen et al. (2007b).

Klein, M.J., Kox, A.J., Renn, J., and R. Schulmann (eds.). 1995. *The collected papers of Albert Einstein, volume 4: The Swiss years: Writings, 1912–1914*. Princeton, NJ: Princeton University Press.

Norton, J.D. 2005. A conjecture on Einstein, the independent reality of spacetime coordinate systems and the disaster of 1913. In *The universe of general relativity, volume 11 of Einstein studies*, ed. A.J. Kox, and J. Eisenstaedt, 67–102. Basel, Boston, Berlin: Birkhäuser.

Norton, J.D. 2007. *What was Einstein's "Fateful Prejudice"?*, 715–83. In Janssen et al. (2007b).

Norton, J.D. 2011. *The hole argument*. http://plato.stanford.edu/entries/spacetime-holearg/.

Räz, T., and T. Sauer. 2015. Outline of a dynamical inferential conception of the application of mathematics. *Studies in History and Philosophy of Modern Physics* 49: 57–72.

Renn, J. 2004. Standing on the shoulders of a dwarf: General relativity: A triumph of Einstein and Grossmann's erroneous "entwurf" theory. In *In the shadow of the relativity revolution*, Preprint 271, 5–20, Berlin: Max Planck Institute for the History of Science.

Renn, J. (ed.). 2007. *The genesis of general relativity (4 vols.)*. Dordrecht: Springer.

Renn, J., and T. Sauer. 2007. *Pathways out of classical physics. Einstein's double strategy in his search for the gravitational field equation*, 113–312. In Janssen et al. (2007a).

Ricci, M., and T. Levi-Civita. 1901. Méthodes de calcul différentiel absolu et leurs applications. *Mathematische Annalen* 54: 125–201.

Sauer, T. 2014. Marcel Grossmann, and his contribution to the general theory of relativity. In *Proceedings of the 13th Marcel Grossmann meeting on recent developments in theoretical and experimental general relativity, gravitation, and relativistic field theory*, ed. Jantzen, R.T., Rosquist, K., and R. Ruffini. Singapore: World Scientific. arXiv:1312.4068.

Stachel, J. 2007. *The first two acts*, 81–112. In Janssen et al. (2007a).

Veblen, O. 1927. *Invariants of quadratic differential forms*. Cambridge: Cambridge University Press.

Wright, J.E. 1908. *Invariants of quadratic differential forms*. New York: Hafner Publishing Co.

Part III
Integration in Practice

Chapter 10
From Discrepancy to Discovery: How Argon Became an Element

Theodore Arabatzis and Kostas Gavroglu

Abstract In this paper, we revisit the discovery of argon by Lord Rayleigh and William Ramsay. We argue that to understand *historically* how argon was detected, conceptualized, and accommodated into chemical knowledge we need to take into account the *philosophical* insight that scientific discovery is often an extended process. One of argon's most intriguing properties was that it did not react with other elements. Reactivity, however, had been a constitutive property of elements. Thus, the discovery of argon could not have been accepted by chemists without a reconceptualization of 'element'. Furthermore, there were difficulties with the accommodation of argon in the Periodic table, because argon appeared to undermine the conception of matter that underlay the Periodic table. The discovery of argon was complete only after those conceptual difficulties had been removed. This is why it has to be understood as an extended process, rather than as an event. Furthermore, we will suggest that some of the factors that complicated the discovery of argon were related to the legitimization of physical techniques of investigation in chemistry and the emergence of physical chemistry.

10.1 Introduction

In the troubled history of integrated history and philosophy of science (&HPS) we can discern two main ways of bringing the two fields together, historical philosophy of science (HPS) and philosophical history of science (PHS). The former explores, in a historically informed manner, general philosophical issues about science; the latter reconstructs, from particular philosophical points of view, specific episodes from

T. Arabatzis (✉) · K. Gavroglu
Department of History and Philosophy of Science, University of Athens,
University Campus, 15771 Athens, Greece
e-mail: tarabatz@phs.uoa.gr

K. Gavroglu
e-mail: kgavro@phs.uoa.gr

T. Sauer and R. Scholl (eds.), *The Philosophy of Historical Case Studies*,
Boston Studies in the Philosophy and History of Science 319,
DOI 10.1007/978-3-319-30229-4_10

the history of science. Thus, the goal of HPS is primarily philosophical, whereas the motivation for PHS is predominantly historiographical. So far &HPS has been dominated by HPS, at the expense of PHS (see Arabatzis 2016). This can be seen most clearly in the discussion on the significance of case studies, which has focused on the value (and limitations) of historical evidence for settling philosophical disputes (see, e.g., Pitt 2001; Burian 2001). In this paper, we want to go beyond that discussion, and investigate the converse (and much neglected) question, namely the "*added value*" (in John Heilbron's words) of taking into account philosophical issues when engaging with episodes in the history of science.

Historical studies of science employ metascientific concepts (e.g., theory, experiment, evidence), which embody particular philosophical points of view (Hanson 1962; Chang 2011; Schickore 2011). As we hope to show below, if we reflect upon those concepts, by engaging with the pertinent philosophical issues, we may clarify and enrich the historical accounts that we produce. The concept of discovery is one of the meta-scientific concepts that scientists, historians, and philosophers have used to make sense of the scientific enterprise. There has been a long tradition of philosophical thinking about discovery, going back to the 17th century (Laudan 1980) and many case studies reconstructing particular scientific discoveries (e.g., Nickles 1980a, b). In this connection, one should note the (in)famous distinction between the context of discovery and the context of justification and the enormous literature that it has engendered. In the discussions on the two contexts, "discovery" is understood as the generation of novel ideas (hypotheses, theories). Thus, these discussions have little relevance for understanding observational or experimental discoveries (Arabatzis 2006b), which are often understood as events about which the following W-questions can be posed and unambiguously answered:

What was discovered?
Who made the discovery?
When was the discovery made?
Where was the discovery made?

This conception of scientific discovery is particularly popular among scientists and science writers. Actually, it underlies the reward system of scientific institutions. From now on we will call it the received view of scientific discovery (RVSD). The RVSD has the following presuppositions:

The discovered object(s) can be unambiguously identified.
Discoveries are happenings that can be attributed to individuals.
Discoveries are events that can be precisely dated.
Discoveries are events that can be located in space.

More than half a century ago, these presuppositions were thrown into doubt by Thomas Kuhn, who argued that many scientific discoveries are not events. Rather they are extended processes that involve several scientists and cannot be precisely dated. According to Kuhn, the reason that scientific discoveries often have a historical structure is that they involve not only the observation of a novel entity but also its conceptualization. Discovery comprises not only the recognition of something novel, but also its (correct?) identification. The latter may require the revision of established concepts and/or the formation of new ones. These are collective achievements that inevitably take time and are extended in space (Kuhn 1962). This is particularly so when a discovery is unexpected or incompatible with entrenched beliefs, concepts, or ways of understanding nature. In such cases a discovery is not complete before a restructuring of the conceptual framework associated with existing knowledge.

Kuhn's critique of the RVSD showed in a striking manner that the uncritical use of a meta-scientific concept (discovery) may mislead historical investigation by creating deadlocks or blocking the discussion of a number of questions. Thus, rethinking the notion of discovery, by taking into consideration the philosophical difficulties pointed out by Kuhn and others, promises to bring about a fresh historiographical perspective on many cases from the history of science that have been already studied—some of them in a detailed manner as well. In recent literature, Kuhn's insights have been borne out by historical studies of scientific discoveries (e.g., Arabatzis 2006a; Caneva 2005; Dick 2013; Frercks et al. 2009) and there are now strong arguments for thinking of discovery as an extended process. It has been argued, for instance, that justification is an essential part of the discovery process and that the mere observation of a novel phenomenon (or object) is not sufficient for establishing its discovery. Furthermore, justification is a process that transcends the technicalities involved in the observation of a phenomenon; it involves rhetoric, persuasion and many other aspects which are endemic to the culture of the specific scientific community.

In this paper, we will discuss the "discovery of argon" by taking into consideration philosophical issues that have been neglected in various narratives about argon which have been loyal to the RVSD. We will argue that to understand historically how argon was detected and conceptualized and how it became a new element we need to take into account the philosophical insight that scientific discovery is often an extended process.

There have been a number of scholarly discussions of the argon story (for example, Giunta 1998, 2001; Hiebert 1963; Hirsh 1981; Matyshev 2005; Scerri 2007, pp. 151–156; Scerri and Worrall 2001; Spanos 2010; Wolfenden 1969). None of them, however, can answer unambiguously all the W-questions concerning the discovery of argon. There are difficulties in attributing the discovery of argon to any particular scientist (Cavendish? Rayleigh? Ramsay?); to pinpoint its date (1892? 1895? later?); to specify its location (which particular laboratory?); and, more importantly, to identify the object of discovery (what was argon? Or, to put it less provocatively, how did chemists and physicists conceptualize the new gas that they had detected?).[1]

[1]Cf. Gordin (2012, p.59): "I have no idea who discovered the periodic system of chemical elements, and I am going to tell you why." The main reason he gives is that there is no way to answer the "What?" question.

10.2 A Discrepancy and Its Discontents

At the end of the 19th century, many physicists, including Lord Kelvin, believed that the state of physics was such that the only viable prospect for its progress and new discoveries involved the "next-decimal-place."[2] This attitude, whereby the only new surprises would come from ever more accurate measurements, came to be known as the "next-decimal-place view of physics." A widespread view concerning the historical interpretation of the discovery of argon is that it was a triumph of that culture of exact measurements. As we will argue below, this view conflates the discovery of argon with its original detection and neglects other essential aspects of the discovery process, such as the identification and assimilation of argon within the conceptual framework of late 19th century chemistry.

Another problem to be understood is the rather violent reactions of some chemists against William Ramsey and Lord Rayleigh (R & R) and the way they chose to deal with the discovery. These reactions are considered as a conservative dismissal of the new discovery—something rather common in many discoveries. Unless one questions the RVSD one cannot properly assess the arguments of those chemists, and especially James Dewar and Henry Armstrong, in the context of the period. What they were reacting against was the conceptual reorientation that had to take place in chemistry if argon was what R & R claimed it to be. They were, in effect, reacting against the prospect of a new kind of chemistry, a chemistry that would accommodate the possibility of inert elements and would, thus, be at odds with the long cherished, almost axiomatic, belief that elements are entities that react with other elements. Apart from the technical issues involved, such a reconceptualization was a major factor in completing the discovery of argon.

Rayleigh's measurements for the exact determination of the densities of gases had started in 1882 while he had succeeded James Clerk Maxwell as the Cavendish Professor of Experimental Physics at the University of Cambridge.[3] He continued them in 1888, having by then left Cambridge and having become the Professor of Natural Philosophy at the Royal Institution in London. This program of exact measurements was a program aimed to test Prout's hypothesis by finding the atomic weights of gases and observing the extent to which they were multiples of the atomic weight of hydrogen. By 1892 Rayleigh found a curious discrepancy. In a letter to *Nature* he noted that the density of nitrogen depended on the method used to isolate the gas. The nitrogen he derived from the two different methods he called "physical" nitrogen and "chemical" nitrogen. The former, physical nitrogen was isolated by passing atmospheric air over red-hot copper. Chemical nitrogen could be derived by

---

[2] Apart from the published sources, the following discussion takes into consideration the notebooks of Lord Rayleigh, William Ramsay and James Dewar. These notebooks have not been explored by the historians who have written about the "discovery of argon" and though they do not add anything substantially new, they do help to clarify a number of issues. Rayleigh's Notebooks are in the Lord Rayleigh Papers in Hanscomb Air Base, Massachusetts; Ramsay's in the William Ramsay Papers at University College, London; and Dewar's in the Sir James Dewar Papers at the Royal Institution, London. We thank the administrative officers of these archives for their permission to use them.

[3] For the precision culture that Rayleigh fostered at the Cavendish see Schaffer (1995).

a method communicated to Rayleigh by William Ramsay, Professor of Chemistry at University College, London, whereby a mixture of dry air and ammonia was passed over red-hot copper. It was found that physical nitrogen was heavier than chemical nitrogen by about 1/1000:

> I read your letter in *Nature* some weeks ago about the discordance between the densities of nitrogen prepared by the old way and by the way i told you. I could think of no reason for the difference and i can't now, but it has just struck me that you perhaps do not know that ordinary ammonia contains traces of amines ... But I really do not see what such an impurity could do ... So I give it up. But I thought it worthwhile to tell you of the possible impurity.[4]

Thus, Ramsay suspected that the "discordance" could be attributed to the presence of impurities in ammonia, without however being able to specify how.

Rayleigh's next step was to find ways to exaggerate this difference:

> One's instinct at first is to try to get rid of a discrepancy, but I believe that experience shows such an endeavour to be a mistake. What one ought to do is to magnify a small discrepancy with a view of finding out the explanation (Rayleigh 1895, p. 525).

Further improvements showed that chemical nitrogen was about 0.5 % lighter than physical nitrogen.

Though the detection of that discrepancy was the initial step towards the discovery of argon, a similar discrepancy had already been noticed more than a century earlier, by Cavendish (1785). Cavendish noticed, while attempting to remove all the nitrogen from a jar, that there was a residue (less than one hundredth of the original sample) which he could not remove. However, he did not attempt to account for that residue. It was Rayleigh (and, then, Rayleigh together with Ramsay) who eventually attributed the discrepancy to a new gas, eliminating various possible explanations of the weight difference between chemical and physical nitrogen. The first two alternatives Rayleigh entertained were either that atmospheric nitrogen was too heavy because of the imperfect removal of oxygen from atmospheric air or that chemical nitrogen was too light because, upon its removal from ammonia, it was contaminated with gases which were lighter than nitrogen. Further experiments by Rayleigh excluded both possibilities. Another possible explanation was that the discrepancy might be due to the dissociation of nitrogen molecules and their subsequent formation into N_3, much like the situation with the production of ozone from oxygen by silent discharge. Rayleigh ruled out this possibility too, by showing that sparking had no appreciable effect in altering the densities of the two kinds of nitrogen. The elimination of those alternative explanations was a significant aspect of the discovery process.

By the beginning of 1894 Rayleigh was convinced that the atmosphere contained a new hitherto unknown constituent.

The experimental procedure used by Rayleigh to isolate the new constituent was, in effect, very similar to the experiments performed by Cavendish in 1785. Rayleigh tried first to remove the oxygen from the atmospheric air, then the nitrogen and then the carbon dioxide and the other known gases. The difficulty, of course, was in the

[4] William Ramsay to Lord Rayleigh, 20 November 1892.

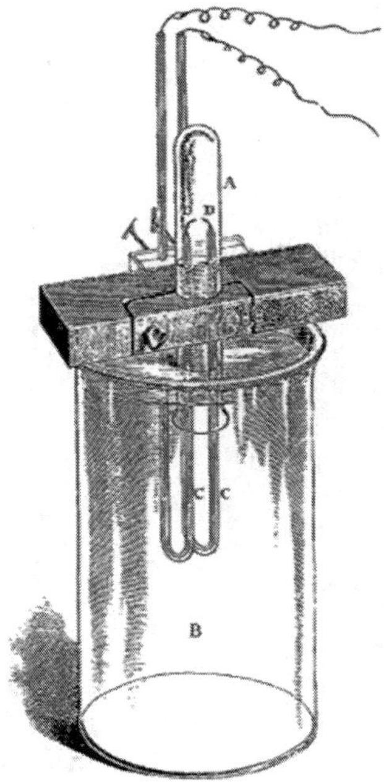

Fig. 10.1 The apparatus used by Lord Rayleigh in 1894 to isolate the residual gas, which remained after removing all nitrogen. Figure taken from Rayleigh and Ramsay (1895, p. 198)

removal of nitrogen since it chemically combines only with certain elements and under specific conditions. Rayleigh used the apparatus in Fig. 10.1. It consisted of a Ruhmkorff coil connected to a battery and five elements of a Grove cell. The gases were in the test tube A, placed upside down in a container with a large amount of light alcalines B. The current went through the wires which passed inside two U-shaped glass tubes (CC in the figure). The platinum ends were secured by being "glued" onto the test tubes. A short spark of about 5 mm was found to be more efficient than a longer one. When the proportions of the gases were the right ones then the absorption was about 30 cc/h—thirty times better than in Cavendish's experiment.

A characteristic run is found in the very first page of Rayleigh's *Notebooks*. He started with 50 cc of air and continuously added oxygen. With the help of the sparks he could have a union of oxygen with nitrogen. The addition of oxygen continued until there was no noticeable contraction of the volume of the gas inside the test tube after sparking for 1 h. What remained was transferred to another tube and found to be 1 cc. The residue was passed over alkaline pyrogallate and the final product was 0.32 cc. This could not have been nitrogen since it did not decrease after continuous sparking; nor could it have come from somewhere else since repeated measurements had shown that it was proportional to the mass of the original intake of atmospheric air. Rayleigh called it the "residue."

Ramsay's method for isolating the new constituent was rather different (see Fig. 10.2). In tube A there was magnesium and in tube B copper oxide. A and B were united with India rubber. The tube CD had preheated lime soda, to ensure that it would not contain water vapour. The other half contained phosphoric anhydride.

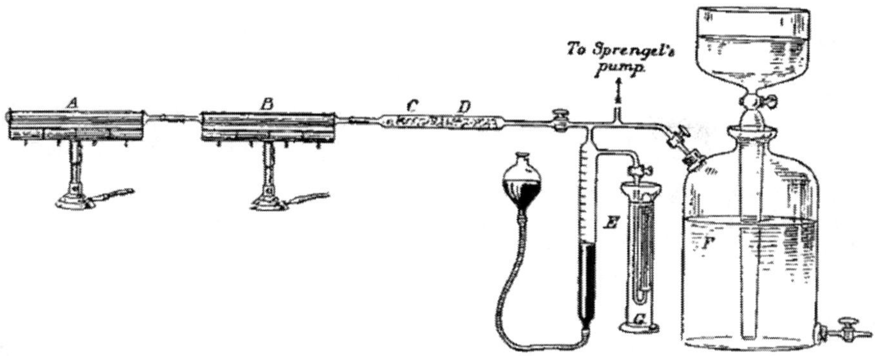

Fig. 10.2 William Ramsay's method for isolating the residue (Rayleigh and Ramsay 1895, p. 200)

E was a measuring vessel and F contained atmospheric nitrogen. The heated magnesium absorbed the nitrogen and the rest was collected in G. Ramsay, then, changed tube A and repeated the same process. In this manner, by starting with 1094 cc of nitrogen he was left with a residue of 50 cc which, however, contained rather large quantities of nitrogen. By May 1894 Ramsay was suspecting that he was on to something.

> I intended to ask you today what is probably quite unnecessary, not to say anything about the gas which I think I have got. It may turn out a mare's nest and it would be well that no one should know of its existence.[5]

This is quite a remarkable letter. Ramsay informed Rayleigh that he had gotten a large amount of nitride of magnesium, which when treated with water gave ammonia. He could get nitrogen out of this ammonia and was willing to send it to Rayleigh. Ramsay suggested to send to Rayleigh the ammonia as chloride of ammonium. Then Rayleigh could liberate the ammonia, mix it with oxygen and pass it over red hot copper. The end of the letter is quite revealing:

> Has it occurred to you that there is room for gaseous elements at the end of the first column of the periodic table? ... Such an element should have the density 20 or thereabouts and 0.8 % (1/120th about) of the nitrogen of the air would so raise the density of nitrogen that it could stand to pure nitrogen in the ratio 230/231 (Ibid.).

Ramsay had already started to think about placing the new gas in the periodic table.

By the beginning of August of 1894 they were both convinced that atmospheric air contained either a new element or a new compound. On August 4, a triumphant Ramsay wrote to Rayleigh:

> I have isolated the gas. Its density is 19.075 and it is not absorbed by magnesium. ...The nitrogen prepared from magnesium nitride is chemical nitrogen; i.e. it has a density 1/230 below that from air (your experiments). The value of the chemical N_2 is identical with yours.

[5]William Ramsay to Lord Rayleigh, 27 May 1894.

> I have been watching the density of X creep up as absorption proceeds; so you see this is no chance determination with a possible source of error.[6]

Rayleigh answered immediately

> I believe that I too have isolated the gas, though in miserably small quantity ...[7]

Up to now they were both working separately and communicating the results to each other. At this point, Rayleigh suggested that they should proceed to a joint publication since so much that has been done was complementary work. Almost by return mail, Ramsay responded stating that a joint publication was indeed the best course to follow.[8]

The results were first presented during the meeting of the BAAS in Oxford on August 13, 1894. In a brief announcement read by Rayleigh it was reported that atmospheric nitrogen, when purified from all the other known constituents of air, was found to contain another gas to the extent of about 1 % which was even more inert than nitrogen. It is interesting to note that the relative inertness of the gas was announced during this first public presentation. The density of this gas was found to be between 18.9 and 20 and preliminary observations of its spectrum had found a characteristic line.

At the Oxford meeting of the BAAS, H.G. Madan (1838–1901), an Oxford chemist who was in the audience, proposed that the new gas, because of its inertness, should be called "argon", a word meaning idle in Greek (see Strutt 1968, p. 205; and Travers 1956, p. 122). Apparently, the extraordinarily careful R & R did not adopt the name immediately, since there were still tests to be conducted about the reactivity of the gas. And it was only later, on November 7, 1894 that Ramsay wrote to Rayleigh: "Seeing that X is very inactive, what do you think of argon, ἁ-εργόν for a name?" In that letter Ramsay did not mention Madan. However, Madan's prior naming of the new gas was pointed out by Morris Travers, a long-time associate of Ramsay's and a great admirer of the master (Travers 1956, p. 122). Thus, we can safely surmise that Ramsay simply reiterated Madan's suggestion, without mentioning him.

The announcement of a new constituent of atmospheric air deeply upset many distinguished British chemists. Shortly after the announcement, a critic wrote:

> Whatever may be the ultimate verdict of chemists—whether it be that a new element or a new form of nitrogen has been discovered in the atmosphere; or whether the verdict may be, as we fear, that a sad blunder has been committed, the method of announcement was most unhappy ... There was no adequate discussion in the Chemical Section, and a manifestly exaggerated importance was at once assigned to the preliminary announcement and to a certain degree the public has been misled.[9]

The two most vocal critics were the President of the Chemical Society, Henry Armstrong, and James Dewar, the Jacksonian Professor of Experimental Philosophy at

---

[6] William Ramsay to Lord Rayleigh, August 4, 1894.

[7] Lord Rayleigh to William Ramsay, August 7, 1894.

[8] William Ramsay to Lord Rayleigh, August 7, 1894.

[9] *British Medical Journal*, September 1, 1894, 508.

the University of Cambridge and the Fullerian Professor of Chemistry at the Royal Institution. He held both posts until his death and he was also the Director of the Davy-Faraday Research Laboratory of the Royal Institution.

Right after the BAAS meeting in Oxford, James Dewar wrote two letters to the *Times* claiming that what had been found by R & R was the triatomic form of nitrogen. Dewar suggested that the allotropic form of nitrogen could produce spectra which were distinct from nitrogen and in the case of R & R "the new substance is being manufactured by the respective experiments, and not separated from ordinary air" (Dewar 1894a). Dewar's suggestion could have undermined R & R's achievement. If he were right, the gas detected by R & R would be an artifact of their experimental setup and, furthermore, an artifact of a familiar kind. Their presumed discovery would evaporate into thin air!

Looking into Dewar's *Laboratory Notebooks*,[10] we notice that right before the meeting of the BAAS he was busy performing a series of experiments on the low temperature behavior of chemical nitrogen and atmospheric nitrogen and found that both kinds of nitrogen liquefied at the same temperature. Dewar was well versed in low temperature experiments. Though his crowning achievement, the liquefaction of hydrogen, would take place in 1898, by the time he performed these experiments with nitrogen he had invented and improved what came to be known as the Dewar Flask. Those results showed, according to Dewar, that the new gas was N_3.[11] Dewar reached his conclusions through characteristic chemical thinking. He suggested that the theoretical density of the new nitrogen, compared to hydrogen, should be 21, while the experimental numbers were between 19 and 20. He surmised that "chemists would infer" that such a substance ought to be characterized by great inertness, because phosphorus, the element most nearly allied to nitrogen, easily passes into an allotropic form known as red-phosphorus, which, relative to yellow phosphorus, was inert. If, therefore, such an active body as phosphorus could become, in condensed form, far less active chemically, then, "by analogy, nitrogen, so inert to start with, must in the new form, become exceedingly active."

On December 6, 1894, Dewar presented to the Chemical Society his experiments concerning the liquefaction of nitrogen. In his talk he claimed that chemical and physical nitrogen liquefied at the same temperature and boiled off at the same rate (Dewar 1894b).[12] From this he inferred that the assumed new substance in atmospheric nitrogen did not condense at those temperatures when oxygen and various gaseous compounds condensed and behaved in exactly the same manner as nitrogen. In an unsigned piece at *The Times* the next day, reporting about the meeting at the Chemical Society, it was remarked that "Chemists will appreciate the extreme singularity of a substance with the assigned density which fulfills either condition."

[10]Dewar's Laboratory Notebooks are in the Dewar Archives at the Royal Institution, London. See note 2.

[11]Dewar Notebooks: Entries for August 9; August 14; November 21; November 27; November 29; December 3; December 14; December 20, 1894.

[12]See also Dewar's Laboratory Notebooks, entries throughout November 1894.

The summary of the discussion to Dewar's paper was most probably written by his most fanatic ally, Henry Armstrong.

Neither Rayleigh nor Ramsay had attended the meeting. "It is useless to deny that special interest attached to the communication to which they had just listened, but, unfortunately, in the absence of Lord Rayleigh and Professor Ramsay, they were left in the position of having to play "Hamlet" with only the ghost present, and, under such circumstances, the play obviously could not be continued to a successful issue".

Dewar's announcement gave Armstrong the opportunity to express the skepticism of his fellow chemists:

> Lord Rayleigh and Prof. Ramsay now could not hope to keep so remarkable a discovery to themselves much longer. After having been told so much, chemists could not be expected to remain quite under the imputation that they had been eyeless during a whole century. Indeed. Although no one would seek to take the discovery out of the hands of those who had announced it, chemists unquestionably had the right, not only to exercise entire freedom of judgement, but also critically to examine the statements which had been made (Armstrong 1894).

Thus, R & R's discovery implied that a ubiquitous constituent of the atmosphere had escaped the chemists' attention for nearly a century. It is small wonder that chemists were skeptical of R & R's accomplishment.

Having completed the various tests, R & R proceeded with the presentation of the characteristics of the new constituent of the atmosphere, at a meeting of the Royal Society at the Theatre of University College London on January 31, 1895. The paper was presented by Ramsay. Lord Kelvin was chairing the meeting where the Councils of both the Chemical and the Physical Society were invited. There were 800 people present. Michael Foster, the other secretary of the Royal Society besides Rayleigh, was there. A.S. Balfour, Lyon Playfair, Henry Roscoe, George Stokes, James Paget, B.W. Richardson, Henry Gilbert, Philip Magnus, Henry Armstrong, Carey Foster, Arthur Rucker, Henry Odling, William Perkin, William Frankland, William Crookes, William Tilden, Sylvanus Thompson, Sydney Young, Ralph Meldola were all there. Dewar was absent.

In the beginning of their paper R & R quoted a passage from De Morgan's *A Budget of Paradoxes*, to highlight their conception of discovery:

> Modern discoveries have not been made by large collections of facts, with subsequent discussion, separation and resulting deduction of a truth thus rendered perceptible. A few facts have suggested an hypothesis, which means a supposition, proper to explain them. The necessary results of this supposition are worked out, and then, and not until then, other facts are examined to see if their ulterior results are found in nature (Rayleigh and Ramsay 1895, p. 187).

Interestingly, according to De Morgan (and R & R) and in contrast to most of those who later analyzed the "discovery of argon," discovery is an extended process involving the generation of a hypothesis and the testing of its consequences.

Ramsay described all the different methods used to isolate atmospheric nitrogen and chemical nitrogen and the difference of less than 1 % in the measured densities of the two kinds of nitrogen. Then he presented the methods for removing the nitrogen and the different methods to induce chemical combinations with nitrogen. There was always a remaining residue which could not be gotten rid of. Ramsay, then, discussed a number of ways to isolate the new gas and to obtain it in relatively large quantities. Having achieved that, William Crookes and Arthur Schuster examined its spectrum and found that it did contain certain lines which were not contained in the nitrogen spectrum. This was one piece of convincing evidence that what had been found was not N_3. The other was the extreme inertness of argon, whereas most of the chemical evidence implied that it would be almost explosive. Ramsay continued describing the solubility of argon in water and its liquefaction; a more detailed account was presented at the same meeting by Karol Stanistaw Olszewski. By measuring the velocity of sound in argon, R & R managed to determine the ratio of its specific heats. It was found to be 1.66. This implied that argon was monatomic and, hence, quite impossible to accommodate in the periodic table as that table was structured at the time.

In the conclusion of their paper, R & R used very careful language: they presented their results as a novel discovery, but they also expressed their doubts as to whether argon was an element or a mixture of elements and pointed out that there was evidence for and against both possibilities. The physical methods they had employed could not unambiguously identify argon as a unique element or a mixture of elements. Furthermore, the evidence for its being chemically inactive derived from a number of failures to induce reactions with other empirically chosen elements, and there was no theoretical justification for its inertness.

After the end of the presentation, Armstrong and Arthur Rucker, professor of physics at the Royal College in London, spoke. Though not as vitriolic as in his remarks after the Chemical Society meeting, Armstrong made a long speech questioning in effect the conclusiveness of the evidence presented by R & R as to the inertness and the monatomicity of argon. Rayleigh said a few words at the end:

> I am not without experience of experimental difficulties, but certainly I have never encountered them in anything like so severe and aggravating a form as in this investigation.[13]

After the formal announcement of the discovery of argon, *Nature* carried a detailed report of the meeting with various comments, most probably written by Rucker. The report remarked that

> All that is known of argon was told to all. ... As has been well said, the result is "the triumph of the last place decimals," that is, of work done so well that the worker knew he could not be wrong ... [and concerning the disagreements about the monatomicity of argon it was added that] The courts of science are always open and every litigant has an unrestricted right of moving for a writ of error (Ibid., p. 337).

It seems then that by early 1895 the discrepancy detected by R & R had been established, to the satisfaction of many chemists, and attributed to a new constituent of the atmosphere. However, the discovery of argon was by no means complete. One of its

[13]From the report in *Nature*, volume 51, number 1319, February 7, 1895, 338.

crucial components was still missing: the accommodation of the new gas within the theoretical framework of late 19th century chemical knowledge, that is, its placement in the periodic table.

10.3 Accommodating Argon

In his Presidential address at the Royal Society at the end of 1894, Lord Kelvin praised the new discovery of "the hitherto unknown and still anonymous fifth constituent of our atmosphere" as "the greatest scientific event of the past year." And then he reminded the audience of the comments he had made 23 years earlier.

> Accurate and minute measurement seems to the non-scientific imagination a less lofty and dignified work than looking for something new. But nearly all the greatest discoveries of science have been but the rewards of accurate measurement and patient long-continued labour in the minute sifting of numerical results (Lord Kelvin 1894, pp.291–292).

It is not uncommon, especially among scientists and historians who have studied the argon case, to come across strong views where the discovery of argon is considered as a paradigmatic achievement of the late 19th century culture of exact measurements. This, however, is a rather misplaced assessment. By considering the discovery of argon exclusively in such a context we lose sight of one of its crucial aspects, namely the reconceptualization that was required for the legitimation and assimilation of the new element. One of the problems that had to be overcome, to that effect, was the inert character of argon. To come to terms with that problem, chemists had to reconceptualize the very notion of a chemical element.

The historical accounts of the discovery discuss the difficulties of placing argon in the Periodic Table, but neglect to note how fundamental those difficulties were. From about the time of the construction of the Periodic Table, the identification of a substance as an element and its "correct" placement in that Table, involved the determination of its atomic weight and the kinds of chemical reactions that it entered into. Thus, more or less by definition, an inert substance would not qualify as an element. To accommodate the possibility of inert elements it was necessary to revise the concept of an element. In fact, in 1899 there were chemists who would argue that the inert "elements" are, in effect, no elements at all and should not be part of the periodic table (see Wolfenden 1969). The complexity of the discovery of argon is not evident unless we realize that it concerned the legitimacy of a new chemical element, the inertness of which was negating the very notion of a chemical element! Argon forced chemists to re-appraise one of the core notions of their discipline, by compelling them to rethink what a chemical element is. Interestingly, J.J. Thomson, some 40 years later, noted that the discovery of argon involved a new chemical element that "had no chemical properties," that "formed no compound with any other element" and "would have nothing to do with the most tempting brides that chemists put before it; as this is the trap on which chemists rely for catching a new element, it is no wonder that argon eluded them" (Thomson 1936, p.400).

The difficulties of accommodating argon within the "fabric" of chemistry are evident from Ramsay's correspondence with other chemists and physicists. Before the announcement of the discovery to the Royal Society, Ramsay had sought the opinion of Wilhelm Ostwald and George Francis FitzGerald. Ostwald's response was that he would gladly publish Ramsay's paper in the *Zeitschrift für Physikalische Chemie*. "The fact is that I do not care very much for the new elements. But one so unexpected and almost impossible as that which you have found is something totally different from the trivial discoveries amongst the rare earths."[14] FitzGerald proposed that Ramsay make a determination of the specific heat at constant volume and a calculation of it from the value of γ and the P, v, T relation and thus decide whether it obeyed the Dulong Petit law. Ramsay was seeking FitzGerald's opinion about the peculiarities of the ratio of specific heats he had found for argon.[15] The latter was convinced that such a calculation would lead to an atomic weight of 40. "This is certainly very mysterious." FitzGerald suggested that this could imply that the two atoms may have little or no independent motion and so the molecule behaved like a single atom. "I make this in the interests of chemistry because physically there can be no objection to an atomic weight of 40."[16] Ramsay had suggested to FitzGerald the possibility of a system of elements with zero affinity and the latter, though very enthusiastic about the suggestion, warned Ramsay that the "Chemists will never believe in an element with no chemical affinity."[17] And Ramsay felt no scruples in telling Arthur Smithells, Professor of Chemistry in the University of Leeds, that the implications of argon were such that "the whole fabric of chemistry is going to receive a shake."[18]

In view of these difficulties, the resistance of many chemists to R & R's discovery was amply justified. Ramsay, however, thought otherwise:

> [Professor Ramsay] more than once allowed it to be guessed that he was not altogether delighted by the way in which the world of science greeted the achievement of "the Argonauts"... He now enunciates the sum of his discontent in a perfectly distinct, but unmelodious fashion. The discovery of argon was first announced in a somewhat bold manner whilst much of the corroborative detail still remained to be worked out. Many chemists of high standing immediately submitted the announcement to free criticism and implied their right to test it by all the experimental methods at their disposal. Mr. Ramsay now asks whether such a course was "consistent with the highest view of scientific morality" and shows that he holds that it was not. He singles out for condemnation the words of the late President of the Chemical Society, Dr. Armstrong, who laid down a doctrine with which few unprejudiced persons will disagree ... From a scientific point of view, Mr. Ramsay's contention is even less to be praised. It has often been declared by the opponents of modern science that it is in reality as obscurantist in its tendencies as was medieval theology. This is a sweeping libel; but such

[14] Wilhelm Ostwald to William Ramsay, December 24, 1894 (our emphasis). William Ramsay Papers, University College Library, London.

[15] George Francis FitzGerald to William Ramsay, December 14 and December 20, 1894. See note 2.

[16] George Francis FitzGerald to William Ramsay, December 28, 1894. See note 2.

[17] George Francis FitzGerald to William Ramsay, January 8, 1895. See note 2.

[18] Wiiliam Ramsay to Arthur Smithells, March 11, 1895. See note 2.

> plaintive appeals as Mr. Ramsay has now issued would go far to substantiate it, if it were common.[19]

Be that as it may, there was another factor that complicated further the consolidation of R & R's discovery, namely the emergence of physical chemistry. The articulation and legitimation of the techniques and principles of physical chemistry got in the way of (and were helped by) the discovery of argon. The end of the 19th century was a rather precarious period for British chemists. The renowned group of scientists who referred to their discipline as experimental science were faced with a number of challenges: physicists were increasingly meddling with their affairs and, most importantly, they were appropriating the atom; some chemists were increasingly showing sympathy to chemical thermodynamics, the only proper theory in chemistry; and the new technique of spectroscopy was slowly becoming a new tool for chemistry.

In a way, the dramatic conflicts concerning argon express all these developments. This is evident in the difference in outlook about the intermediate region of physical chemistry between Ramsay and Dewar. The contrast in the narrative and the problems being discussed by Ramsay in his address at the International Congress of Arts and Sciences at St. Louis in 1904 or in the Introduction of the series he edited on Physical Chemistry and Dewar's 1902 Presidential address to the British Association for the Advancement of Science is, also, quite striking in this respect (see Ramsay 1904; Dewar 1903). Dewar never transcended the view of physical chemistry as a way of adopting physical techniques in chemistry. At the beginning, this view appeared convenient and promising but, as physical chemistry was becoming a new autonomous sub-discipline, it was soon marginalised. In any case, Dewar's and Ramsay's disagreements around and about the new element have to be assessed within the context of the emergence of physical chemistry as a distinct new field. At the end, after the dust had settled, it appeared that argon "belonged" to those physicists who for a moment felt like chemists and to those chemists who started realizing that physical chemistry was not simply a way of enriching chemistry with techniques borrowed from physics.

Thus, the reaction to the discovery of argon, and the lingering unwillingness to accept what R & R proposed, was shaped by the increasing "physicalisation" of chemistry, the meddling of physicists into the affairs of chemistry and the few but increasing number of chemists who were sympathetic to that development. The reactions came almost exclusively from the quarter of the "pure" chemists. One is reminded of what Henry Armstrong said several years later in his presidential address to the chemistry section of the British Association for the Advancement of Science:

> Now that physical inquiry is largely chemical, now that physicists are regular excursionists into our territory, it is essential that our methods and our criteria be understood by them. I make this remark advisedly, as it appears to me that, of late years, while affecting almost to dictate a policy to us, physicists have taken less and less pain to make themselves acquainted with the subject matter of chemistry, especially with our methods of arriving at the root conceptions of structure and the properties as conditioned by structure. It is a serious matter that chemistry should be so neglected by physicists (Armstrong 1909, p.423).

---

[19] *Glasgow Herald*, February 22, 1896. See also G.G. Stokes to Kelvin, March 11, 1896: "I quite agree with the Glasgow Herald. I should think it expresses the feeling of scientific men in general."

The public debate was between different groups of chemists, and it was, also, suggestive of the way each group chose to map and delineate the undefined middle ground that some called physical chemistry and some called molecular physics. In a manner analogous to the situation in spectroscopy, the question was asked as to whose domain physical chemistry was. Was it an activity for physicists or chemists? Or was it a subject for a new field, physical chemistry or molecular physics? How would the boundaries of this newly emerging area be drawn? What would be the character and extent of the practitioners' allegiances to physics and chemistry? These issues, which bore an immediate relation to the whole question of the *status* of physical chemistry, would be discussed and disputed well into the interwar years, and even after the successes of quantum mechanics in chemical problems. The discovery of argon happened during a time when physical chemistry was articulating its own autonomous language, charting its own theoretical agenda and formulating its own theoretical framework.

By the time they presented their results to the Royal Society, Rayleigh and Ramsay were convinced that the new gas they called argon could not be N_3: The densities appeared to be different, the ratio of the specific heats could not be that of a triatomic molecule, and the prominent lines shown in the spectrum of argon were in sharp contrast with the diffuse nitrogen lines. Furthermore, they noted that the gas they isolated was not manufactured in the process of isolating nitrogen, thus, excluding the possibility of it being N_3. Yet, Dmitri Mendeleev, in his congratulatory message to Rayleigh and Ramsay, expressed his belief that "the molecules [of the gas they had discovered] ... contain[ed] three nitrogens."[20] He insisted on the matter since he thought that there was no place in the Periodic Table for an element with atomic weight about 40, especially since argon's chemical characteristics had nothing in common with those of Potassium (atomic weight 39) or Calcium (atomic weight 40) (Mendeleev 1902, pp. 496–497). And he suggested a number of other tests in order to investigate the characteristics of the gas: to subject it to very high temperatures and see whether it "breaks up" (in case it is a triatomic element), to find out whether there are lines in the spectrum reminiscent of those of nitrogen, etc.

Mendeleev's objections were echoed by those of Armstrong, who with typical "chemical thinking" could not envisage a place for argon in the Periodic Table. There was a strong counterargument from Rucker, who thought that the ratio of the specific heats should be taken as the dominant criterion for identifying argon as an element, since that ratio was based on a sound physical theory. Faced with the dilemma to choose between the constraints of the Periodic Table, which he considered as a generalization of empirical data with no dynamical theory to support it, and the kinetic theory, which if shown to be problematic would upset "the whole of our fundamental notions of science,"[21] Rucker chose the latter. In this approach, it appeared that the crucial matter was the monatomicity of argon. This exchange was, again, symptomatic of the differences concerning the theoretical as well as conceptual framework within which argon would be legitimized. Not all chemists,

[20]Dmitri Mendeleev to William Ramsay, 12 February 1895; quoted in Matyshev (2005, p. 1283).

[21]*Chemical News*, February 1, 1895, 61.

of course, were as insistent on the conceptual framework underlying the Periodic Table. Ralph Meldola, for example, was open to accepting the existence of inert elements, expressing his belief in the monatomicity of argon. Interestingly, he acquiesced after the discovery of helium, another event strengthening our suggestion about the protracted nature of the discovery of argon.

In fact, until about 1900, Mendeleev continued to reject the view that argon, with an atomic weight of 40, ratio of specific heats 1.66, and without any reactions with other elements, could be accommodated in the Periodic Table. He regarded the Periodic Table as something that had the status of "law": all atoms of each element were identical, atoms of different elements had nothing in common, strength of affinity depended on mass, and thus the Periodic Table could be constructed by elements with increasing atomic weight and in groups which showed similar chemical reactivity.[22] Though an increasing number of chemists were subscribing to this view, there were, also, some who considered the Periodic Table as simply a systematic generalization of empirical facts.

Mendeleev's understanding of the Periodic Table was based on a specific view about the structure of matter itself. It was predicated on three principles: the atomicity of matter, the immutability of atoms, and the valency of each element as a measure of its reactivity with others. Thus, any anomaly for the Periodic Table would have undermined that view. And that was a much more serious problem than the inability to find an appropriate place for a possible new element. When in the mid– to late 1890s all three principles were seen to be strongly undermined by new phenomena (the electron, radioactivity, and argon), what was threatened was not only the Periodic Table, but the view of matter that was such an integral part of its construction. In fact, Mendeleev's view about the character of the Periodic Table predisposed him negatively towards a number of new developments: he denied the theory of electrolytic dissociation, he rejected the electron ("the incomprehensible hypothesis of the electron"), and rebuffed radioactivity by insisting on the "individual originality of chemical elements."[23] As Michael Gordin aptly remarks,

> these assaults threatened both the borders of chemical knowledge and the stability of the entire discipline. Mendeleev felt he had to preserve the integrity of the chemical worldview to which his periodic system had contributed so substantially. ... The integrity of Mendeleev's chemical vision was at stake in each (Gordin 2004, p. 209).

A combination of new data on densities and spectra, as well as the finding of new gases with "zero valency", brought about the re-conceptualizations that were needed in order to open up space for accommodating the inert gases. The crucial step was to make a conceptual shift: "inert element" ceased to be a self-contradictory notion. Then argon and the other noble gases turned out to be an asset rather than a liability for the Periodic Table. That, of course, was not without any cost. Mendeleev had to reshape his views about matter.

---

[22] For an illuminating discussion of Mendeleev's concept of elements and its significance for the Periodic Table see Scerri (2007, pp. 112ff).

[23] See Matyshev (2005). The quotes are from p. 1284, the source being Mendeleev's book, *Periodic Law* (the Russian edition of 1958).

Mendeleev's reactions to argon indicate an additional aspect of the protracted character of its discovery. Its accommodation in the Periodic Table involved something more fundamental than the technical issue of how to find the characteristics of a new element and, eventually, would strengthen the validity of Mendeleev's understanding of the Periodic Table as a law-like framework. The discovery of argon was completed when Mendeleev, finally, included a new column for the inert gases in 1905, by which time there was a wide consensus as to the status of the new gases.[24]

Reactions against the characterization of argon as an element stopped when the isolation of neon and helium during the next 2 years "imposed" the necessity of a new column in the Periodic Table. Thus the Periodic Table was not to be composed only of elements with high and low chemical reactivity, but also with no chemical reactivity as well. The importance of atomic weight was enhanced until a few years later it was substituted by the principle of atomic number. We have a situation here where the discovery of the inert gases has been unequivocally accepted when it was found that there were many such gases. The discovery of the "second batch" reinforced the status of argon and the other way around. The placing of the Group 0 in the Periodic Table did not occur before 1898 and not after 1902. In a way, the discovery of argon was "completed" by the discovery of helium and neon. So, as it were, all three discoveries were simultaneously realized. As Wolfenden has pointed out,

> Although the debate over "this kind of chemical monster brought unexpected and unwelcome, like the cuckoo, into the previously happy family of the elements" was spirited, long, and confused, two simplifying factors should be recorded. The data on argon provided by the original paper of Rayleigh and Ramsay were essentially accurate and, apart from the refractive index of the gas, omitted no information about argon that was relevant to the problem of its position in the Periodic Table. Second, helium was discovered in cleveite by Ramsay less than two months after the argon paper so that the problem almost immediately became that of finding places for two new elements in the Periodic Table; more than three years were to elapse between the discovery of helium and the period (June–September, 1898) during which Ramsay isolated krypton, neon, and xenon in that order. Third, it is broadly true to say that all remotely plausible hypotheses for nesting the cuckoo had been made within six months of the Rayleigh-Ramsay report to the Royal Society (Wolfenden 1969, p.572).

The acceptance of argon by the chemical community came in late, because it brought about a public dispute concerning issues in the very core of the chemists' ontological and cultural framework. It undermined a regulative aspect of those frameworks (i.e., that elements are necessarily reactive) and was established more on physical rather than chemical methods. Eventually, chemists started thinking of this particular gas within a different framework, which included the possibility of an inert element, and the re-consideration of the techniques used for its identification (and the identification of other "similar" gases). It was within that framework that chemists became convinced that argon was an element, since it now fitted the Periodic Table. No additional experimental data for argon itself were needed in order to convince the community that it was a "proper" element. Thus, the consensus among chemists

[24] For more detailed information about others' attempts to accommodate argon in the periodic table, see Giunta (2001).

was the result of a conceptual shift. The physicists, on the other hand, were careful not to altogether disturb the basic (empirical) tenet of the Periodic table, that is, the periodicity of chemical properties with increasing atomic weight.

10.4 Conclusion

The above case-study shows that the RVSD cannot do justice to the discovery of argon, which has to be understood as an extended process. As we have shown, that process had several stages: first, Rayleigh's detection of a discrepancy between "physical" and "chemical" nitrogen; second, R & R's experimental demonstration that the discrepancy was not an artefact of the experimental situation; third, the identification of the entity "responsible" for the discrepancy. The third stage was a collective undertaking of several physicists and chemists. Besides R & R, the spectroscopists Crookes and Schuster, the low temperature physicist Olszewski, and R & R's critics Dewar and Armstrong played an important role in the identification of argon. That, in turn, involved several steps: first, the exclusion of the possibility that the new gas was a species of a familiar substance (i.e., N_3); second, the employment of (chemically) controversial techniques from spectroscopy, necessitated by the chemical inertness of the new gas; third, the conceptual accommodation of the new gas within the chemical taxonomy (i.e., the enrichment and, possibly, the modification of the periodic table). All these steps were precarious, met resistance by chemists and had the potential of undermining the discovery and turning it into an insignificant discrepancy.

Furthermore, our analysis indicates that scientific discoveries are extended processes for reasons that go beyond those envisaged by Kuhn. He attributed the extended nature of scientific discoveries to the conceptual adjustment that they require. That was definitely an aspect of the discovery of argon. However, Kuhn neglected another aspect of scientific discoveries that contributes to their being protracted episodes, namely that they may involve controversial experimental techniques and methods. When that happens, the establishment of a discovery goes part and parcel with the acceptance of the techniques and methods by means of which it came about. As we have seen, some of the factors that complicated the discovery of argon were related to the legitimization of physical techniques of investigation in chemistry and the concomitant emergence of physical chemistry.

An enrichment of Kuhn's account of scientific discovery along the above lines provides an expansive framework for understanding particular discoveries. It has definitely helped us to make sense of the collective, multi-dimensional, and conflictual character of the discovery of argon. What more could we ask from an adequate philosophy of historical case studies?

Acknowledgments We would like to thank John Heilbron for asking us to spell out the "added value" of our philosophical approach to the history of science. John suggested in conversation the first part of the title of this paper ("From Discrepancy to Discovery"). Earlier versions of this paper were presented at the 2014 Meeting of the History of Science Society in Chicago, and at "Knowledge, Technologies, and Mediation: A Workshop in Honor of Norton Wise" (UCLA, October 2015). We are indebted to the audiences for helpful discussion. Moreover, we are grateful to the editors for their constructive comments. Finally, Theodore Arabatzis's work for this paper was supported by European Union (European Social Fund—ESF) and Greek national funds through the Operational Program "Education and Lifelong Learning" of the National Strategic Reference Framework (NSRF)—Research Funding Program: THALIS—UOA—Aspects and Prospects of Realism in the Philosophy of Science and Mathematics.

References

Arabatzis, T. 2006a. On the inextricability of the context of discovery and the context of justification. In *Revisiting discovery and justification: Historical and philosophical perspectives on the context distinction*, ed. J. Schickore, and F. Steinle, 215–230. Dordrecht: Springer.

Arabatzis, T. 2006b. *Representing electrons: A biographical approach to theoretical entities*. Chicago: The University of Chicago Press.

Arabatzis, T. 2016. The structure of scientific revolutions and history and philosophy of science in historical perspective. In *Shifting paradigms: Thomas S. Kuhn and the history of science*, ed. Blum, A. et al. Berlin: Edition Open Access, Max Planck Institute for the History of Science.

Armstrong, H. 1894. [No Title]. December 6, 1894; as reported by the *Chemical News*, December 21, 1894, 301.

Armstrong, H. 1909. Presidential address to section B-Chemistry. *Proceedings of the British Association for the Advancement of Science:* 420–454.

Burian, R. 2001. The dilemma of case studies resolved: The virtues of using case studies in the history and philosophy of science. *Perspectives on Science* 9(4): 383–404.

Caneva, K. 2005. 'Discovery' as a site for the collective construction of scientific knowledge. *Historical Studies in the Physical Sciences* 35(2): 175–291.

Cavendish, H. 1785. Experiments on air. *Philosophical Transactions of the Royal Society of London* 75: 372–384.

Chang, H. 2011. Beyond case-studies: History as philosophy. In *Integrating history and philosophy of science: Problems and prospects*, ed. S. Mauskopf, and T. Schmaltz, 109–124. Dordrecht: Springer.

Dewar, J. 1894a. [No Title]. *The Times* (London), August 18, 1894.

Dewar, J. 1894b. The relative behaviour of chemically prepared nitrogen and of atmospheric nitrogen in the liquid state. *Proceedings of the Chemical Society* 10(144): 222–225.

Dewar, J. 1903. Presidential address. In *Report of the 72nd meeting of the British Association for the Advancement of Science held in Belfast in September 1902*. London: John Murray.

Dick, S. 2013. *Discovery and classification in astronomy: Controversy and consensus*. Cambridge: Cambridge University Press.

Frercks, J., H. Weber, and G. Wiesenfeldt. 2009. Reception and discovery: The nature of Johann Wilhelm Ritter's invisible rays. *Studies in History and Philosophy of Science* 40: 143–156.

Giunta, C. 1998. Using history to teach scientific method: The case of argon. *Journal of Chemical Education* 75(10): 1322–1325.

Giunta, C. 2001. Argon and the periodic system: The piece that would not fit. *Foundations of Chemistry* 3: 105–128.

Gordin, M. 2004. *A well-ordered thing: Dmitrii Mendeleev and the shadow of the periodic table*. New York: Basic Books.

Gordin, M. 2012. The textbook case of a priority dispute: D.I. Mendeleev, Lothar Meyer, and the periodic system. In *Nature engaged: Science in practice from the renaissance to the present*, ed. M. Biagioli, and J. Riskin, 59–82. New York: Palgrave Macmillan.

Hanson, N.R. 1962. The irrelevance of history of science to philosophy of science. *The Journal of Philosophy* 59(21): 5745–5886.
Hiebert, E.N. 1963. Historical remarks on the discovery of argon, the first noble gas. In *Noble-gas compounds*, ed. H. Hyman, 3–20. Chicago: The University of Chicago Press.
Hirsh, R. 1981. A conflict of principles: The discovery of argon and the debate over its existence. *Ambix* 28(3): 121–130.
Kuhn, T.S. 1962. Historical structure of scientific discovery. *Science* 136(3518): 760–764.
Laudan, L.L. 1980. Why was the logic of discovery abandoned? In *Scientific discovery, logic, and rationality*, ed. T. Nickles, 173–183. Dordrecht: Reidel.
Lord Kelvin. 1894. Anniversary Address. *Chemical News*, December 14: 288–292.
Matyshev, A. 2005. 'Prout's law' and the discovery of argon. *Physics-Uspekhi* 48(12): 1265–1287.
Mendeleev, D. 1902. *The principles of chemistry. Trans. from the sixth*, Russian ed. New York: P.F. Collier & Son. Part IV.
Nickles, T. 1980a. *Scientific discovery: Case studies*. Boston Studies in the Philosophy of Science, vol. 60. Dordrecht: Reidel.
Nickles, T. 1980b. *Scientific discovery, logic, and rationality*. Boston Studies in the Philosophy of Science, vol. 56. Dordrecht: Reidel.
Pitt, J. 2001. The dilemma of case studies: Toward a Heraclitian philosophy of science. *Perspectives on Science* 9(4): 373–382.
Ramsay, W. 1904. The present problem of inorganic chemistry. In *International congress of arts and science*, vol. IV, ed. H. Rogers, 258–275. London: University Alliance.
Rayleigh, L., and W. Ramsay. 1895. Argon, a new constituent of the atmosphere. *Philosophical Transactions of the Royal Society of London* 186A: 187–241.
Rayleigh, L. 1895. *Argon. Proceedings of the Royal Institution* 14: 524–538.
Scerri, E.R. 2007. *The periodic table: Its story and its significance*. Oxford: Oxford University Press.
Scerri, E.R., and J. Worrall. 2001. Prediction and the periodic table. *Studies in History and Philosophy of Science* 32(3): 407–452.
Schickore, J. 2011. More thoughts on HPS: Another 20 years later. *Perspectives on Science* 19(4): 453–481.
Schaffer, S. 1995. Accurate measurement is an English science. In *The values of precision*, ed. M.N. Wise, 135–172. Princeton: Princeton University Press.
Spanos, A. 2010. The discovery of argon: A case for learning from data? *Philosophy of Science* 77: 359–380.
Strutt, R.J. 1968. (Fourth Baron Rayleigh). *Life of John William Strutt, Third Baron Rayleigh*. Madison: The University of Wisconsin Press.
Thomson, J.J. 1936. *Recollections and reflections*. London: G. Bell and Sons.
Travers, M.W. 1956. *A life of Sir William Ramsay*. London: Edward Arnold.
Wolfenden, J. 1969. The noble gases and the periodic table, telling it like it was. *Journal of Chemical Education* 46: 569–576.

Chapter 11
"So How Do We Know that the Moon Is Mountainous?" Problems of Seeing in Galileo's Reflections on Observing the Moon

Simone De Angelis

Abstract In the debate on integrated history and philosophy of science, the idea of re-evaluating historical episodes in terms of how they contribute to illustrating the history-philosophy relationship is very intriguing. This position implies that historical episodes contain philosophical concepts or arguments that can be investigated on a basic historical level, i.e. by analysing historical data. This article reconstructs the historical episode of Galileo's observation of the moon surface by analysing additional historical data that were neglected by scholars who treated the same episode. However, I focus also on the question on how an historical episode is constructed from historical data. I suggest to consider an historical episode on a more abstract level than concrete historical data. Thus, an historical episode can be conceived as an abstract scheme or model in which the historical data are collocated and interconnected. The starting point of my historical narrative was Galileo's emphasis on the perception problem which clearly emerged only after the publication of 'The Sidereal Messenger' in 1610. The historical episode of Galileo's observation of the moon contains a nuanced concept of seeing and observing, a proposition on how to determine the height of a lunar mountain, as well as a detailed argumentation regarding the mountainous surface of the moon. Thus, the historical episode illustrates a 'style of scientific work', a 'mode of argumentation' or a form of scientific explanation.

11.1 Constructing an Historical Episode

In the ongoing debate on how the history and the philosophy of science (&HPS) could, or should, be integrated, much work still remains to be done. However, the idea to re-evalute the role of historical episodes in terms of how they contribute to illustrating the history-philosophy-relation is worth considering. According to Richard Burian "case studies, properly deployed, illustrate styles of scientific work and modes of

S. De Angelis (✉)
Zentrum für Wissenschaftsgeschichte, Universität Graz, Graz, Austria
e-mail: simone.de-angelis@uni-graz.at

T. Sauer and R. Scholl (eds.), *The Philosophy of Historical Case Studies*, Boston Studies in the Philosophy and History of Science 319,
DOI 10.1007/978-3-319-30229-4_11

argumentation that are not well handled by currently standard philosophical analyses" (2001, p. 383). Burian's position implies the claim that historical episodes *contain* philosophical concepts or arguments that can be investigated on a basic historical level, i.e. by analysing historical data. I consider this a very stimulating approach to &HPS which deserves and requires further detailed conceptual, methodological and historical work.

The historical episode of Galileo's observation of the moon surface has recently been treated by two eminent scholars: Heilbron (2010) and Graßhoff (2010b). John L. Heilbron, on the one hand, maintains that Galileo "understood the significance of what he had seen" (2010, p. 151f.). He evidently underestimates the perception problem created by those observations. Gerd Graßhoff, on the other hand, uses the episode as an opportunity to illustrate a scientific model, neglecting some epistemological problems which clearly emerged only after the publication of *Sidereus Nuncius*. Graßhoff's procedure is indeed legitimate, but it is unsatisfactory from a historical point of view. In my paper I would like to reconsider the historical episode by analysing additional historical data.

The historical episode of Galileo's observation of the moon (1609–1611) contains, among other things, a nuanced concept of seeing and observing, a proposition for how to determine the height of a lunar mountain, as well as a detailed argumentation regarding the mountainous surface of the moon. These philosophical concepts—vision, observation, and argument—are reflected and developed by the dynamics of the historical episode. It is the aim of this article to reconstruct the exact dynamics of the historical episode. However, my focus is also on the question of how an historical episode is constructed from historical data, given that historical data do not possess an explicative force from their own. This means that the historical episode is to be considered on a more abstract level than concrete historical data. I select a series of historical data according to a philosophical position, which, for example, holds that the perception problem is relevant to science, as Galileo himself explicitly underlines in the episode. This was the starting point of my historical narrative.

A historical episode can be thus conceived as an abstract scheme or model in which the elements—i.e. the historical data—are collocated and interconnected. The analysis of how these data combine and interconnect determines an *epistemic situation*. In such a situation, individuals negotiate *knowledge claims* in the respective historical and social context. According to Lutz Danneberg, in the history of science it is advisable to not speak of knowledge *tout court*, but of knowledge claims ("Wissensansprüche"): "What ought to be understood as a knowledge claim of a specific period are cognitive entities about which there could be a dispute or about which there has been a dispute at a certain time" (Danneberg 2006, p. 213). The concept of epistemic situation allows me to grasp the complexity of the knowledge claims negotiated by the protagonists of the historical episode. I would like to show that the way I (re)construct the historical episode takes the complex epistemic situation better into account than it has been done before.

What exactly does this mean for the historian (or philosopher) who is concerned with historical data, and how does he or she actually determine an epistemic situation? Let us consider the case that one starts by determining certain core elements of an episode, for example of the history of a scientific instrument such as Galileo's telescope. Then one adds to the analysis further historical material such as papers, books, images, diagrams, notebooks, protocols, letters, objects, etc. Finally, one puts the collected historical records into a chronological order and thus reconstructs the

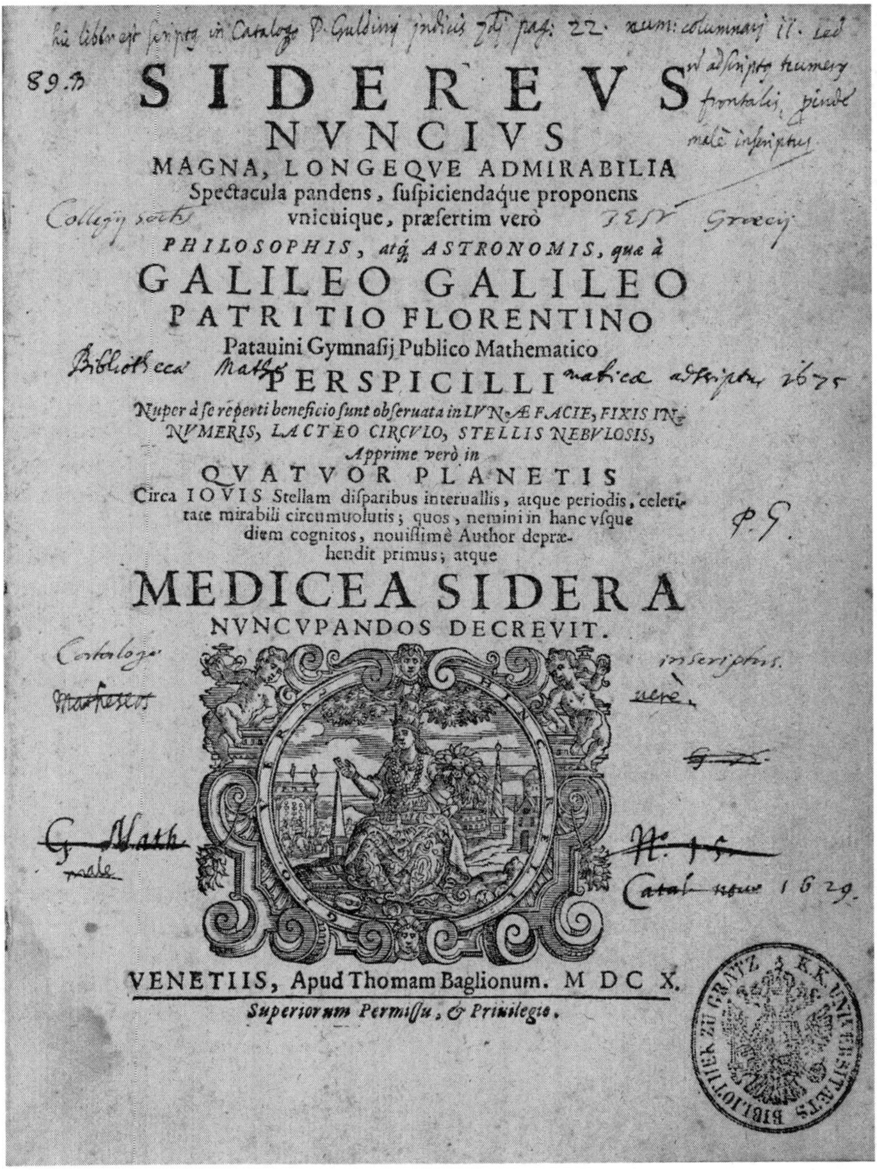

SIDEREVS
NVNCIVS
MAGNA, LONGEQVE ADMIRABILIA
Spectacula pandens, suspiciendaque proponens
vnicuique, præsertim verò
PHILOSOPHIS, atq; ASTRONOMIS, quæ à
GALILEO GALILEO
PATRITIO FLORENTINO
Patauini Gymnasij Publico Mathematico
PERSPICILLI
Nuper à se reperti beneficio sunt obseruata in LVNÆ FACIE, FIXIS INNVMERIS, LACTEO CIRCVLO, STELLIS NEBVLOSIS,
Apprime verò in
QVATVOR PLANETIS
Circa IOVIS Stellam disparibus interuallis, atque periodis, celeritate mirabili circumuolutis; quos, nemini in hanc vsque diem cognitos, nouissimè Author deprædendit primus; atque
MEDICEA SIDERA
NVNCVPANDOS DECREVIT.

VENETIIS, Apud Thomam Baglionum. M D C X.
Superiorum Permissu, & Priuilegio.

Fig. 11.1 Title page of the Graz original edition of *Sidereus Nuncius*, Venice 1610 (with kind permission of the University Library of Graz)

historical episode. The case of the telescope which Galileo built in the fall of 1609 is particularly interesting, since a private document demonstrates that Galileo's was mainly a practical approach to optics. As Giorgio Strano has recently shown, Galileo's shopping list "is direct evidence" of the fact that he wanted to "surpass the ordinary spectacle quality of lenses available" and that "the list reveals Galileo to have been an individual with very good knowledge of glass and lens making" (Strano 2012, p. 18). Galileo was particularly interested in a method of producing sheets of glass which displayed but few distortions. Those he referred to as 'German glass' (ibid., p. 13; Dupré 2014). By the end of the year 1609, Galileo had successfully constructed his most powerful telescope "which was capable of magnifying objects more than 30 times" (Strano 2012, p. 18). Being familiar with the material, cultural, and practical conditions under which the telescope was constructed is certainly an important element of the epistemic situation, as it was this very telescope which enabled Galileo to perform the outstanding observations of the firmament he presented in *Sidereus Nuncius* in 1610 (Fig. 11.1). *Sidereus Nuncius* is an early seventeenth-century text written in Latin, which requires of the reader a high level of linguistic and philological competence to translate and comprehend it. Yet Galileo's text also contains images which illustrate both his observations of the moon and the problems he encountered in their course.

The point I am making is that in the historical episode of Galileo's lunar observations conceptual and argumentative elements are intrinsically connected to the physical, practical, textual, and pictorial elements, and that it is a *combination* of all said elements which determines the epistemic situation this article focuses on. In the following I will explicate the dynamics of the episode and show how its elements provoked a debate that forced Galileo to further specify his concepts and arguments. I will start with the analysis of two historical documents: one of Galileo's drawings of the moon and a diagram which visualises his geometrical argument.

11.2 Galileo's Argument for Determining the Elevation of the Lunar Mountains

In one of his drawings Galileo identified and depicted spots of light on the dark side of the moon (Fig. 11.2). On the half-moon he observed the dividing line between the moon's day side and its night side, also referred to as *terminator*. Galileo, moreover, noted that said line was not straight but irregular, from which he later deduced that such a distorted line could not appear on a spherical surface. Furthermore, he was able to detect spots of light on the night side of the moon which he estimated to be removed from the terminator by one-twentieth of the lunar diameter. The moon's diameter was an empirical datum, a figure which had been amazingly well known since antiquity. However, Galileo used rounded figures to facilitate his calculations. He assumed that the spots of light were mountain tops illuminated by sun light. The foot of the mountains, however, remained in the shadow and was not touched

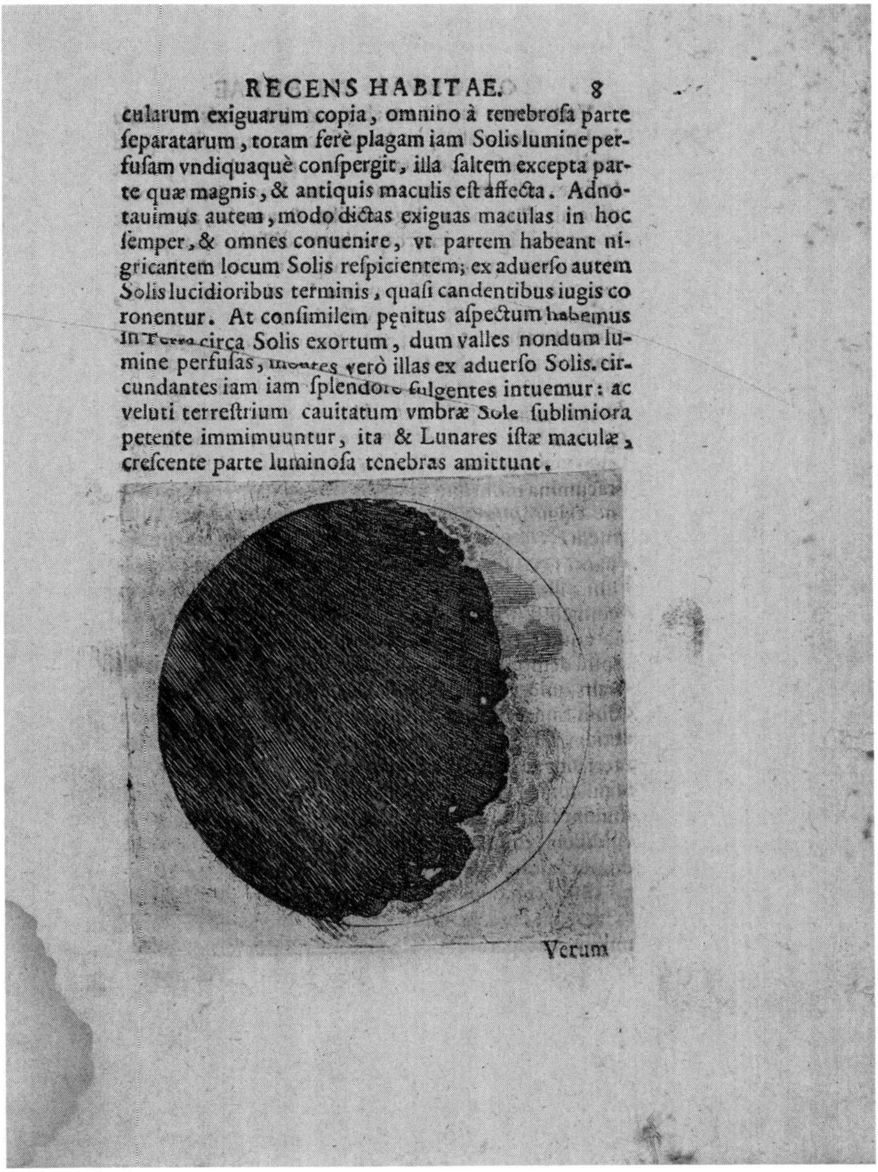

RECENS HABITAE. 8

cularum exiguarum copia, omnino à tenebroſa parte ſeparatarum, totam ferè plagam iam Solis lumine perfuſam vndiquaquè conſpergit, illa ſaltem excepta parte quæ magnis, & antiquis maculis eſt affecta. Adnotauimus autem, modo dictas exiguas maculas in hoc ſemper, & omnes conuenire, vt partem habeant nigricantem locum Solis reſpicientem; ex aduerſo autem Solis lucidioribus terminis, quaſi candentibus iugis coronentur. At conſimilem penitus aſpectum habemus in Terra circa Solis exortum, dum valles nondum lumine perfuſas, montes verò illas ex aduerſo Solis circundantes iam iam ſplendore fulgentes intuemur: ac veluti terreſtrium cauitatum vmbræ Sole ſublimiora petente imminuuntur, ita & Lunares iſtæ maculæ, creſcente parte luminoſa tenebras amittunt.

Verum

Fig. 11.2 Spots of light on the night side of the moon. Galileo Galilei, *Sidereus Nuncius*, Venice 1610, p. 8

by the sun's rays because of the moon's curvature. Assuming this enabled Galileo to construct his argument for determining the height of a lunar mountain, which is well known among historians of science (Graßhoff 2010b; Heilbron 2010). As I will show in more detail later, Galileo was quite aware that it was not obvious why he

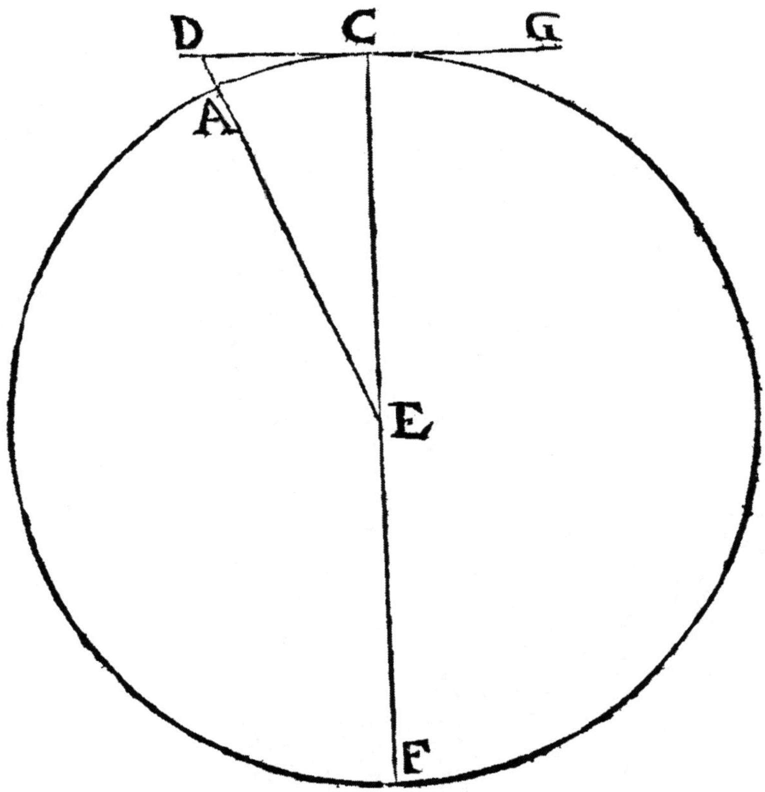

Fig. 11.3 Galileo's geometrical model for determining the height AD of a moon mountain. Galileo Galilei, *Sidereus Nuncius*, Venice 1610, p. 13 v

would identify the spots of light as mountain tops and that his assumption had to be explained, for the spots could have also been caused by other lunar phenomena such as bright, reflecting material.

Galileo's reasoning is as follows: Fig. 11.3 shows that to represent the three-dimensional moon, he used a simple two-dimensional geometrical model in the form of a perfect circle CAF (although the moon is not actually a perfect sphere). The axis CF represents the terminator, and the line GCD designates a ray which emanates from the sun in G and touches the moon as a tangent in C, obliquely illuminating the mountain AD on the night side of the moon. D indicates the mountain top illuminated by the sun, which hence appears as a spot of light (see Fig. 11.2). Thus it follows that if the mountain AD is high enough to be illuminated by the ray GC, the mountain AD and the tangent GC intersect in the top D. As Gerd Graßhoff and John Heilbron have shown, Galileo calculated the height of the lunar mountains by means of this model and by the Pythagorean theorem ($ED = \sqrt{EC^2 + DC^2}$), and he found that the mountains were almost five miles high (Graßhoff 2010b, p. 30; Heilbron 2010,

p. 152f.). Galileo also determined that the lunar mountains were higher than those on earth which, according to him, were about one Italian mile high. When we consider that one Italian mile equals 1.851 km, this estimation is too low, even when we take into consideration that Galileo only knew the European mountains. Elsewhere Galileo stated that the lunar mountains were 3–4 miles high.

Despite the fact that Galileo simplifies the actual circumstances and that his assumptions are partly incorrect, his argument for determining the height of the lunar mountain is still convincing, and his achievement was remarkable for his day. At the same time, it has to be made plain that it was not Galileo's aim to determine the exact height of the lunar mountains, but rather to estimate it so that he could compare the lunar mountains with the ones on earth. However, as I will show in more detail later, even the mere supposition that there were mountains on the moon was not easily accepted by Galileo's contemporaries. For the sake of the argument it is acceptable to disregard Galileo's simplifications and the inaccurately calculated elevations of the lunar and the terrestrial mountains. As long as it serves its purpose, a theorem may also entail wrong assumptions (Graßhoff 2010b, p. 30f.). As it was not Galileo's aim to be exact, he was fully aware of his own simplifications. Besides, he also assumed that rays of light travel in straight lines. He obviously did not know that light is reflected by heavy bodies, which is of significance on the earth, where rays of light are reflected by the atmosphere, but not on the moon. This implies that Galileo must have assumed that there is no atmosphere on the moon. Moreover, his theory only applies to a half moon. A geometrical construction of the situation during a new moon would be substantially more complicated. In that case, especially the Pythagorean theorem would be rendered inoperative, as there would no longer be a right-angle triangle (Graßhoff 2010b, p. 31). Putting it simply, we can say that Galileo was able to approximate the height of the lunar mountains despite all the simplifications.

However, these findings raise the question of *why* Galileo used such an idealised model in his lunar observations, a question to which he himself provided the following, somewhat surprising answer: the reason was that in truth we do not actually see the lunar mountains from the earth. From such a distance it is impossible to see them even through a telescope (despite the fact that they are observable in principle). Thus Galileo raised an interesting *epistemological problem* concerning observation in a more general sense, which deserves to be examined carefully, as it (at least in part) explains his decision to introduce a geometrical argument, i.e. an idealised model. However, Galileo did not raise this observation-related problem in *Sidereus Nuncius* but in his famous letter to the Jesuit priest Christoph Grienberger in Rome, of 1 September 1611, to which scholarship has thus far not paid sufficient attention. This letter, which is a significant component of the historical episode studied in this article, also contributes to shedding light on the complex epistemic situation in which Galileo was immersed after the publication of *Sidereus Nuncius*.

11.3 Outline of the Problem at Hand

When Galileo posed that there are mountains on the moon, most Peripatetics and astronomers considered this an impertinence. In *Dialogo sopra i due massimi sistemi* (1632) Galileo depicted this conflict in dialogue form in a number of concise scenes: With regard to the mountainous qualities of the moon, Galileo, for example, puts into *Simplicio's* mouth the following standard argument of the Peripatetics: that one could 'save this phenomenon', that is to say explain it, by considering it an optical illusion, stemming from the fact that parts of the moon are unevenly opaque and transparent.[1] What was to be 'saved', according to the language of the Peripatetics, was the perception of the moon as a perfectly round sphere, which, like any other perfect sphere in the universe, was regarded as immutable, durable, and eternal. Galileo's observations of the moon and the stars from 1609 and 1610 made it more difficult to maintain these metaphysical perceptions, even though the argument that the moon's mountainous appearance was an optical illusion or even a hallucination persisted for a long time, especially in relation to Galileo's observations of the moons of Jupiter. In *Dialogo*, *Salviati* hence tries to argue for a new perception of the moon's surface in a persuasive and vivid manner: We should imagine the face of the moon covered in high mountains: upon doing so, would we not see how the mountainsides and ridges rise above the curvature of the perfect sphere, how they receive the sun's rays and glow like the rest of the moon?[2] The meaning of the word "existence"—or *exsistere* from Latin *ex-sisto* in the sense of rising up, or standing out, from a place—could hardly be described more graphically. *Salvati* thus *in nuce* already encapsulates the problem of viewing the lunar mountains which an observer on earth has to face. This is the problem Galileo confronted intensely, especially after the publication of *Sidereus Nuncius* (1610).

In the last decade, Galileo research has come to the fore with a multitude of specialised studies and accounts, whose primary feature is to identify, contextualise, and comment on specific types of knowledge, which are reflected in the texts of the Pisa scholar. This is particularly true of *Sidereus Nuncius*, the "Sidereal Messenger", which Galileo published in Venice in 1610 (see Fig. 11.1).

As, for instance, Reeves (2011) notes in her review of the state of research, the historiography of the last decade has substantially changed our perception of this book, its author and of the instrument, i.e. the telescope, itself. Contributions in this field could be divided into four areas of investigation: 1. the telescope before Galileo,

---

[1] "SIMPLICIO.[...] e circa questo particulare della montuosità della Luna, resta ancora in piede la causa che io addussi di tale apparenza, potendosi benissimo salvare con dir ch'ella sia un'illusione procedente dall'esser le parti della Luna inegualmente opache e perspicue" (Galilei 1998, p. I/263, 93). Unless indicated otherwise, all translations and paraphrases from Italian and/or Latin in this contribution are mine (S. De A.).

[2] "SALVIATI [...] Figuratevi ora la faccia della Luna piena di montagne ben alte: non vedete voi come le piagge e i dorsi loro, elevandosi sopra la convessità della perfetta superficie sferica, vengono esposti alla vista del Sole, ed accomodati a ricevere i raggi, assai meno obliquamente, e perciò a mostrarsi illuminati quanto il resto?" (Galilei 1998, p. I/241, 88).

2. Galileo before the telescope, 3. the emergence of *Sidereus Nuncius* and 4. the book's impact and dissemination (ibid., p. 39). This categorisation makes it possible to systematically gather the newly determined knowledge about *Sidereus Nuncius*, as the following two examples illustrate: Sven Dupré's studies have shown what *optical knowledge* existed in theory and praxis in the mid- and late 16th century. The studies particularly indicate which information Galileo and other contemporaries gleaned from Ettore Ausonios' theory of concave mirrors from 1592 onward, and how the situation changed once again in 1609, when the emergence of telescopes with two lenses rendered the lens-mirror combination obsolete, at least temporarily (Dupré 2005, p. 173f.; Reeves 2011, p. 50). Furthermore, what needs to be taken into account are *Theories of Perspective*, or rather the knowledge of *viewing in perspective*. This was especially the expertise of artists and consequently also codified in contemporary treatises on art. Hence, we are looking at a form of artists' knowledge, which, as Filippo Camerota has shown, was brought to bear in Galileo's observations concerning the moon (Camerota 2004b, 2006, pp. 210–220). The interesting point here is that Galileo (as will have to be shown) used this artists' knowledge as part of an argumentative strategy in the dispute with the Jesuit astronomers of the Collegio Romano, who felt constrained to demonstrate a reaction after the publication of *Sidereus Nuncius*.

However, this reaction came, as Galileo felt, slightly belatedly, that is to say in the spring of 1611, when Galileo travelled to Rome after a long illness and came into direct contact with the Jesuit astronomers of the Collegio Romano, for example with their head Christoph Clavius, who had already attested to his appreciation of Galileo's discoveries in a letter of 1610 (Camerota 2004a, 202; Wallace 2003, p. 108).[3] In fact, Clavius went even further and arranged for Galileo's telescope observation to be included in his Sacrobosco commentary; in 1611 this commentary was published in the third volume of his works in Mainz: The Jesuit professor of mathematics Otto Cattenius, for example, already drew upon Clavius' wording, when he referred to Galileo's new observations of the firmament during his lectures on mathematics in Mainz at the beginning of June 1611 (Krayer 1991, p. 50, 156f.).[4] This shows just how rapid the reception of *Sidereus Nuncius* was in the circles of Jesuit mathematicians at the time.

However, as Camerota (2004b, pp. 200–218) shows in detail, the judgement of the Jesuit mathematicians about the new observations of *Sidereus Nuncius* was characterised by ambivalence from the beginning: Four renowned mathematicians of the Collegio Romano's "Accademia di mathematica"—Christoph Clavius, Christoph Grienberger, Odo van Mealcote and Paolo Lembo—confirmed Galileo's observations

[3]"The Jesuit astronomers there [sc. at the Collegio Romano] had built a telescope and, after several efforts, had succeeded in confirming the main results he [sc. Galileo] had presented in his 'Sidereal Messenger'. Clavius wrote to Galileo, diagramming in the letter the positions he had observed for the satellites of Jupiter" (Wallace 2003, p. 108).

[4]The *Sphaera* by Johannes von Sacrobosco (13th century) "remained decisive for the elementary education at the universities until far into the sixteenth century" (Krayer 1991, p. 57) and was constantly supplied with newer and more exhaustive commentaries.

about the shape of Saturn, the phases of Venus, and the moons of Jupiter; however, whether the entire milky way consisted of nothing but small stars was uncertain; the mathematicians were also in disagreement on the issue of the moon's uneven surface: While Clavius ascribed this phenomenon to "more or less dense parts" on the moon, and thus did not deviate from the traditional theory of the completely smooth globe, "others thought that the surface of the moon was truly (*veramente*) uneven" (I Matematici del Collegio Romano a R. Bellarmino, 24 aprile 1611, qtd. in Camerota 2004b, p. 208). This ambivalent judgement was, however, not only due to the inferior quality of the telescopes used by the Jesuits, as the latter were able to also see what Galileo saw, at least since mid-November 1610, when they started using improved optical instruments (Camerota 2004b, p. 201f.; Reeves 2011, p. 68f.). Galileo's discovery of the mountainous surface of the moon, which according to Galileo himself resembled the earth's surface, was excessively harsh or strong, because it, for instance, endangered the traditional distinction between sub- and supra-lunar worlds (Reeves 2011, p. 67). Yet also on this issue the Jesuits had meanwhile slightly relaxed the strict metaphysical distinction between a transient sub-lunar and an eternal, everlasting supra-lunar sphere of the world: The Jesuit astronomer Christoph Scheiner writes in his letter of 10 November 1612 to the Swiss Jesuit Paul Guldin in Rome that he considered the sky to be, if not transient (*corruptibilis*), then certainly fluid (*fluidum*)[5] and therefore changeable,[6] while on the other hand he by no means considered the solar spots an inherent part of the sun; quite the opposite: he wanted to point out Galileo's blatant misjudgement.[7] At the same time, in February and March 1612, Scheiner minutely observed and described the moons of Jupiter, the satellites of Saturn and the phases of Venus with the help of small sketches, which he included in his letter.[8] The ambivalent assessment of Galileo's observations would therefore appear to have been a widespread fact among the Jesuits even in the years following Galileo's stay in Rome. The majority of Jesuits seem to have remained convinced that Galileo's observations were supposed to serve to correct the Ptolemaic world system in so far as the new telescopic data could be integrated and the "phenomenon [could be] saved" (Camerota 2004b, p. 211).

Thus, one the one hand, the Jesuits approved Galileo's observations, on the other, they disapproved and attacked them. However, in returning to the question of why Galileo used contemporary knowledge in his astronomical texts, the following issue

---

[5]"[...] quid enim ego afferam de Maculis solaribus, quid de Faculis et rebus similibus, si ea proferre mihi possim, quin caelum si non corruptibile, certé fluidum faciam? Nam penetrationem corporum non dari evidens est". (Scheiner 1612; With kind permission of Universitätsbibliothek Graz, specifically Herrn Manfred Mayer; Signatory of the Letter Convolute: Ms III 159).

[6]"By the opening decades of the seventeenth century at last some astronomers had abandoned solid celestial globes for a fluid heavens" (Barker 2011, p. 25).

[7]"[...] incoeptas perficiam Maculas, et D. Galileo una satisfaciam, eiusque errores craßissimos ostendam" (Scheiner 1612).

[8]"Joviales planetas saepe contemplo; nuper mensis Februarij 10.11.12. vidj Saturnum comites hoc modo; ° O ∘ [...] Venus nunc es talis [...] circiter; cornua, quod miror hebetia semp[er] exhibet: eius ad Lunam quasi plenam in diametro proportio est, ut 1 ad 20. die hoc 22. Martij unde si diametrum Lunae faciamque 34" (Scheiner 1612).

still appears more important to me: Galileo used his knowledge of perspective not exclusively, but especially in the context of an argument in which he defended certain claims of knowledge, namely his discovery of the lunar mountains. Therefore the fact that Galileo was per se familiar with knowledge concerning perspective is less crucial than the questions of how and under which circumstances he used this knowledge, and of what his objectives were. Because even though his observations about celestial bodies were generally acknowledged and to some extent also admired, they were still in the spring of 1611 the object of criticism, a fact which was hard on Galileo. To me the situation that Galileo on his part reacted to being attacked by the Jesuit astronomers, his accepting the challenge, and the fact that he reconsidered his observations of the moon seem to be of pivotal importance. Galileo's oft-quoted letter of 1 September 1611 to the Jesuit priest and astronomer Christoph Grienberger—Christoph Clavius' successor in Rome—bears eloquent witness to this situation: there Galileo states that he is trying for a clearer and more elaborate argumentation than he presented in *Sidereus Nuncius*.[9]

The reason for this letter to have been written is generally known as the "Mantua Problem", which likewise emerged in the spring of 1611. At a conference in Mantua, a Jesuit of the Collegio di Parma had not so much questioned Galileo's conclusions with respect to the existence of high mountains on the moon, but rather the fact that Galileo took account of the mountains when calculating the circumference of the moon. That is to say, as the Jesuit mathematician Giuseppe Biancani put it, that "mountains existed on the periphery of the moon [, so that] the moon's periphery also runs through the mountain tops" (qtd. in Camerota 2004b, 213). Whereas for Biancani the "periphery of the moon [remained] entirely luminous, without shadow and irregularities" (Ibid.; on the "Mantua Problem" see also Pantin 2005). This was the problem on which Galileo took a stand in his letter to Grienberger of 1 September 1611, where he took the opportunity to reflect and elaborate more precisely on the diverse implications of his lunar observations than he had done to date. Galileo had already met the mathematician Biancani during his time in Padua. He also shows the latter respect in his epistle to Grienberger and offers him 'eternal gratitude' for the possibility of a reply (Galilei 1901, p. 180; cf. also Wallace 2003, p. 108f.).

At first glance, the subject under consideration seems confusing and contradictory, for Galileo, at this point in the letter, suggests that the angular or uneven mountain tops on the periphery of the lunar hemisphere are not visible, nor do they have to be (Galilei 1901, p. 186). While the periphery of the lunar hemisphere is visible, the tops of the mountains under discussion apparently are not. This already addresses an important aspect of the problem of seeing or visual perception, with which Galileo confronts the reader, and which I would like to discuss in this contribution.

Another prominent passage in this letter clarifies the issue even further: the question of the Jesuit priest Giuseppe Biancani, who had raised objections to Galileo, was: "Are there elevations to be seen on that side of the moon which faces the earth?"

[9]"[…] et poi, che io di nuovo mi affatichi in dichiarare più lucidamente et diffusamente che non feci nel mio Nunzio Sidereo" (Galilei 1901, p. 186).

Galileo: "No, I reply, and I even say it is not only that we do not, and cannot, see the heights and elevations of the moon from such a distance, but that they would not even be visible from a vicinity of 100 miles, just as one would be unable to see our hills and highest mountains from an altitude and a distance of 50 miles or less."[10] And then Galileo asks the crucial question, which I have adopted in the title of this contribution: "So how do we know that the moon is mountainous?" His answer: "We do not know it by simple sensory perception (*senso*), but by combining and linking rational argumentation (*discorso*) with observations (*osservationi*) and with that which appears to the senses (*apparenze sensate*)."[11] There are, in fact, detailed studies about the development of Galileo's concept of irradiation (Lat. 'radiation'), which concerned him in connection with his observations of the light of the moon (Dupré 2003). However, notwithstanding this, the epistemological implications of Galileo's statement—and especially the implications relating to the theory of perception, which to me seem fundamental to an understanding of Galileo's reflections on observing the moon—have not yet been sufficiently illustrated (yet see also Piccolino and Wade 2008b).[12] It will be necessary to clarify these implications in detail in a first step. A second step ought to further elucidate how Galileo's problem is reflected in epistemological terms and in terms of the theory of perception within the argumentations he propounds in detail in the subsequent pages of his long letter; in other words, how Galileo actually achieves the postulated linkage between *senso* and *discorso*.

11.4 Considering the Problem in Light of Perception Theory

So how can we talk about an epistemological problem or about a problem concerning the theory of perception? If we follow the exposition of Dominik Perler and Markus Wild in their book on *Wahrnehmungstheorie in der frühen Neuzeit* ('Theories of Perception in the Early Modern Times', 2008), we find the following phrasing: "The problem of perception consists in finding a way to explain the relationship between the sensory experience and the material objects" (Perler and Wild 2008, p. 48). I will not address in detail Galileo's basic distinction between primary and

[10]"Rispondo io di no, et dico che i tumori et eminenze della [Luna] (come eminenze) non solamente non si veggono o possono vedere da tanta distanza, ma non si scorgerebbero nè anco dalla vicinanza di 100 miglia; sì come i nostri colli et le maggiori montagne niente si discernerebbero sorgere da i piani, da un'altezza e lontanza di 50 miglia et di meno ancora" (Galilei 1901, p. 183).

[11]"[…] non col semplice senso, ma coll'accoppiare e congiungere il discorso coll'osservationi et apparenze sensate" (Galilei 1901, p. 183).

[12]Even though in his contribution the biologist Piccolino focuses on sensory perception, he still puts an emphasis on regarding Galileo as 'precursor' of 19th-century optics and physiology of the senses (Johannes Müller and Hermann von Helmholtz) as well as of present-day neurosciences (Piccolino and Wade 2008a).

secondary qualities, which actually underlies the problem of perception, for this would go too far here, and Galileo himself deals with it much later, in *Saggiatore* of 1623 (on Galileo's positions cf. ibid., p. 39f.). Yet we can still say that Galileo's argumentation in the letter to Grienberger is essentially concerned with giving an explanation of the phenomena and appearances of the moon as we perceive them from earth. This is because Galileo obviously distinguishes between that which we see, what the visual sense allows us to perceive of the lunar surface (also with help of a telescope), i.e. that which lies within the *subject* of perception on the one hand, and that which really is, i.e. how the *object* of perception is actually constituted on the other hand. However, the problem of perception addressed here can not only be contextualised in an abstract manner, i.e. within the framework of the theory of perception of early modern times, but also in the immediate context of the reception of *Sidereus Nuncius* via the Jesuit mathematicians of the Collegio Romano. As Paul Guldin reports in a letter from 13 February 1611 from Rome to his fellow Brother Johann Lanz, Clavius and his colleagues at that point still considered Galileo's observations of the moons of Jupiter the effect of an optical illusion, perhaps even a hallucination: "hallucinationem potius deceptionem visus esse existimabunt, quam veras observationes Astronomicas" (P. Guldin to J. Lanz, 13 February 1611 qtd in Camerota 2004b, p. 201); the likely reason for this is the fact that they were still using a rudimentary telescope (Ibid.). This circumstance would at least explain why it was a matter of concern to Galileo to travel to Rome himself and to convince the Jesuit mathematicians that his observations were real and true (Ibid.). In any case, the acceptance or non-acceptance of Galileo's observations was linked to questions regarding perception from the very beginning.

11.5 Light-and-Dark Effects

I would now like to use an example to explain how the above-mentioned problem of perception ought to be understood. For this purpose it is useful to focus on an illustration from *Sidereus Nuncius* (Figs. 11.2 and 11.4).[13] The starting point of Galileo's deliberation is the dividing line or arc, also called the *terminator*, which separates the dark from the light part of the moon, and which, according to Galileo, appears to the eye as winding and rather uneven ("si vede crestata, sinuosa"; Galilei 1901, p. 183); thus this line could not be the boundary of light on the surface of a smooth and

[13]Claus Zittel discusses the conclusion that an image does not per se determine "how [… it] is perceived and what we expect from it" but rather "the respective theoretical setting" (2012, p. 283f.) Therefore, Galileo's images of the moon in *Sidereus Nuncius* are not so much meant to create a specific 'realistic' effect, but rather (according to the thesis I support) to explain problems relating to theory of perception in the context of his lunar observations. As will be shown, this is even more true in the case of the sketches and diagrams which Galileo included in the letter to Grienberger of 1 September 1611.

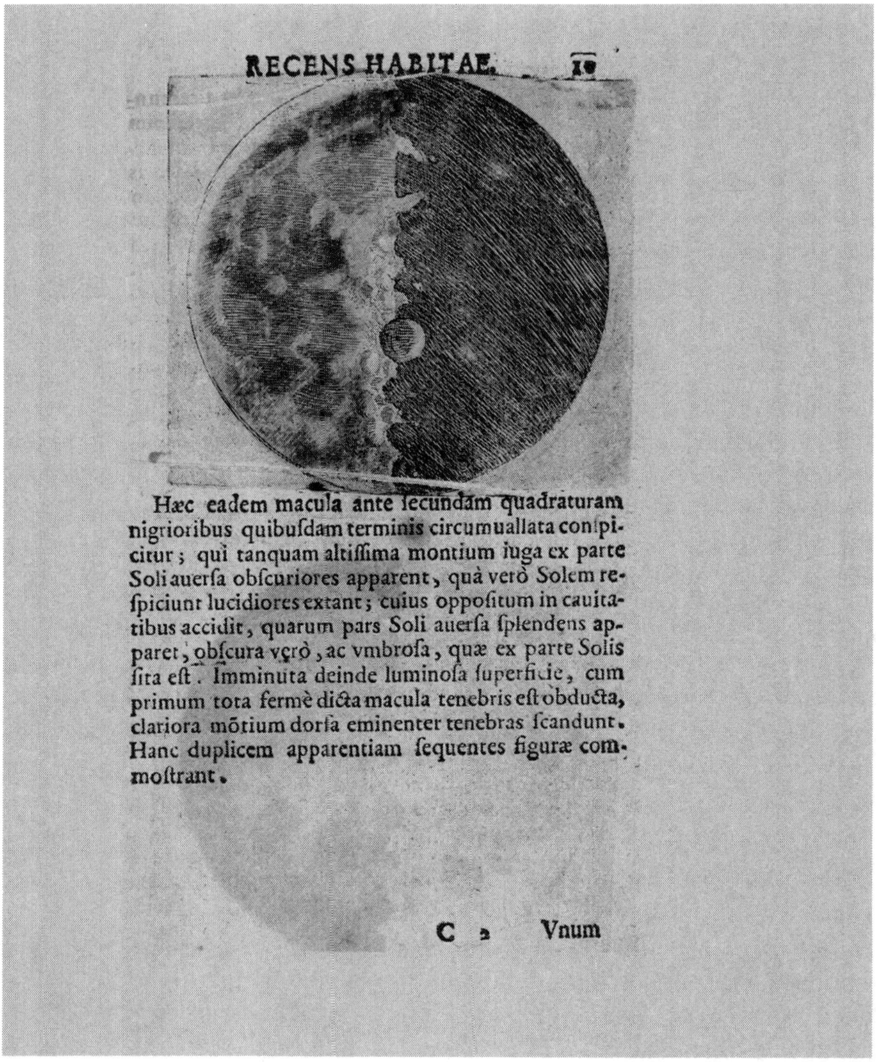
RECENS HABITAE. 10

Hæc eadem macula ante ſecundam quadraturam nigrioribus quibuſdam terminis circumuallata conſpicitur; qui tanquam altiſſima montium iuga ex parte Soli auerſa obſcuriores apparent, quà verò Solem reſpiciunt lucidiores extant; cuius oppoſitum in cauitatibus accidit, quarum pars Soli auerſa ſplendens apparet, obſcura verò, ac vmbroſa, quæ ex parte Solis ſita eſt. Imminuta deinde luminoſa ſuperficie, cum primum tota fermè dicta macula tenebris eſt obducta, clariora mõtium dorſa eminenter tenebras ſcandunt. Hanc duplicem apparentiam ſequentes figuræ commoſtrant.

C 2 Vnum

Fig. 11.4 The dividing line or arc, also called the *terminator*, which separates the *dark* from the light part of the moon. Galileo Galilei, *Sidereus Nuncius*, Venice 1610, p. 10

even sphere, but rather on a mountainous and uneven one.[14] Galileo now provides a number of minute descriptions of dark and light spots which are visible either on one side of the dividing line or on the other, whereby the darker spots disappear entirely

[14]"[...] adunque ella [sc. the line] non può esser termine dell'illuminatione in una superficie sferica, tersa et eguale, ma sì bene di una montuosa et ineguale" (Galilei 1901, p. 183).

in the direction of the light parts of the moon. (Ibid.) Hence these spots are visible solely because of the contrast between light and dark, since larger, lighter parts are situated next to a dark spot and vice versa (see Fig. 11.5). Galileo therefore describes *light-and-dark effects*. According to him, such phenomena could never occur on the surface of an even and smooth sphere. Hence, the necessary conclusion must be that the surface of the moon is covered in heights and depths.[15] Thus Galileo reasons from the effects he observed—the light-and-dark effect—to the physical reason for these effects: the heights and depths upon the surface of the moon.[16]

Now Galileo also reflects on the epistemological status of his observations. Galileo effectively does not see the heights and depths, but the appearances and phenomena ("queste sono le apparenze e fenomeni"; Galilei 1901, p. 184). Even though Galileo designates them 'facts', these appearances and phenomena count, according to him, as suppositions and hypotheses of rational argumentation ("suppositioni et ipotesi del discorso"; ibid.). This means that Galileo further defines the 'facts' ("i quali fatti") or the appearances through the epithet *suppositioni e ipotesi del discorso* in respect of their *epistemic* qualities. Thus, *hypothetical* arguments are derived from the phenomena. Although the appearances are convincing, they still remain hypothetical, because we cannot really see the heights and depths, i.e. the properties of the *object* of perception. Galileo promptly supplements the following statement which is relevant in terms of the theory of perception: "For that such mountains and heights might become visible to us by protruding and growing in front of our eyes is completely impossible."[17] However, for Galileo it is nonetheless beyond dispute that these heights and depths 'exist' in reality, exactly in the sense of 'protruding' or 'elevating' in a specific place.

Galileo expands his explanation further as to also include the case in which we do not even perceive the light-and-dark effects on the lunar surface. This affects, e.g., the zones of the moon which are removed from the boundary of light, especially on the periphery of the lunar hemisphere, where the sun's rays render all shadows invisible. So even though Galileo assumes that light-and-dark effects would exist on the periphery of the moon's hemisphere—as well as in the internal zones of the moon—we cannot possibly see them. In this case, according to Galileo, we cannot derive conjectures, evidence, and arguments in favour of a mountainous lunar surface

[15]"Ma tali accidenti et apparenze in niun modo possono accadere in una superficie sferica, che sia liscia et eguale; [...] adunque con necessaria dimostrazione si conclude, la superficie lunare esser piena di eminenze e bassure" (Galilei 1901, p. 184).

[16]On the topic of the pre-modern hypothetical-deductive method employed by Galileo cf. Graßhoff (2010a). Among other things, he draws attention to the deficit of this model in Galileo's later reasoning which was in favour of heliocentricity and based on the theory of tides. The latter may have also been the reason for his argumentative weakness towards the critics of the church.

[17]"Ma che simili montuosità et prominenze fossero a noi visibili (rimosse le narrate mutationi di ombre e di lumi) mediante il loro sporgere et rigonfiare verso la vista nostra, è del tutto impossibile" (Galilei 1901, p. 184).

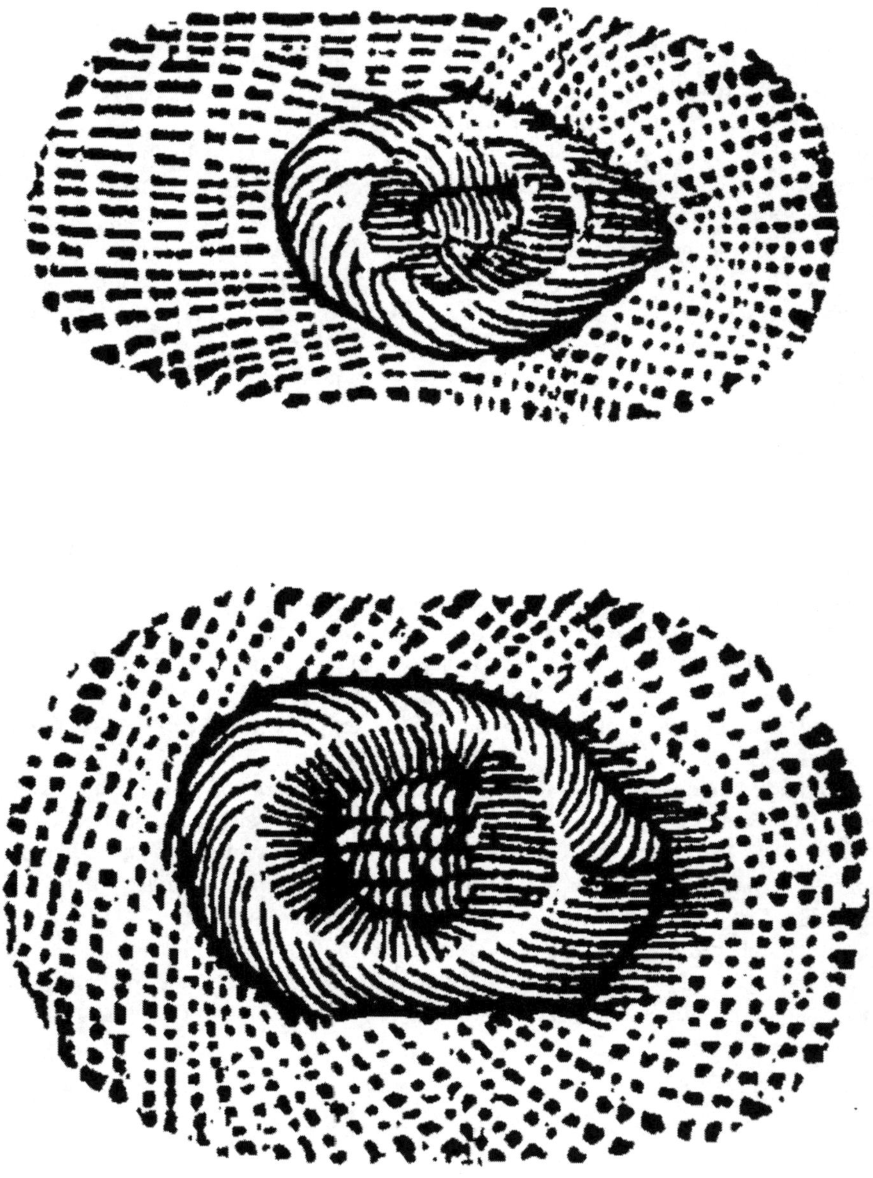

Fig. 11.5 Visualisation of the light-and-dark contrast as it occurs at the so-called *terminator*, from Galileo's letter to Christoph Grienberger, 1 September 1611, p. 183

based on the light-and-dark effects.[18] This fact is important, since it allowed the author of the "Mantua Problem" to come to the conclusion that no mountains exist on the periphery of the lunar hemisphere, and, if they did exist, other phenomena than the light-and-dark effects ought to be visible (Galilei 1901, p. 186). Galileo therefore has to explain why there are no light-and-dark effects visible on the periphery of the lunar hemisphere. In doing so, he illustrates the problem of perception under discussion with a drawing that he includes in the letter.

11.6 Seeing in Perspective

The drawing in Fig. 11.6 exemplifies the situation in which sun rays fall horizontally upon a mountainous surface BC seen from a point O.

In this case, according to Galileo, the sun, or someone standing in the sun ("il sole, o chi nel sole gosse collocato"; ibid. p. 185) can only see the illuminated parts of the surface and not a single dark side. This is because in such an instance the eyesight's and the sun's rays run in parallel lines (Ibid.).

In order to see the dark side, says Galileo, it is necessary for the eye to rise above the surface, as can be seen in Fig. 11.6 (Ibid.). There, the eye is located in a position A at an angle above the illuminated surface. Yet, Galileo also regarded a third constellation, which emerges when the luminary is positioned higher than the illuminated surface, and the eye is positioned lower, as if (in the same diagram) the eye were in O and the sun in A (Ibid.). In this case, the side of the surface enveloped in shadow would also remain hidden from the eye (Ibid.). This is exactly the situation of an observer on earth when the rays of his or her eyesight touch the periphery of the moon's illuminated hemisphere. Since we are positioned lower than the surface we observe, the sun can be positioned in any location, yet we will never be able to see the dark sides of the mountain depths.[19] Furthermore: Since the mountain depths are obscured all around by illuminated mountain heights, nothing will be visible but a completely illuminated continuum.[20] This last constellation of the sun, the moon, and the observer on earth therefore forms the basis of Galileo's considerations regarding *seeing in perspective*: This is because even though the boundary of the light is situated on the periphery of the lunar hemisphere—as for examples just after a new moon,

[18]"Hora, perchè delle sopranarrate apparenze di lumi et ombre, quanto bene, sicome io assolutamente credo, siano ancora circa l'estrema circonferenza non meno che nelle parti più interne, niuna può in modo alcuno da noi scorgersi e distinguersi; però niuna coniettura, inditio ed argomento ci possono elle somministrare dell'essere o non essere la detta circonferenza montuosa" (Galilei 1901, p. 184).

[19]"Hora, perchè i raggi visivi che abbracciano l'estrema visibil circonferenza del corpo lunare, non hanno elevazione alcuna sopra essa, ma toccano in lei la superficie della luna, manifestatamente si scorge come, costituito il sole in qualsivoglia luogo, mai non potranno da noi esser vedute le ombre delle bassure alla detta circonferenza vicinissime" (Galilei 1901, p. 185).

[20]"[...] anzi, restando tali parti oscure celate tra le eminenze circonvicine illuminate, altro non si scorgerà che una continuazione tutta luminosa" (Galilei 1901, p. 185f.).

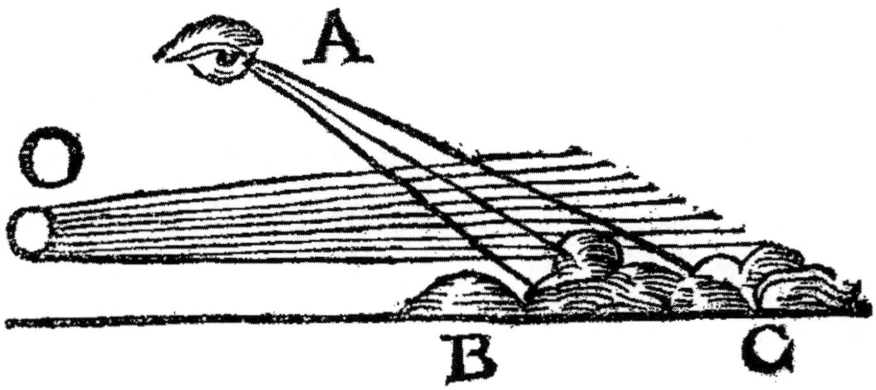

Fig. 11.6 Sketch of Galileo's considerations concerning seeing in perspective, from Galileo's letter to Christoph Grienberger, 1 September 1611, p. 185. Galileo demonstrates where the observer should be located in order to see the dark side of the illuminated mountainous surface BC

when the latter appears as a sickle—we still cannot see the surface of the moon, because we perceive this location *foreshortened* or side-on ("si veggono in scorcio et quasi in profilo"; ibid., p. 187f.). The shadings and darkened areas appear distorted or disappear entirely, especially during a full moon, which Galileo shows with Fig. 11.7. Thus it becomes obvious that the reference to knowledge concerning perspective in Galileo's texts about the observation of the moon is very closely linked to the problem of perception which he addresses.

At this point, it is possible to glean the first results of Galileo's argumentative strategy. In the first part of the letter to Grienberger, Galileo *first of all* concentrated on appearances—on the *sensate apparenze*—which play a central role in the observation of the moon. This is particularly true for the light-and-dark effects, from which findings about the mountainous structure of the moon can be derived in the form of hypotheses. Formulating hypotheses is the *second* part of the rational argumentation (*ragione*), among which Galileo also counts the knowledge of the terminology and of the effects concerning perspective (*termini et gl'effetti di prospettiva*). He emphasises that he did not assert his knowledge claims without reasoning ("non senza momento alcuno di ragione") and he explicitly accuses his critics of not having taken seeing in perspective sufficiently into account.[21] However, the analysis of the *sensate apparenze* (the light-and-dark effects) and the knowledge of the effects of perspective are not the only means of his argumentative strategy. Galileo's argumentation much rather rests on a bundle of arguments, or on a series of arguments respectively, which take into account various levels of the perception of reality. It is

[21]"Da quanto sin qui ho narrato credo che ciascheduno che mediocremente intenda i termini et gl'effetti di prospettiva, haverà sentito che non senza momento alcuno di ragione, come assai resolutamente pronunzia l'autore del Problema, ma spinto e forzato da manifeste apparenze et necessarie conietture, ho affermato, le montuosità lunari distendersi fino all'ultima visibil circonferenza" (Galilei 1901, p. 190).

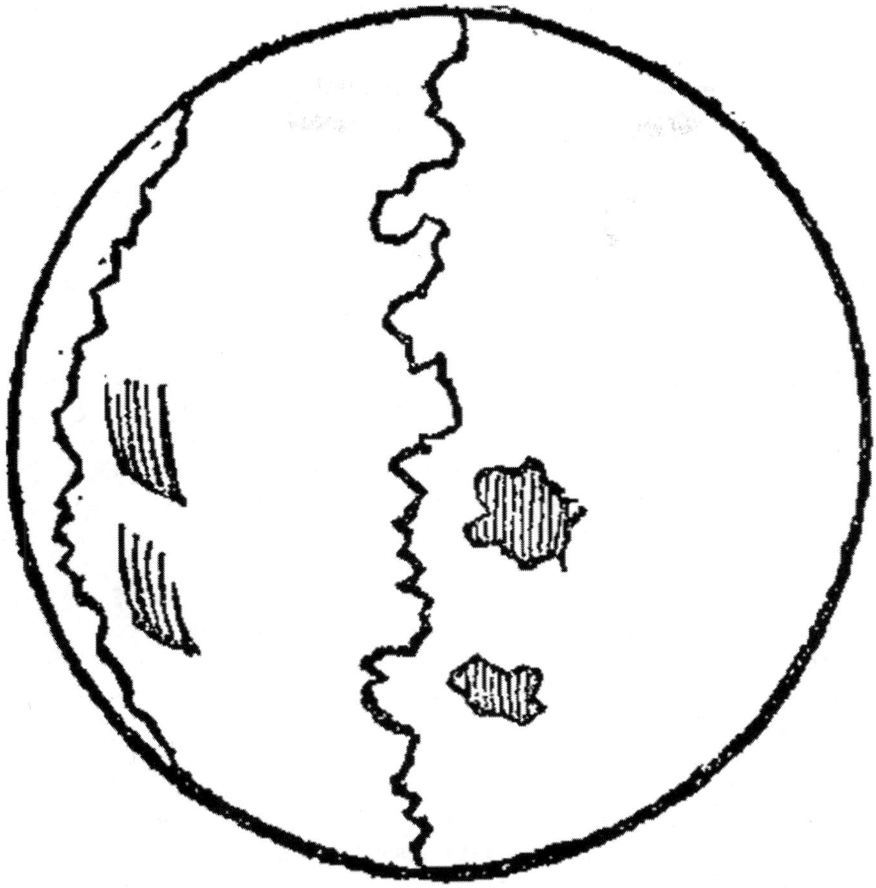

Fig. 11.7 Distortions and shading effects on the moon's surface at a full moon from a perspectival point of view, from Galileo's letter to Christoph Grienberger, 1 September 1611, p. 187. Galileo shows that the shadings and darkened areas on the moon surface at a full moon are perceived distorted or disappear completely from a perspectival point of view

as though Galileo wanted to show that the visual perception of lunar phenomena is not that different from the perception we experience in everyday life, and that we can hence easily comprehend lunar phenomena in the context of our everyday lives. Therefore, Galileo *thirdly* adds further arguments regarding the visual perception of lunar phenomena which are based on imagination. Furthermore, he *fourthly* characterises a so-called *esperienza sensata* as well as *fifthly* and finally an *esperienza*, i.e. an actual experiment.

11.7 Visual Perspectives and Irradiation Effects

As part of further considerations I would like to roughly outline Galileo's argumentation in the second part of his letter to Grienberger and then arrive at a number of conclusions. The main problem, which obviously concerned Galileo for some time, was to explain why we cannot see the lunar mountains and especially the heights on the periphery of the lunar hemisphere. Galileo contemplates three-dimensional solid bodies and notes that the lunar mountain ranges are in comparison hundreds of miles long, 50–60 miles wide, but only 3–4 miles high (Galilei 1901, p. 190). As Heilbron (2010, p. 152f.) has shown, Galileo calculated the height of the lunar mountains in *Sidereus Nuncius* according to Pythagoras and came to the conclusions that the mountains on the moon must be higher than those on earth. In this letter too, Galileo sustains his considerations with geometric drawings and diagrams (see Fig. 11.8). So the axis DAE represents the course of the light boundary, while CAN is the plane which represents a lunar spot, which is evenly dissected by the boundary border. The lunar spot is surrounded by large mountain ranges, whose long and wide ridges ABC reach on to the even darker side of the moon. As these mountain ranges are very large, illuminated, and surrounded by dark areas, they become clearly visible to us. Once again, light-and-dark effects play a decisive role here. However, Galileo also allows the readers to entertain a different variant in their imagination: If we imagine that the same mountain ranges were to be shifted to the periphery of the lunar hemisphere DFG, we would only see their elevations FG; but since they are only four miles high, i.e. a fiftieth of the lunar diameter, they still remain completely invisible (Galilei 1901, p. 191). The objection, which Galileo raises himself, is this: Why do the mountain tops on the dark side of the moon appear as light spots, while on the opposite light side the elevations FG are not visible, especially when we consider that these elevations are surrounded by the pitch black night sky (Ibid., p. 192). The light-and-dark contrast described here can also be illustrated by means of Fig. 11.9.

Galileo's answer to the question he posed is as follows: Just because I do not know the reason why the surface of the moon is invisible on the periphery of the lunar hemisphere does not mean that there are no such reasons, because so many of them may be unknown. Yet Galileo says that there still is a difference as to whether I observe the illuminated mountain tops in the central zone of the moon or at the periphery. This is because the former are larger than the latter simply due to where they are positioned, since I view the former from the front (*in faccia*), but the latter side-on (*in profilo*) (Galilei 1901, p. 193). Galileo shows the following illustration (Fig. 11.10): Here, one ought to observe the surface of the sphere which is encased inside a polar circle. When looking at the pole vertically from above or below, we will behold a perfect circle. However, on casting ones eye over the polar circle in profile only a small fraction of the same circle will be visible. The first and the second point of view *grosso modo* correspond to the difference in size between the circle ABCE and the segment of the circle ADC (Ibid.). Transferred to the lunar observations, this means that we see the mountain tops in the moon's central zones in front view (*in maestà*), similar to the circle BAEC. However, on the periphery we see them in

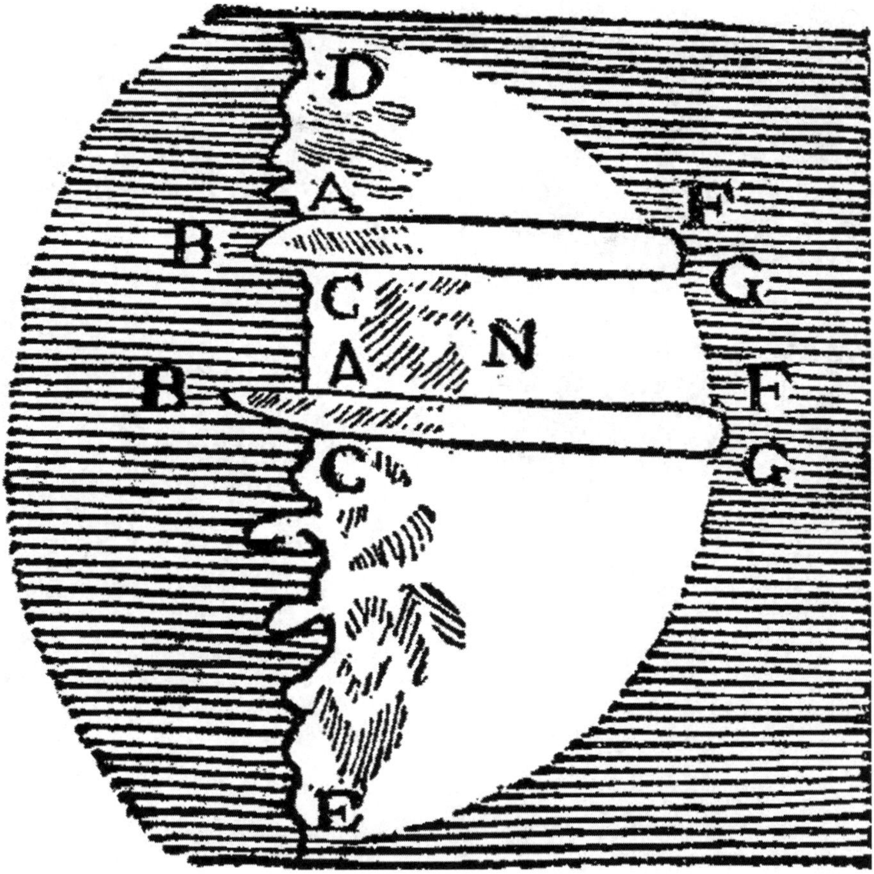

Fig. 11.8 Representation of the heights and of the conditions of illumination on the periphery of the lunar hemisphere, from Galileo's letter to Christoph Grienberger, 1 September 1611, p. 191. Galileo illustrates the difference between observing the illuminated mountain tops in the central zone of the moon and observing them at the periphery. The former ABC is observed from the front (*in faccia*), the latter FG side-on (*in profilo*). From the side, light-and-dark contrasts disappear

profile (*in profilo*), similar to the circle fragment ADC. Thus, the position (*positura*) of the mountain tops alone determines which area I perceive—the semi-circle ABC or the section ADC (Ibid.).

Although these considerations already address problems of perspective, Galileo is, in fact, concerned with a bigger issue, namely with the basics of the truth of his assertions ("I fondamenti della verità della nostra asserzione"; ibid.). These are very clearly rooted and founded within the theory of perception by way of the following generalising statement:

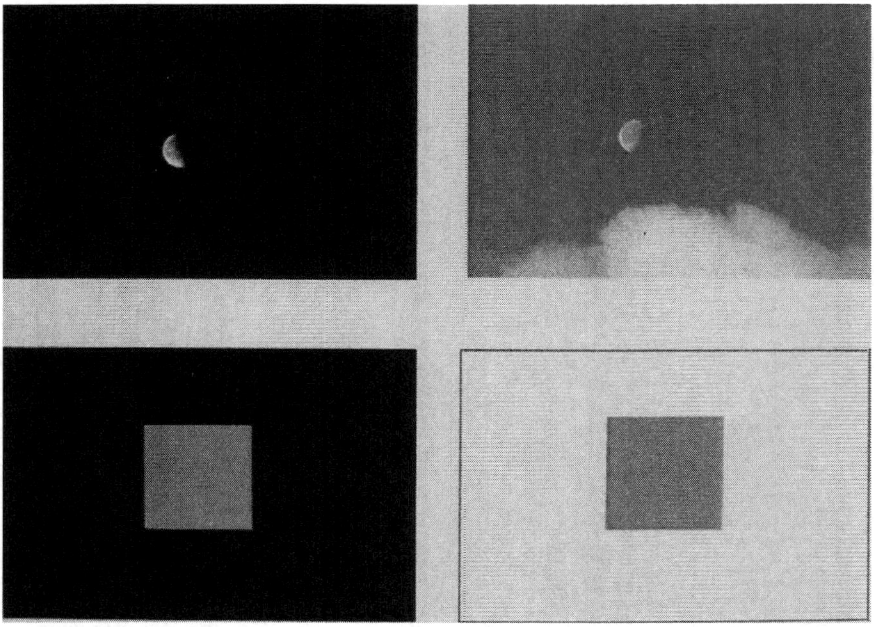

Fig. 11.9 Illustration of the light-and-dark contrasts in the day-time and night-time sky (Piccolino and Wade 2008a, p. 587). The pictures in the lower half show the same light-and-dark-contrast by a psychological experiment: The *square* against the *black* background on the *left* appears to be brighter than the *square* against the *white* background on the *right*

> While every illuminated body shows its true and real appearance when observed from a close proximity, when regarded from afar it seems to surround itself with rays (*s'inghirlandi)* that are not its own, and which cause it to lose its contours and make it appear larger.[22]

This is a *sensata esperienza* which we experience in principle with all luminaries, including the celestial bodies. Galileo illustrates this principle once again and uses the flame of a candle (*fiammella*) as an example: on the one hand, it is the *distance* to the flame which determines our perception: regarded from up close, the flame's profile is clear, while from a distance it seems to us much bigger and loses its form completely due to the rays which surround it, i.e. through the *effect of irradiation*. On the other hand, the *surroundings* amplify the effect of perception: while at sunset the flame is small, it grows in darkness and the rays consume its shape (Ibib., pp. 193–195; Dupré 2003, p. 390).[23] In this context, for Galileo it basically made no difference whether we

[22]"Ogni corpo luminoso, mentre è veduto da vicino, ci si mostra sotto la sua vera et real figura; ma da lontano pare che s'inghirlandi di alcuni raggi ascitizii, tra i quali i termini della sua figura si perdono, et pare che la sua mole s'accresca" (Galilei 1901, p. 193).

[23]"In 1611, irradiation became such a central concept for Galileo that he used it to reinterpret his earlier observations of the light of the Moon" (Dupré 2003, p. 390).

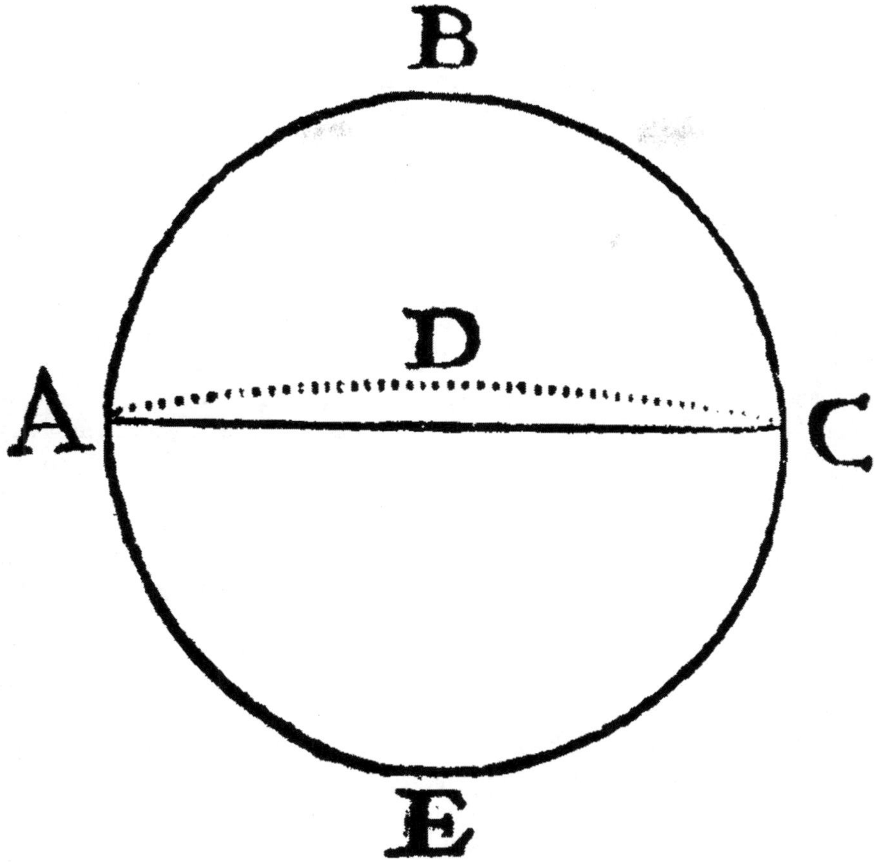

Fig. 11.10 Schematic representation of the frontal and lateral views of mountain elevations in the central and peripheral zones of the moons surface, from Galileo's letter to Christoph Grienberger, 1 September 1611, p. 193

look with the naked eye ("col semplice occhio natural") or through a telescope. This is because the telescope can only partially eliminate irradiation, as e.g. in the case of the planets Mars or Venus, which are close to the sun. More specifically: When Venus is situated above the sun, a telescope is insufficient, according to Galileo, to bring it close enough for us to fully distinguish its sphere and to separate the latter from its irradiation. When Venus is situated below the sun, on the other hand, its contours are very clearly distinguishable (Ibid. pp. 193–195). Of course, the principle of the candle flame also applies to the observations of the moon: Despite the close proximity to earth and despite the telescope, the described light effects cause 'homogeneous perceptions', i.e. they cause us to perceive the moon's surface as *evenly and continuously structured*. In this context, heights and depths are balanced out.[24] This effect in the course of perception still occurs more frequently on the periphery of the lunar hemisphere than in the central zones, as there we do not perceive the large mountain ranges according to height, but according to length and to how wide they are. Thus light effects cannot to balance out the heights and depths.[25]

11.8 'Homogeneous Perceptions'

The experiment (*esperienza*) which Galileo finally describes aims to point out such 'homogeneous perception'.[26] Two gaps are cut out of a thin metal plate, similar to the drawing in Fig. 11.11. When looking at the figure closely, one can see that the two gaps or shapes are not exactly the same. The line of the upper incision runs evenly, that of the lower incision less so, as if it had been drawn with an unsteady hand. Galileo instructs that the metal plate be put into a dark room. Behind the plate, a flame is positioned, which is so big that it entirely covers the area of both gaps. The flame is then shielded from other light sources, so that no other light is visible than that which comes through the two gaps. Thus are the instructions for the experiment.

When we now look at the two openings from up close, according to Galileo, we will be able to clearly differentiate two illuminated strips from one another. One gap is defined by a clear line, the other by an irregular one. If we now look at both gaps from a larger distance—say we take 100 or 150 steps back—it is no longer possible to see these differences, because we now get the impression that the light shines evenly onto both gaps, and the uneven line is rendered indistinct in these rays. Both gaps now look practically equal. We can draw nearer to them with the help of a telescope and see both gaps distinct again. However, if we step back one more time (around 1000 or 1500 yards), the telescope cannot bring the gaps close enough for us to be able to see them differentiated once more (Galilei 1901, p. 198f.). In principle, every astronomer was now able to verify this effect of perception by experiment. Whether Galileo actually conducted the experiment himself is not made explicit in the text. However, the test with the candle light can be undertaken without any effort.

---

[24] "[...] et benchè il telescopio toglia in gran parte la detta irradiazione, col portarci la specie della [Luna] molto vicina, non è però tanta la vicinanza, nè sì poca la irradiatione, che non ve ne avanzi soprabbondantemente più di quello che basterebbe per adeguare la scabrosità delle escrescenze di alcune rupi che in qualche parte soverchiassero le eminenze disposte in molti e lunghissimi ordini intorno al perimetro lunare" (Galilei 1901, p. 196f.).

[25] "[...] et se una tale irradiazione è potente a nasconderci la immensa cavità di Venere, quando è cornicolata, et che noi la rimiriamo con la vista naturale mostrandocela similissima alle altre stelle, ben si può senza un minimo scrupolo ammettere et senza alcuna ombra affermare, che i piccolissimi cavi e colmi dell'immensa circonferenza lunare siano talmente dalle loro scambievoli irradiationi ingombrati, che del tutto si perdino, veduti ancora col telescopio" (Galilei 1901, p. 198).

[26] "At no relevant point does Galileo demand more from an experiment than the artificial reproduction of the phenomenon which is to be explained" (Graßhoff 2010a, p. 31).

Fig. 11.11 Representation of two gaps cut into a metal plate, which form a part of Galileo's instructions on how to conduct the experiment to create homogeneous perception by way of using candlelight, from Galileo's letter to Christoph Grienberger, 1 September 1611, p. 198

11.9 Conclusion

Here are my final conclusions. This is certainly not the place to become embroiled in a debate about Galileo's empiricism. Segre (2005) once called Galileo's empiricism "a historiographical excuse" and endowed it with a question mark. According to Segre, if one were to pose the question as to Galileo's methodology, that is, as to how he worked, one could "state with conviction that he was an empiricist, because his manuscripts prove that he conducted experiments" (Ibid, p. 96). However, if one were to ask about Galileo's epistemology, "i.e. what the origin of his knowledge was, one would find that many of Galileo's remarks point towards the fact that he did not consider experiments and observations to be sources of knowledge. Galileo's epistemology was therefore almost certainly not empirical" (Ibid.). In the end, I think the question as to empiricism has to remain open as regards Galileo, and it can very definitely not be answered in an abstract manner. This is to say, it depends on the *context*. In fact, in the (re)constructed historical episode the experiment is postponed at the end of a long series of arguments. Its aim was basically to illustrate an observation made with the naked eye or the telescope. But also observation cannot be considered as the only source of knowledge. Thus the reconstruction of Galileo's argumentation in the letter to Grienberger has shown that he indeed takes observations as a starting point, but he does not take them at face value. In the specific case of the lunar observations, the point was to uphold a seemingly paradoxical position: Galileo asserts, namely, to have shown that the lunar mountains reach up to the periphery of the lunar hemisphere. At the same time, he believes that it is not necessary for us to actually see these lunar mountain ranges (Galilei 1901, p. 199). I have tried to argue in favour of the idea that these statements only make sense if we try to understand them from the point of view of the theory of perception. This is because in regard to the lunar observations all we really perceive are light-and-dark effects, according to Galileo. This would then again mean that Galileo was ultimately trying to find an explanation for a range of *problems* of perception in his argumentation, i.e. that he tried to explain what constituted the relationship between visual perception and material objects. And as Galileo's extremely sophisticated argumentation shows, there were a multitude of ways to explain this relationship.

It should have become clear that this article focuses on one particular historical episode and shows how the latter can be (re)constructed and conceived in terms of an *epistemic situation*. I have concentrated on the analysis and the interconnection of historical data which supposedly contains philosophical concepts and arguments. By way of summary we can say that the episode of Galileo's lunar observation can only be adequately comprehended if we consider how the various types of historical data combine: 1. Galileo's use of different *types of texts* and *forms of representation* (scientific treatises, letters, illustrations, diagrams, abstract models, etc.; 2. the *chronology* of the episode (from the construction of the telescope in November 1609 to Galileo's letter to Grienberger in September 1611); 3. the different instances of how Galileo's knowledge claims were *received* (from Jesuit criticism in 1611 to the reproduction of Galileo's observations by Jesuit astronomers, as e.g. in the case of Christoph Scheiner's letter from November 1612); 4. the even more important fact that *the geometrical argument must not be considered in isolation*, at least in retrospect. Its significance only becomes apparent in conjunction with Galileo's reaction to the Jesuit criticism he had to face after the publication of *Sidereus Nuncius*, and which forced him to expose more accurately the problem of seeing and observing astronomical phenomena; 5. it was notably in the course of exposing and defending his position that he adopted and combined various *epistemic strategies*: instruments (the telescope), observations, empirical data (the lunar diameter), theories (light rays, seeing in perspective), models, arguments and one experiment. 6. Lastly, we have determined that historical episodes are characterised by their being *open for interpretation*, and that we may not regard them as set in stone. In principle, it is always possible to reveal new historical data or to reinterpret existing knowledge in the light of new findings, as did Galileo with his concepts and arguments in the letter to Grienberger. Such changes of perspective may also change our perception of the episode on the whole, and in retrospect this may throw new light on *Sidereus Nuncius* as a fundamental text of the new astronomy. The (re-)construction of the historical episode thus helped me to extract from it some more general or abstract knowledge about observation and perception in science. Certainly, at the end of the day we have a more detailed notion of Galileo's *concept of observation*, especially of his argument for combining *senso* (sense perception) and *discorso* (reasoning) in observing astronomical phenomena. Galileo seems exactly to distinguish between (a) *senso*, sense perception, i.e. the physiological process of seeing, (b) *apparenze sensate*, appearances, i.e. what appears to the observer, (c) observation, which also includes the interpretation of what is seen, and (d) *discorso*, which is the combination of both observation and reflection. The *senso-and-discorso argument* deduced from the perception of light-and-dark effects is particularly subtle. It compels Galileo to reflect on the relationship between sensual perception and (hypothetical) reasoning in observational processes and explains why (hypothetical) reasoning becomes indispensable. At least, Galileo clearly expounded how he dealt with observational data when the phenomena, i.e. the lunar mountains, could not actually be perceived with the naked eye or with the telescope. Thus, his argument is central to determining the character of the epistemic situation under consideration here.

References

Barker, P. 2011. The reality of Peurbach's orbs: Cosmological continuity in fifteenth and sixteenth century astronomy. In *Change and continuity in early modern cosmology*, ed. P.J. Boner, 7–32. Dordrecht: Springer.

Burian, R. 2001. The dilemma of case studies resolved: The virtues of using case studies in the history and philosophy of science. *Perspectives on Science* 9(4): 383–404.

Camerota, F. 2004a. Galileo's eye: Linear perspective and visual astronomy. *Galilaeana* 1: 143–170.

Camerota, F. 2006. *La prospettiva del Rinascimento: arte, architettura, scienza*, vol. 19. Milan: Electa.

Camerota, M. 2004b. *Galileo Galilei e la cultura scientifica nell'età della Controriforma*. Rome: Salerno.

Danneberg, L. 2006. Epistemische Situationen, kognitive Asymmetrien und kontrafaktische Imaginationen. In *Ideen als gesellschaftliche Gestaltungskraft im Europa der Neuzeit. Beiträge für eine erneuerte Geistesgeschichte*, ed. L. Raphael, and H.-E. Tenorth, 193–221. Munich: Oldenbourg Wissenschaftsverlag.

Dupré, S. 2003. Galileo's telescope and celestial light. *Journal for the History of Astronomy* 34: 369–399.

Dupré, S. 2005. Ausonio's mirrors and Galileo's lenses: The telescope and sixteenth-century practical optical knowledge. *Galilaeana* 2: 145–180.

Dupré, S. 2014. Galileo and the culture of glass. In *Tintenfass und Teleskop: Galileo Galilei im Schnittpunkt wissenschaftlicher, literarischer und visueller Kulturen im 17. Jahrhundert*, ed. A. Albrecht, G. Cordibella, and V.R. Remmert, volume 46, 297–319. Berlin: De Gruyter.

Galilei, G. 1901. Galileo a Cristoforo Grienberger [in Roma], Firenze, 1° settembre 1611. In *Le Opere di Galileo Galilei, Edizione Nazionale, volume 11: Carteggio 1611–1613*, ed. A. Favaro, 178–202. Florence: Barbera.

Galilei, G. 1998. *Dialogo sopra i due massimi sistemi del mondo tolemaico e copernicano*, ed. O. Besomi, and M. Helbing, 2 volumes. Padua: Antenore.

Graßhoff, G. 2010a. Mit allen Wassern gewaschen? Galileis Theorie der Gezeiten. In *"Natur", Naturrecht und Geschichte: Aspekte eines fundamentalen Begründungsdiskurses der Neuzeit (1600–1900)*, ed. S. De Angelis, F. Gelzer, and L. Gisi, 23–46. Heidelberg: Winter.

Graßhoff, G. 2010b. Modelle. In: *Einführung in die Wissenschaftsgeschichte und Wissenschaftstheorie*. http://philoscience.unibe.ch/documents/VorlesungenFS10/wtwg/wtwgscript.pdf. Accessed in Feb 2015.

Heilbron, J. 2010. *Galileo*. Oxford: Oxford University Press.

Krayer, A. 1991. *Mathematik im Studienplan der Jesuiten: Die Vorlesung von Otto Cattenius an der Universität Mainz (1610/11)*. Stuttgart: Steiner.

Pantin, I. 2005. Galilée, la lune et les Jésuites: À propos du Nuncius Sidereus Collegii Romani et du 'Problème de Mantue'. *Galilaeana* 2: 19–42.

Perler, D., and M. Wild. 2008. Einführung. In *Sehen und Begreifen: Wahrnehmungstheorien in der frühen Neuzeit*, ed. D. Perler, and M. Wild. Berlin: De Gruyter.

Piccolino, M., and N.J. Wade. 2008a. Galileo Galilei's vision of the senses. *Trends in Neurosciences* 31(11): 585–590.

Piccolino, M., and N.J. Wade. 2008b. Galileo's eye: A new vision of the senses in the work of Galileo Galilei. *Perception* 37(9): 1312–1340.

Reeves, E. 2011. Variable stars: A decade of historiography on the Sidereus Nuncius. *Galilaeana* 8: 37–72.

Scheiner, C. 1612. Letter of Christoph Schreiner to Paul Guldin, 10 November 1612. UB Graz, Ms 139, Letter Nr. 6.

Segre, M. 2005. Galileis Empirismus: Eine historiographische Ausrede? In *Miscellanea Kepleriana: Festschrift für Volker Bialas zum 65. Geburtstag*, ed. F. Bookmann, D. Di Liscia, and H. Kothmann, 89–100. Augsburg: Rauner.

Strano, G. 2012. Galileo's shopping list: An overlooked document about early telescope making. In *From earth-bound to satellite. Telescopes, skills and networks*, ed. A.D. Morrison-Low, S. Dupré, S. Johnston, and G. Strano, 1–19. Brill: Leiden.
Wallace, W.A. 2003. Galileo's Jesuit connections and their influence on his science. In *Jesuit science and the republic of letters*, ed. M. Feingold, 99–126. Cambridge: MIT Press.
Zittel, C. 2012. Die Lunatiker von Aix-en-Provence: Peiresc—Gassendi—Mellan. In Et *in imagine ego: Facetten von Bildakt und Verkörperung. Festschrift für Horst Bredekamp*, ed. U. Feist, and M. Rath, 277–299. Akademie Verlag: Berlin.

Chapter 12
Multiple Perspectives on the Stern-Gerlach Experiment

Tilman Sauer

Abstract Different or conflicting accounts of the same episode in the history of science may arise from viewing that episode from different perspectives. The metaphor suggests that conflicting accounts can be seen as complementary, constructing a multi-dimensional understanding, if the different perspectives can be coordinated. As an example, I discuss different perspectives on the Stern-Gerlach experiment. In a static interpretation, the SGE has been viewed as an experiment that allows the determination of the magnetic moment of silver atoms. Based on the concept of magnetic momentum arising from orbital angular momentum, the original experiment was designed in 1922 as an *experimentum crucis* to decide between Bohr's quantum theory and classical electromagnetic theory, and its outcome was interpreted as a confirmation of the Bohr-Sommerfeld quantum postulates. After the advent of quantum mechanics, the SGE was reinterpreted in terms of magnetic moment arising from the electron's spin angular momentum. In a dynamical interpretation, physicists have asked for the physical mechanism responsible for the quantization of the angular momentum with respect to the direction of the magnetic field. Although different suggestions were explored, none was ever accepted as fully satisfactory. Today this difficulty is seen as a paradigmatic instance of the unsolved quantum measurement problem.

12.1 Introduction

The historical development of scientific knowledge and understanding is an inherently unique sequence of events. Every insight is different. As historians and philosophers of science we try to identify typicalities, regularities, methodologies, and to some extent these efforts prove successful. There are, indeed, cases of scientific inquiry that lend themselves to philosophical generalizations about the nature of science. Our ideas about the nature of science influence our historical narratives about

T. Sauer (✉)
Institute of Mathematics, Johannes Gutenberg University Mainz, Mainz, Germany
e-mail: tsauer@uni-mainz.de

T. Sauer and R. Scholl (eds.), *The Philosophy of Historical Case Studies*,
Boston Studies in the Philosophy and History of Science 319,
DOI 10.1007/978-3-319-30229-4_12

how that science came about. This leads to a problem of underdetermination: two or more historical accounts of the same episode differ as a result of the underlying philosophical assumptions and the difference is such that one needs to decide which one of the accounts is to be accepted. The issue then arises whether and in which way evidence in one way or other may decide between the competing accounts. But such clearcut cases of underdetermination are perhaps rare.

Quite often, it seems, we are faced with the situation that an episode in the history of science gives rise to historical investigation of different sorts, and the resulting accounts are simply different but not necessarily conflicting. They may be conflicting in certain conclusions, and this seems to be an interesting situation, but they may not be conflicting as a result of explicitly competing philosophical stances. Why different historical accounts are conflicting is itself something that needs be looked into in every single instance, and the causes may be very different every time.

But the very fact that historical events and episodes may be seen from different perspectives, and the fact that these different views may result in conflicting accounts of the same aspects of the past, is a fascinating (some say: troubling) feature of the history of science. In this paper, I will take one historical episode, the original Stern-Gerlach experiment, and look at different historical views that it has provoked. My contention will be that diversity and conflict in historical accounts can be turned into richness and depth of understanding, if the conflicting views can be understood as arising from different but complementary perspectives. They then give rise to a multi-dimensional understanding of the scientific enterprise that any single philosophical case study—or, for that matter, any straightforward historical narrative without explicit reflection on its underlying research interest—would not be able to provide on its own.

The metaphor I propose is taken from the theory of visual perception. Three-dimensional features of an object in space can only be perceived and mentally reconstructed if the object is viewed from more than one viewpoint. Any two-dimensional image produces only a projection, and a three-dimensional object may look very different from different perspectives. But the differences are explained by the identity and invariance of the actual object of perception which appears in certain projections to different observers.

12.2 The Stern-Gerlach Experiment

The Stern-Gerlach experiment (SGE) sends individual silver atoms through an inhomogenous magnetic field and observes that the silver atoms are deflected into two distinct directions. The observation is in agreement with the idea that the magnetic moments of the silver atoms align in two distinct projections with respect to the (vertical, say) direction of the magnetic field, and the gradient of the magnetic field therefore exerts a force on the atoms upward or downward depending on the alignment of the magnetic moment.

The SGE has been described as one of the most fundamental experiments of quantum theory, "one of the milestones on the winding road to modern quantum physics, one which offered other-than-spectroscopic evidence that quantum objects (atoms) exhibit behavior incompatible with classical physics" (Toennies et al. 2011, p. 1045).

For one, it is said that the SGE demonstrated for the first time the concept of space quantization, more precisely, angular momentum projection quantization ("Richtungsquantelung"), and in particular showed that individual silver atoms carried an angular momentum of one Bohr magneton in concordance with today's understanding of quantum atomic physics if one assumes that the electron carries an intrinsic angular momentum, the electron spin. Secondly, the SGE is a fundamental experiment because it demonstrated for the first time and in a very manifest way an anticlassical feature of quantum dynamics, the collapse of the wave function in a quantum measurement.

These assessments may be true for "a" SGE performed today, as it is with such fundamental experiments in the context of physics education.[1] But if we transfer the assessment to "the" SGE, i.e. the historical experiment done in 1922 by Otto Stern and Walther Gerlach (Stern 1921; Gerlach and Stern 1921, 1922a, b, 1924),[2] we run all the risks that are associated with backward projection of later successful science. Nevertheless, there was one historical SGE, "the" SGE, and the first question to ask is whether the actual historical experiment indeed shows what we now say that it shows.[3]

Since any experiment is a material constellation (in contrast to theoretical ideas and concepts), the SGE may be reconstructed both on the basis of historical documents (publications, notes, correspondence, etc.) and by replicating certain of its material aspects (Trageser 2011). The replication of the SGE is a difficult task. It uses techniques that are no longer in use or the use of which is no longer permitted, at least in the way they were used then (mercury pumps), and it requires an extensive infrastructure in terms of electricity, vacuum technology, glass blowing, chemical processing, etc. Even if some or all of the original materials and parts are substituted by their modern counterparts, the SGE remains an experiment that is not easily replicable.[4]

[1] The historical SGE is not a good experiment for ad oculos demonstration purposes, as it is difficult to do and the results are not easily transparent for display in a classroom. But its principle can be easily visualized and conveyed in schematic and idealized displays.

[2] For historical accounts of, or comments on, the SGE, see Gerlach (1969), Schütz (1969), Mehra and Rechenberg (1982), Friedrich and Herschbach (1998, 2005), Bernstein (2010), Toennies et al. (2011), Trageser (2011), Schmidt-Böcking and Trageser (2012).

[3] We will keep in the following the distinction between "a" SGE and "the" SGE. This distinction does not preclude that "the" SGE was an extended process of experimentation with various distinct stages, see note 9 below.

[4] The replication of parts of the SGE was done by Trageser (2011) in the context of his larger reconstruction of the genesis and early development of the SGE. Certain aspects of the historical SGE were also put to a replicative test by Friedrich and Herschbach (2003). There exists a large body of literature on replication of historical experiments which cannot be reviewed here.

Nevertheless, on the basis of extensive contemporary documentation, we can still be sure that Stern and Gerlach did perform the SGE experiment and did see the results that we now consider an important datum in the empirical foundation for our modern understanding of quantum theory.

From our modern understanding then, we can distinguish two aspects of the SGE that are to some extent independent and both establish the modern significance of the experiment. The first aspect pertains to the fact that the quantitative analysis of the original SGE and of any of its later incarnations demonstrated that the individual silver atom carries an angular momentum of about 1 Bohr magneton. From today's theory, this finding confirms our current understanding according to which the angular momentum of a silver atom arises from the spin angular momentum of its outer valence electron.

The second aspect pertains to the fact that the quantization that we see in the SGE is described, from our modern understanding, as arising from the projection postulate or from the collapse of the wave function. The projection postulate says that the outcome of an individual measurement of a quantum observable realizes one of the eigenvalues of the operator associated with the physical quantity. But the projection postulate does not tell us which of the eigenvalues of the operator will be realized in any individual measurement. The theory only gives statistical predictions for an ensemble of measurements of observables in systems that were each prepared in the same way. Any individual measurement realizes a stochastic, non-linear dynamics that is incompatible with the deterministic, linear dynamics of the time-dependent Schrödinger equation. This dynamical dualism is known as (one aspect of) the quantum measurement problem. The idealized material constellation[5] of an SGE is often used to illustrate and discuss the intricacies of the quantum measurement problem.

Both these aspects of modern physical theory give us a framework for how to understand what was going on in the original historical SGE. Since there is a continuity between the original historical SGE and modern installations of it,[6] we know how to interpret the actual material constellation that constituted the original SGE. In the following, I will look at the two aspects and discuss different perspectives as well as their mutual relationship.

---

[5]The expression "idealized material constellation" may sound paradoxical. It refers to the fact that "a" SGE can be done in many ways, using different materials, vacuum technology, geometries, etc. Whether any such *material* constellation would be permissible to constitute an SGE depends on the *idea* of the SGE, which tells you, e.g., that the magnetic field has to have a gradient, etc. But any defining aspect of such an SGE must be realized in some way or other materially. The SGE is not a thought experiment.

[6]This continuity was established, e.g., by Otto Stern's successful efforts in establishing a program of molecular beam experimentation in his laboratory in Hamburg, indicated, e.g., by a series of publications from his laboratory which explicitly were called "Untersuchungen zur Molekularstrahlmethode aus dem Institut für physikalische Chemie der Hamburgischen Universität" (Toennies et al. 2011). Walther Gerlach, too, for a while continued to do experiments similar to the original SGE but then changed to other fields, see Friedrich and Herschbach (2005).

12.3 Static Interpretation of the SGE

A standard criticism of writing history in a whiggish way questions that there is any reason to believe that the historical actors saw the state of affairs in the same way as we do now. In the case of the SGE this anti-Whig objection is immediately supported by the observation that neither the concept of electron spin nor the concept of a wave function or of the projection postulate were available to the physicists in 1922. Barring isolated and incomplete precursor conceptions articulated by Arthur Holly Compton and Alfred Landé, the concept of electron spin was introduced into quantum theory only in 1925 by Davisson and Germer. Likewise, the concept of a wave function only arises with the invention of wave mechanics by Erwin Schrödinger in 1925 and its statistical interpretation by Born in 1926, and a clear axiomatic conception of the projection postulate dates even later with Stone and von Neumann.

If we don't take the historical SGE as a mere inspiration of how to do an experiment that demonstrates angular momentum projection quantization or the wave function collapse, we have to look at the historical SGE from a new perspective, from the perspective of the historicity of the material conditions and, more importantly, of the historicity of intellectual resources, intentions, and interpretational horizons available to the actors in 1922.

The biographical perspective reconstructs the individual trajectory of the actors in an intellectual, cultural, and social universe. In the case of the SGE, we have a biography of Otto Stern (Schmidt-Böcking and Reich 2011) but a biography of Walther Gerlach is still missing. To be sure, scientific biographies—and a fortiori biographies of scientists in general—serve many purposes and answer to many questions. Given that Otto Stern had to emigrate Nazi Germany because of his Jewish background while Gerlach stayed on and even became head of the German uranium project in 1944, we may wonder, for example, how two scientists whose later trajectories were so very different after 1933 worked together harmoniously in 1922 and whether and to what extent political and social boundary conditions are or are not relevant for scientific collaboration. But for our purposes, the biographical perspective provides information about the knowledge and specific outlook on the scientific problems of the day. Taking up a terminological distinction by Giora Hon (this volume), the biographical perspective helps us identify the "baseline" and "snapshot" of the knowledge relevant for the historical SGE.

It is relevant for our purposes here that Otto Stern was trained and worked both as a theoretician and an experimentalist. As a theoretician—at one stage in his early career he was an assistant to Einstein in Zürich in 1913—, he was well aware of modern theoretical ideas, in particular, with Bohr's quantum theory of atomic spectra. As many of his contemporaries, he was also very skeptical about Bohr's bold postulates that were at variance with and had no foundation in classical physics. Indeed, Bohr's postulate of radiation free quantum orbits were really only justified by their success in explaining the hydrogen spectrum and confirmed by the successful explanation of the Stark effect in 1916 by Sommerfeld and Epstein. In addition, and more specifically relevant for the SGE, he had come to Frankfurt as a "Privatdozent" and titular professor in 1919 and had been engaged on the premises of the new physics institute in an experiment to test empirically Maxwell's velocity distribution. This experiment involved measuring the radial velocity of silver atoms effusing off a silver coated platinum wire by sending the evaporating silver atoms through a system of diaphragms in a rapidly rotating frame. This experiment was successful and for the first time demonstrated the empirical validity of Maxwell's distribution, after Stern corrected his initial theoretical interpretation in response to a criticism against his initial reasoning raised by Einstein. But this correction only amounted to the adjustment of a numerical factor. This precursor experiment provided Stern with practical experience in dealing with atomic silver beams with respect to (a) the creation and handling of molecular beams, (b) the possibilities and difficulties of capturing and evaluating thin silver deposits on brass plates, and, last not least, (c) it gave him confidence in the calculation of thermal velocities of effusing silver atoms.

Gerlach, too, brought some necessary experience to the table. He had previously done extensive experiments on thermal black-body radiation and was well versed with state-of-the-art electrotechnology, vacuum pumps, chemical processing, and other infrastructure technology. He also knew about experimental methodology with respect to isolating and determining minute effects.

Stern then knew about Bohr's quantum postulates and the refined versions that Arnold Sommerfeld and others had introduced to explain the Zeeman and Stark effects, and he was sceptical about this theory. One reason for his scepticism was that he thought angular momentum projection quantization should manifest itself in some kind of double refraction for light rays passing through a gas exposed to an external magnetic field. More specifically, he argued as follows. The angular momentum of the hydrogen atom's electron should be quantized with respect to an external magnetic field, even if the field is very weak. A light ray passing through the hydrogen gas orthogonal to the external magnetic field should exhibit double refraction because its polarization components orthogonal and parallel to the magnetic field should affect the electron orbits in very different ways. But such dispersion anomaly had never been observed. When he then heard about deflection experiments of atoms carrying an electric dipole in inhomogeneous electric fields, he put the various elements together and devised the idea for the SGE. In his proposal, he gave some numerical estimates about the feasibility of observing a splitting of the silver beam on the basis of the Bohr-Sommerfeld theory.

As Weinert (1995) emphasized, the theoretical model that motivated and guided the design of the historical SGE was not the model that we now associate with the historical SGE. Weinert pointedly expressed this fact by saying that for the historical SGE, the theory was "wrong" but the experiment was "right."

Stern had devised the SGE setup as an *experimentum crucis*. The idea was that the observation of the behavior of a beam of silver atoms in its transit through an inhomogeneous magnetic field would decide between classical and quantum theoretical conceptions because the two theoretical frameworks predicted different outcomes of the experiment.

According to classical Maxwell-Boltzmann theory, the magnetic moments of the thermal silver atoms would be distributed evenly in all directions before entering the inhomogeneous magnetic field. The effect of the magnetic field gradient would be to exert a force on the silver atoms proportional to the component of the magnetic moment along the gradient direction. The individual silver atoms would then be deflected according to the projection of their magnetic moment before entering the field, and the effect on the overall beam of silver atoms would be a simple broadening of the beam's profile. According to the angular momentum projection quantization hypothesis of quantum theory, however, the angular momentum, and hence the magnetic moment of the silver atoms would have to align in discrete quantized projections along the direction of the magnetic field. The theory therefore predicted a splitting of the beam into several discrete components.

There is a little complication here that Weinert did not take into account in his discussion because there was an ambiguity in the quantum theoretical prediction. According to contemporary understanding, the angular momentum of the silver atom arose from the angular momentum of its outer valence electron and was quantized to be one Bohr magneton. The angular momentum projection quantization now predicted that there were *three* different discrete projections for the angular momentum vector corresponding to electron orbits in a plane orthogonal to the magnetic field vector and parallel to the magnetic field vector. But Bohr had hypothesized that the latter case, in which the angular momentum vector would be orthogonal to the magnetic field and therefore the magnetic field vector would lie in the plane of the orbit, was dynamically impossible—or to be forbidden—because, Bohr argued, the electron's trajectory in this orbit would inevitably collide sooner or later with the atom's nucleus. The latter assumption clearly was an additional feature of the general model of quantized electron orbits and was subject to theoretical debate even if the overall quantum theoretical framework was shared. It amounted to an additional postulate of excluding some quantized orbits as dynamically forbidden. Thus, it seems that there were speculations about whether to expect a bipartite splitting of the beam of silver atoms without a central component according to Bohr's hypothesis of forbidden orbits or a three-partite splitting of the beam of silver atoms in the SGE with a non-vanishing central component.

As Weinert pointed out, the SGE was designed by Stern as an *experimentum crucis* to decide between two alternatives, a broadening of the beam according to classical physics and a splitting of the beam into two components according to Bohr's quantum theory. But in effect, a third option was being discussed before decisive results were obtained. According to this third option the beam would split up, as suggested by quantum theory, but into three components, keeping a central, undeflected component. The viability of this third option made the demands on the experimental accuracy much more stringent since the third option was much more difficult to distinguish from the classically expected continuous broadening of the beam.

In a way, the outcome of the SGE meant good luck for Bohr's quantum theory. Had the outcome been a three-partite splitting, it would have been much harder to interpret the SGE as deciding between theoretical alternatives, at least without further experimental refinements. As it came out, however, the SGE clearly supported the second of the three options, the splitting of the beam in two discrete components with no central component.[7] Naturally, the outcome was interpreted as a highly non-trivial confirmation of Bohr's theory. It convinced even the unbelieving Stern.

Returning to our theme of multiple perspectives on the SGE, we have to note that we now have two quite different perspectives of the same material constellation. According to our modern understanding, the historical SGE demonstrated for the first time angular momentum projection quantization of the silver atom's electron spin. According to the perspective of the historical actors, the historical SGE decided between distinct explanatory alternatives and confirmed Bohr's theory. In terms of the symmetry principle, as discussed, e.g., by Michael Bycroft (in this volume), we are dealing here with a case of rational explanation of a false belief. What do we do about this? Of course, one option is to fall for negative meta-induction and deny that today's explanation captures any truth about the state of affairs, either. As we have seen, the same experiment looks very different from two different perspectives. The perspectival metaphor, however, suggests a strategy to resolve this apparent discontinuity. Just as different views of the same object can be transformed into one another in a continuous way by a continuous change of local position of the observer, we can now ask how the interpretation of the SGE changed over time along with a change in the theoretical framework of quantum theory. Unfortunately, just as in the perspectival metaphor two different view points may be kept apart by occluding obstacles, we may not have sufficient historical documentation of how the interpretation of the SGE changed in the transition from Bohr's quantum theory to modern quantum mechanics.[8]

The perspectival metaphor suggests that the experiment as a material constellation remains relatively stable with respect to its changing theoretical interpretations. In the case of the SGE this implication of the perspectival metaphor is confirmed by interpreting the SGE from a causal theoretic point of view. Indeed, the SGE can be seen as a clear instance of a difference test of causal relevance. Quite independent of specifics of quantum-theoretical interpretation, the SGE demonstrated an empirically robust feature of quantum reality. It showed that an inhomogeneous magnetic field is causally relevant for a discrete deflection of individual silver atoms in their transit through that field.[9]

---

[7]One may be reminded here of the confirmation of gravitational light bending by the British eclipse expedition of 1919. Here the theoretical alternatives were also threefold: no deflection according to Newtonian gravitation and classical wave theory or according to Nordström's relativistic theory, deflection of 0.85″ according to Newtonian gravity and light corpuscles, or calculations based only on the equivalence hypothesis, and a deflection of 1.7″ for Einstein's general theory of relativity. The observations decided in favor of the third alternative. In this case, not only the experimental result proved to be robust, Einstein's theory of general relativity, too, remained valid to this day.

[8]I still have to find the place where the historical SGE is explicitly reinterpreted in terms of electron spin versus orbital momentum quantization. Perhaps there is a dark period here in which the SGE was not interpreted at all, and when it was interpreted again it was done so in the new framework without reference to the old Bohr theory.

[9]More precisely, the SGE developed with ever increasing accuracy. The first paper (Gerlach and Stern 1921) only reports a broadening of the silver deposits in the presence of a magnetic field, demonstrating the causal relevance of the field for some kind of broadening of the beam. The magnetic field affected the motion of the silver atoms, at least in some way. This first result was

12.4 Dynamical Interpretation of the SGE

Let me now look at the SGE from yet another perspective. As I have mentioned in the beginning, the SGE is seen today as a significant experiment not only for its demonstration of angular momentum projection quantization and as a manifestation of electron spin. It is also discussed routinely in debates about the quantum measurement problem.

In Stern's and Gerlach's as well as later *static* interpretations, the outcome of the SGE was analyzed with a view of determining the numerical value of the silver atom's magnetic moment. It was assumed that each silver atom moves along its path through the magnetic field with a magnetic moment that is aligned according to one of the quantum theoretically possible projections, and that this alignment stays fixed from the very first moment of its entrance into the magnetic field region. But the furnace from which the silver atoms effuse was shielded from any magnetic field to a very good approximation. The natural assumption therefore is that the magnetic moments of the silver atoms are directed randomly in all directions when they leave the furnace and hit the magnetic field region. This state of affairs immediately raises the question of a *dynamical* explanation for the alignment of the silver atoms at the moment of entering the region with a magnetic field. The problem was raised already by Stern in his proposal paper for the SGE. There he called it a "difficulty for the quantum conception" that had been noted already "by various parties" ("von verschiedenen Seiten bemerkt"). The difficulty was that one could not imagine how the silver atoms would "manage" ("es fertig bringen") to align their magnetic moments when they are sent into a magnetic field.

Immediately after the successful observation of a beam splitting had been published by Stern and Gerlach, Einstein and Ehrenfest (1922) took up this question in what was probably the first published reaction to the SGE. They, too, had asked themselves how the alignment could be explained dynamically. They stated that there are in principle, according to classical physics, only two dynamical mechanisms that might explain such an alignment of magnetic moments. The alignment could happen, somehow, by collisions between atoms, or by an interaction of the electron's orbital magnetic moment with the magnetic field, specifically by sending out Larmor radiation. The first mechanism was excluded since by quantitative estimates of the vacuum and the mean free path, the silver atoms in the SGE were clearly too far away from each other and effectively moved as isolated particles. Einstein and Ehrenfest then did "a little calculation" by applying Larmor's radiation formula to the case in

(Footnote 9 continued)
interpreted as demonstrating that the silver atoms do indeed carry an angular momentum. Only after further instrumental refinements was it possible to see that the silver deposits on the plate were showing the characteristic bipartite splitting, at which stage the causal inference was that the inhomogeneous magnetic field was causally relevant for a splitting of the beam (Gerlach and Stern 1922b). Still further refinements finally also made a numerical evaluation of the hypothesized magnetic moment possible (Gerlach and Stern 1922a).

hand. They found that with the numerical values of the SGE, the time needed by an individual atom to align its magnetic moment by sending out Larmor radiation was of the order of 100 years, whereas the time of flight through the magnetic field in the SGE was only of the order of a few microseconds. In their published paper, Einstein and Ehrenfest discuss a few theoretical implications of this dynamical impossibility of alignment by any classical mechanism. They conclude that any conceivable theoretical interpretation was at variance with fundamental tenets of classical physics (Unna and Sauer 2013).

Einstein's and Ehrenfest's remarks may be regarded as "prescient," if we take this expression as a descriptive category in the sense that their conclusions were made before but remained valid after the advent of today's accepted theoretical interpretation of the SGE.

The first dynamical interpretation of the SGE in terms of the time-dependent Schrödinger equations was given, to the best of my knowledge, by David Bohm in his 1951 textbook on *Quantum Theory*.[10] In the sixth and last part of that book, Bohm offered a discussion of the "Quantum Theory of the Process of Measurement." The account that Bohm gives is a very clear, explicit, and illuminating discussion, although he does not yet state the difficulty we now refer to as the quantum measurement problem in such terms. He distinguishes between the quantum system that we wish to obtain information about and the observing apparatus by which we interact with the system. After a general discussion of the role of observers, Bohm quickly proceeds to a more technical discussion in which he associates three different Hamiltonians with the system alone, the apparatus alone, and the interaction between system and apparatus, respectively. The example by which he exemplifies his theory is none other than the SGE. Here the system Hamiltonian describes the spin of the silver atom's electron, the apparatus Hamiltonian describes the motion of the center-of-mass coordinate of the atom. The interaction Hamiltonian H_{I} arises from an interaction energy of the form $H_{\mathrm{I}} = \mu(\sigma \cdot \mathscr{H})$ where μ denotes the magnetic moment, σ the vector of Pauli matrices representing the electron's spin, and $\mathscr{H}$ the magnetic field. Bohm introduces the notion of what he calls an impulsive measurement by which he means a measurement where the interaction happens on such short time scales that the change of the observables without the interaction would be negligible. This allows him to distinguish between a premeasurement state, an interaction state, and a postmeasurement state. In the premeasurement state the wave function can then be expanded in eigenfunctions to the operator associated with the system and apparatus observables, and in the interaction state, the two variables are linked together. In the SGE, the interaction energy is expanded as $H_{\mathrm{I}} \approx \mu(\mathscr{H}_0 + z\mathscr{H}_0')\sigma_z$, where the index 0 denotes values at the position of the silver atom, and the gradient of the magnetic field points in the z-direction. The second term in the expansion

[10]Note that this textbook was written from the point of view of the standard Copenhagen interpretation i.e., before Bohm began to question that by then canonical understanding and to investigate the alternative interpretations of quantum theory for which he is best known today.

couples the z-component of the center-of-mass coordinate with the z-component of the spin vector through the gradient along z of the magnetic field. The time-dependent Schrödinger equation can now be invoked to describe the evolution of the eigenfunctions of system and observable before the interaction sets in and results in a phase shift of those wave functions. If the initial state is prepared from the eigenfunctions as a wave packet localized with a certain width around the center-of-mass coordinate, the resulting state turns out to be a state of two wave packets associated with the two spin eigenfunctions with a certain width, and the peaks of the two wave packets run away from each other so that after some distance they are spatially distinct. Of course, the post measurement system is still an entangled system that turns into a mixed state after actually observing where the silver atom is located on the screen.

From the point of view of the quantum measurement problem, it must be said that Bohm's discussion very nicely explicates how the motion of the silver atoms in the SGE are to be described from the point of view of the dynamical Schrödinger equations. But it remains unaccounted for at which position an individual silver atom will hit the screen. Bohm gives a dynamical explanation for the splitting of the wave packet but he cannot give a dynamical explanation of the individual atom's place of detection on the screen. The quantum measurement problem is just this impossibility. The actual measurement of the value of a quantum observable introduces a second independent dynamics, a non-linear, stochastic collapse dynamics that is independent of the linear, deterministic dynamics of the time-dependent Schrödinger dynamics. The latter describes the smooth motion of the initial localized wave packet through the SGE magnet and its splitting into a double-peaked wave packet, the former describes the collapse of the wave function to one particular place on the screen where the atom is caught.

To the extent that the dualism between these two dynamical laws are considered unsatisfactory, as the term "quantum measurement problem" suggests, a satisfactory dynamical explanation of the SGE is still missing. In that respect, Stern's original difficulty, Einstein's and Ehrenfest's pointed analysis of the impossibility to account for the SGE dynamically, and Bohm's later critique of the standard Copenhagen interpretation all point to the same desideratum of providing a dynamical explanation of the motion of the silver atoms in the SGE.

If we compare this conclusion with the previous discussion of the determination and explanation of the silver atom's angular momentum, we see that the theoretical interpretation of the SGE does not change in its assessment. It remains a dynamical mystery to this day. What a historical view shows is a variety of different possibilities to conceptualize the dynamics of the SGE within different explanatory frameworks. But, in contrast to the static interpretation, this state of affairs cannot be adduced to justify negative meta-induction. In the former case, a belief in a true explanation of the effect was overturned. In the latter case, none of the attempted explanations were ever accepted as valid by the physicists. The difference is one between falsifying a false belief and stating that a true explanation has not yet been found.

12.5 Concluding Remarks

In this paper, I have taken the historical experiment by Stern and Gerlach of 1922 and looked at it from different perspectives. The starting point is the fact that the SGE is seen from today's perspective as a foundational experiment for quantum theory, demonstrating several key features of modern quantum mechanics. I have contrasted this modern view with the view of the original experiment by Stern and Gerlach as well as other contemporaries like Einstein and Ehrenfest. As regards, the question of identifying the quantized angular momentum projection, it was observed that the view of the historical actors differed substantially from our modern view but the experiment as a material constellation proved stable and its observed features proved robust. As a valid difference test, it demonstrated a robust causal relevance of the inhomogeneous magnetic field for the splitting of the atomic beam.

The metaphor of different perspectives suggests the task to account for a continuity between both the historical and the modern understanding. In the case of the SGE, this task remains to be done. With respect to a dynamical interpretation of the SGE, it was seen that the essential impossibility of giving a causal dynamical account of the SGE has remained unchanged from the very beginnings to this day, even if very different conceptual frameworks were invoked for tentative explanations. In this second aspect, the various historical interpretations also provide different perspectives although not necessarily contradictory ones but only those that are all false. In either case, the interpretation of the SGE from different perspectives, modern ones and historical ones, allows for an understanding of the significance of the SGE that is not possible by any one perspective alone nor by a mere additive sequence of different interpretations.

Acknowledgments My understanding of the SGE effect has profited a lot from discussions with Horst Schmidt-Böcking and Wolfgang Trageser. I also thank Tim Räz, Raphael Scholl, and Adrian Wüthrich for helpful criticism of an earlier version of this paper.

References

Bernstein, J. 2010. The Stern-Gerlach experiment. arXiv:1007.2435.

Bohm, D. 1951. *Quantum theory*. Englewood Cliffs, NJ: Prentice-Hall.

Einstein, A., and P. Ehrenfest. 1922. Quantentheoretische Bemerkungen zum Experiment von Stern und Gerlach. *Zeitschrift für Physik* 11:31–34. Reprinted in Kormos Buchwald et al., 2012, Doc. 315d.

Friedrich, B., and D. Herschbach. 1998. Otto Stern's lucky star. *Daedalus* 127(1): 165–191.

Friedrich, B., and D. Herschbach. 2003. Stern and Gerlach: How a bad cigar helped reorient atomic physics. *Physics Today* 56(12): 53–59.

Friedrich, B., and D. Herschbach. 2005. Stern and Gerlach at Frankfurt: Experimental proof of space quantization. In *Stern-Stunden. Höhepunkte Frankfurter Physik*, ed. W. Trageser. Frankfurt: University of Frankfurt, Fachbereich Physik.

Gerlach, W. 1969. *Otto Stern zum Gedenken. Physikalische Blätter* 25(9): 412–413.

Gerlach, W., and O. Stern. 1921. Der experimentelle Nachweis des magnetischen Moments des Silberatoms. *Zeitschrift für Physik* 8: 110–111.

Gerlach, W., and O. Stern. 1922a. Das magnetische Moment des Silberatoms. *Zeitschrift für Physik* 9: 353–355.

Gerlach, W., and O. Stern. 1922b. Der experimentelle Nachweis der Richtungsquantelung im Magnetfeld. *Zeitschrift für Physik* 9: 349–352.

Gerlach, W., and O. Stern. 1924. Über die Richtungsquantelung im Magnetfeld. *Annalen der Physik* 74: 673–699.

Kormos Buchwald, D., Illy, J., Rosenkranz, Z., and T. Sauer, T. (eds.). 2012. *The Collected Papers of Albert Einstein, Vol. 13. The Berlin Years: Writings & Correspondence, January 1922–March 1923*. Princeton: Princeton University Press.

Mehra, J., and Rechenberg (eds.). 1982. *The historical development of quantum theory, Vol. 1 in 2 parts. The quantum theory of Planck, Einstein and Sommerfeld: Its foundation and the rise of its difficulties 1900–1925*. New York, Heidelberg, Berlin: Springer Verlag.

Schmidt-Böcking, H., and K. Reich. 2011. Otto Stern. Physiker, Querdenker, Nobelpreisträger. Frankfurt/Main: Societäts-Verlag.

Schmidt-Böcking, H., and W. Trageser. 2012. *Ein fast vergessener Pionier. Physik Journal* 11(3): 47–51.

Schütz, W. 1969. Persönliche Erinnerungen an die Entdeckung des Stern-Gerlach-Effektes. *Physikalische Blätter* 25(8): 343–345.

Stern, O. 1921. Ein Weg zur experimentellen Prüfung der Richtungsquantelung. *Zeitschrift für Physik* 7: 249–253.

Toennies, J.P., H. Schmidt-Böcking, B. Friedrich, and J.C. Lower. 2011. Otto Stern (1888–1969): The founding father of experimental atomic physics. *Annalen der Physik* 523(12): 1045–1070.

Trageser, W. 2011. Der Stern-Gerlach-Effekt. Genese, Entwicklung und Rekonstruktion eines Grundexperimentes der Quantentheorie 1916–1926. Ph.D. Thesis, Johann Wolfgang Goethe-Universität Frankfurt.

Unna, I., and T. Sauer. 2013. Einstein, Ehrenfest, and the quantum measurement problem. *Annalen der Physik* 525(1–2): A15–A19.

Weinert, F. 1995. Wrong theory—right experiment: The significance of the Stern-Gerlach experiments. *Studies in the History and Philosophy of Modern Physics* 26(1):75–86.

Chapter 13
From Zymes to Germs: Discarding the Realist/Anti-Realist Framework

Dana Tulodziecki

Abstract I argue that neither realist nor anti-realist accounts of theory-change can account for the transition from zymotic views of disease to germ views. The trouble with realism is its focus on stable and continuous elements that get retained in the transition from one theory to the next; the trouble with anti-realism is its focus on the radical discontinuity between theories and their successors. I show that neither of these approaches works for the transition from zymes to germs: there is neither continuity nor discontinuity, but, instead, a gradual evolution from zyme to germ views, during which germ elements are slowly incorporated into zymotic views until, eventually, none of the original zymotic constituents are left. I argue that the problem with both realism and anti-realism is that they rest on the unwarranted assumption that there are clearly delineated zymotic and germ theories as well as arguments for and against these theories, an assumption that does not hold.

13.1 Introduction

One of the most popular areas for case studies in philosophy of science has been the discussion surrounding the pessimistic meta-induction in the debate about scientific realism. According to the pessimistic meta-induction, we have reason to believe that our current theories are just as false as were their predecessors. Proponents of this argument draw attention to theories that were once regarded as highly successful, yet ended up being discarded and replaced by radically different ones. This line of reasoning has a long history, but the most discussed version in the realism-debate is the one developed by Laudan in 1981. Laudan puts forward a by now famous list of theories that supposedly fit the pessimistic meta-induction's pattern, a list that includes the phlogiston theory, the theory of the electromagnetic aether, the caloric theory of heat, and theories of vital force and spontaneous generation (1981, p. 33; cf. also Laudan 1984, p. 121). All of these theories, according to Laudan,

D. Tulodziecki (✉)
Department of Philosophy, Purdue University, West Lafayette, USA
e-mail: tulodziecki@purdue.edu

T. Sauer and R. Scholl (eds.), *The Philosophy of Historical Case Studies*, Boston Studies in the Philosophy and History of Science 319,
DOI 10.1007/978-3-319-30229-4_13

were "successful and well confirmed, but ... contained central terms which (we now believe) were non-referring" (1981, p. 33). Realists, in response, have argued that Laudan's list is too permissive and ought to be restricted only to examples of theories that enjoyed 'genuine' success. This, according to realists, consists in a theory's ability to make novel predictions, predictions that played no role in the generation of the original theory.[1] In dealing with the remainder of the so diminished list, realists have proposed and endorsed a variety of selective realisms which emphasise the carrying over of stable and continuous elements from earlier to later theories. For example, Kitcher (1993, p. 149) argues that a distinction between working and idle, presuppositional posits ought to be made and that the pessimistic meta-induction is a worry for realists only if our theories' working posits are systematically found not to refer. Worrall (1989, 1994) suggests that our theories' mathematical structure is retained, and Psillos offers the *divide et impera* move, proposing "that if it turns out that the theoretical constituents that were responsible for the empirical success of otherwise abandoned theories are those that have been retained in our current scientific image, then a substantive version of scientific realism can still be defended" (1999, p. 103; cf. also his 1996).[2]

In this paper, I will show that none of these selective realisms can account for the transition from zymes to germs. However, anti-realists don't do any better: there was also no radical discontinuity between zyme and germ views. Instead, we see various germ aspects slowly being incorporated into zymotic views, while the original constituents of these views were, bit by bit, discarded and, eventually, completely eliminated. As we will see, however, the problem lies not with specific realist or anti-realist proposals, but, rather, an unwarranted assumption they share, namely the assumption that there are well-delineated theories that can be assessed on the terms of the realism-debate in the first place.

In Sect. 13.2, to set the scene for the zymotic 'theory', I will provide important background on mid-19th century views on contagionism and anti-contagionism. Zymotism itself is explained in some detail in Sect. 13.3. In Sect. 13.4, I show, first of all, that zymotic views were highly successful; second, I examine how various realist and anti-realist proposals might try to deal with the zymotic case. Through an analysis of the transition from zyme to germ views, I show that none of these proposals work. Section 13.5 traces both the realist and anti-realist failure to provide workable accounts of the zymotic case to their shared, and faulty, assumption that there are specific zyme and germ theories to work with. I provide a brief summary of the main points in Sect. 13.6.

---

[1] See, for example, Psillos (1999, p. 100ff) and Douglas and Magnus (2013). The role and impact of novel predictions, is, however, debated. For a number of recent discussions, see Votsis et al. (2014).

[2] Chakravartty (1998, 2007) proposes drawing a distinction between detection and auxiliary properties and argues that we ought to be (semi-)realists about those parts of our theories that involve detection properties. It would be interesting to see how this proposal, which is not as closely tied to the notion of theory as some of the others, would account for the zymotic case; however, it is not clear to me, for any of the views discussed, what the best candidates for detection and auxiliary properties are.

13.2 Contagionism and Anti-Contagionism

The debate about disease theories in the mid-1800s is often characterised as a staunch debate between contagionists and anti-contagionists, with the contagionists—and reason—eventually winning out. This view is misleading in several respects, however.[3] First, it has to be noted that the contagionism that finally won the day—the contagionism of the germ theory—is not the contagionism that was under debate in the early to mid-1800s. Contagionism was not new, but had a long history and, importantly, many well-known problems, and it was largely in response to the problems of contagionism that anti-contagionist elements became popular, as we will see. Second, there was no sharp dividing line between contagionism and anti-contagionism and most people did not fall straightforwardly into either camp. It is thus misleading to think of the two positions in general opposition to each other, since everybody, including "such radical anticontagionists as Ch. Maclean or J.A. Rochoux admit[ted] the existence of such contagious diseases as syphilis, gonorrhea, smallpox, measles, and the itch" (Ackerknecht 2009, p. 9). What was disputed was not the general question of contagion as such, but, rather, the nature of the traditionally epidemic diseases, such as plague, typhus, cholera, scarlet fever, influenza, and yellow fever. These diseases were tied closely to certain localities, regions, and seasons, and, notably, could be contracted more than once. With respect to these diseases, the big question was the extent to which the atmosphere was involved both in their initial bringing about and also their transmission. And, just as both contagionists and anti-contagionists agreed that there were some clearly contagious diseases, they also agreed that, in general, the role of the atmosphere was significant. According to both, it could transmit diseases, but where anti-contagionists thought that epidemics could originate from environmental conditions alone, contagionists held that an original, ancestral case of the disease was required (cf. Eyler 1971, p. 204). This highlights not just the common elements among contagionists and anti-contagionists, but also the fact that there was no opposition between contagionists and miasmatists, as is often claimed. Instead, people could be, and usually were, contagionists and miasmatists at the same time.

Despite its long history, however, contagionism—the transmission of disease from person to person by some sort of material substance—faced problems: it had trouble explaining the initial origin of epidemics and also lacked an explanation for the generally accepted fact that diseases clearly were tied to certain environmental conditions, seasons, and localities. In this vein, William Farr, Statistical Superintendent of the General Register Office, for example, complained that contagionists lacked an explanation for the fact "that digestive diseases dominated in the summer months while pulmonary diseases ruled the winter death registers" (Eyler 1971, p. 210).[4] Even worse, contagionism was thought to fail in its own domain, since it could not adequately account for phenomena associated even with the contagious diseases themselves. Here, the problem was that these diseases would sometimes

[3]For details, see Eyler (1979, p. 97) or Pelling (1978, Chap. 2 and p. 297ff.).

[4]For an excellent account of Farr's many and diverse achievements, see Eyler (1979).

take the form of an epidemic, yet, at other times, only a small number of people would fall sick. Since, however, these diseases were held to be equally contagious at all times, it was objected that contagion alone could not account for this disparity in disease incidence (cf. Eyler 1971, p. 209). Lastly, there was an abundance of self-experiments with diseases such as plague, cholera, and yellow fever, almost all of which (miraculously) failed to produce the disease (for details, see Ackerknecht 2009, p. 9).

The situation is aptly summed up by Winslow (1980, p. 182; cf. also Ackerknecht 2009, p. 8):

> We cannot dismiss the resistance of the medical profession to the doctrine of contagion as merely an evidence of hidebound conservatism. There were sound reasons for this attitude. The layman perceived the broad truth of contagion as he watched the plague spread from country to country and from seaport to seaport; but the physician knowing the facts more intimately realized that no *existing* theory of contagion taken by itself could possibly explain those facts. Contagion, before the germ theory, was visualized as the direct passage of some chemical or physical influence from a sick person to a susceptible victim by contact or fomites or, for a relatively short distance, through the atmosphere. The physician knew that such a theory was clearly inadequate. Cases occurred without any possibility of such a direct influence. Cases failed to occur when such a direct influence was present. Epidemics broke out without the introduction into the locality of any recognizable cases from without; and within the city or country they raged in a particular section and failed completely to spread beyond the border of that area. Outbreaks began and outbreaks ceased without any causes that would be directly related to the presence or the absence of the sick.

Anti-contagionism thus arose partly in response to the many problems of contagionism. And, even though in popular accounts one often sees contagionism associated with reason and science and anti-contagionism (and the miasma theory) with the opposite, the situation in the mid-1800s was, in fact, reversed. Ackerknecht, for example, points out that "[i]t is no accident that so many leading anti-contagionists were outstanding scientists. To them this was a fight for science, against outdated authorities and medieval mysticism; for observation and research against systems and speculation" (2009, p. 9). And the anti-contagionist list was an illustrious one indeed, including, among many others, Alexander von Humboldt, Villermé, Magendie, Broussais, and Liebig (Eyler 1971, p. 202).

Like contagionism, the core of the miasma theory also had a long history: the idea of the atmosphere being involved in disease causation can be traced back to Hippocrates and his so-called 'epidemic constitutions'. However, the miasmatic views of the 19th century were a long way both from Hippocrates and even from the later incarnations this idea found during the 17th century in people like Sydenham (see Pelling 1993, 2001, p. 18). Instead of making vague pronouncements about the atmosphere, 19th century views were heavily focused on local sources of miasma, emphasizing the role of organic environmental pollutants (Eyler 1971, p. 204), which were thought to be responsible for causing epidemics in certain places and at certain times, thus providing the much sought-after explanation for why epidemics occurred where they did and when they did.

Here as elsewhere, however, the situation was complicated, since there was no one miasma theory, but, rather, a group of views united by the idea that atmosphere and environment played a crucial role both in disease causation and transmission. Shared by these views was the idea that decomposing animal and vegetable matter would give off toxic odours that would be transmitted through the atmosphere and so cause various diseases in people. What disease would be caused, and also its severity, was thought to be due to various environmental factors and the specific degree of organic environmental pollution. Again, there was no real dividing line between contagionists and anti-contagionists. Instead, most people espoused so-called 'contingent contagionism' according to which diseases, non-contagious on the whole, came about as a result of decomposing materials, but, under adverse conditions, could turn into contagious diseases (for more detail, see Hamlin 2009, Chap. 3). What was crucial to this view was the fact that a previously sick individual was not required for disease to be passed on: instead, if the circumstances were sufficiently unfavourable, previously healthy bodies could generate their own disease poison. This was thought to happen when environmental pollution in the form of decomposing organic matter (especially sewage and decomposing animals) was extreme, and the meteorological conditions were conducive. This poison, in conjunction with other disadvantageous factors such as overcrowding and lack of ventilation, could then be transmitted to a healthy person through high concentrations of bodily exhalations that would be breathed in. In addition to these bodily exhalations—and, in the case of diseases such as syphilis or small-pox, the actual disease material itself—disease could also be transmitted through fomites: clothes or other articles or objects that were thought to absorb the noxious disease gas and so pass it on to other people. The importance of predisposition was also retained in this account, to explain why, even during epidemics, only certain people, but not everybody, fell sick. All of these aspects of the miasma theory were wide-spread and, for certain diseases such as cholera, held by virtually everyone (Eyler 1971, p. 208).

13.3 The Zymotic Theory of Disease

One of the most sophisticated and popular versions of the miasma theory was the zymotic theory of disease, so named by William Farr to highlight that disease-causing materials could be thought of as similar to ferments. The term 'zymotic' was chosen by Farr in reference to "the kind of pathological process which the name is intended to indicate" and to communicate that disease processes "are of a chemical nature, and analogous to fermentation; by which they are moreover to a certain extent explained", yet also so that "persons which have not made themselves acquainted with the researches of modern chemistry can scarcely fall into the gross error of considering this peculiar kind of diseased action and vinous fermentation absolutely identical" (1842, p. 201).

According to the zymotic theory, diseases occur as a result of introducing into the body various zymotic materials. These were thought to be "organic matter in a state of pathological transformation" and, through transmission of this matter, zymotic diseases had "the property of communicating their action, and effecting analogous transformations in other bodies" (Eyler 1971, p. 213; Farr 1842, p. 202). The zymotic materials were the 'exciters' of the various diseases and "in the blood corresponding bodies exist, which are destroyed, and by the transformation of which the exciters are generated or reproduced" (1842, p. 199). Zymotic material was thought to be able to enter the body either through direct inoculation or through inhalation after being dispensed in the air. Farr also explicitly recognised the importance of water supplies; however, he did not think (until much later in life) that zymotic materials could be transmitted directly through water. Rather, he thought that polluted water would evaporate and so contribute to a higher concentration of zymotic materials in the air, which could then be inhaled. Moreover, according to Farr "[t]he blood is probably, in the greater number of them [zymotic diseases], the primary seat of disease; and they may be considered, by hypothesis, the results of specific poisons, of organic origin, either derived from without, or generated within the body" (Farr 1842, p. 147). Thus, Farr already believed that different zymotic materials would cause different diseases. In the absence of more detailed knowledge about the zymotic substances themselves, Farr named them after the diseases they were assumed to cause: the zymotic material for small-pox was named 'varioline', the one for cholera 'cholerine' (or, sometimes, 'cholrine'), the one for syphilis 'syphiline', and so on (1842, pp. 199–200). However, while Farr did have a notion of disease specificity, he also thought that the various zymotic materials were related and, under the right circumstances, could transform into each other, more easily so in the case of related diseases such as small- and cow-pox (1842, p. 201). In addition, the same zymotic material could sometimes cause different diseases (ibid.).

A version of the miasma theory, the zymotic theory unsurprisingly also featured miasmas, with Farr explaining that "[t]he miasma which excites intermittent fever may be designated *pyretine*" (1842, p. 199). More generally, just as other miasmatists, he retained a crucial role both for the atmosphere and for the environment. According to Eyler:

> [Farr's] zymotic theory supported the view that a polluted atmosphere or squalid living conditions were responsible for local outbreaks. These physical conditions were believed not only to aid the transference of the zymotic material but also in extreme cases to permit the disease-causing material to be generated spontaneously (Eyler 1971, p. 215, cf. also Farr 1842, p. 200).

The environmental conditions associated with producing disease were tied to the processes of decomposition and putrefaction. The link between disease theory and decomposition was an old one and, as Hamlin points out, "[t]o Victorian sanitarians these processes were not parts of the normal workings of the world but instead represented the essence of morbidity" (1982, p. 90, cf. also Hamlin 1985). Decomposition was thought to transfer its morbid influence to its surroundings, thus explaining, for example, the well-known connection between malaria and marshes, with their abundance of decomposing vegetable matter. Because of the centrality of decomposition, mid-19th century disease theorists such as Farr drew heavily on contemporary chemical theories, especially those of Liebig, who had both a comprehensive system for explaining the various morbid processes of decomposition, putrefaction, and fermentation, but also his own specific zymotic pathology (see Hamlin 1982, p. 93ff.).

Liebig's chemical theories were popular, highly respectable, and had already had great successes, and so the zymotic theory may be seen as drawing on some of the most successful science at the time.[5] Liebig's goal was to provide an underlying scientific account for the basic miasmatic idea that tied decomposition to disease, and Liebig thought that, through "the recognition of the cause of the origin and propagation of putrefaction in complex organic atoms, the question of the nature of many contagions and miasma is rendered capable of a simple solution" (1852, p. 137). This simple solution consisted in his contact theory of decomposition,[6] according to which people who are in direct contact with or inhale decomposing materials would absorb these into the blood. These substances could then communicate their decomposing state to healthy people whose blood would undergo transformations excited by the absorption of the zymotic material. Liebig thought that zymotic materials in the blood were like ferments, volatile chemical substances that could transfer their volatility to other materials. Just as ferments produced fermentation, zymotic materials produced disease. However, ferment was not a specific substance as such, but rather "the carrier of the activity of fermentation or decomposition" (Eyler 1971, p. 217; see also Liebig 1852). Zymotic material, similarly, was "an animal substance in the act of decomposition" (1842, p. 381). Only a small amount of zymotic material needed to be introduced into the blood of a healthy person; there, it would be reproduced in the blood like "yeast is reproduced from wort" (1842, p. 378; see especially pp. 377–378 for more details on the process and on the analogy with fermentation). This in turn would initiate a process of transformation and it was this state of transformation that was "communicated to a constituent of the blood; and in consequence of the transformation suffered by this substance, a body identical with or similar to the exciting or contagious matter will be produced from another constituent substance of the blood" (1842, p. 378). This state of transformation was then communicated to other particles of blood until it ran out of susceptible particles to contaminate, and it was also the blood that would transmit the disease to various organs and other body parts. Moreover, "as long as the decomposition has not completed itself, the disease will be capable of being transferred to a second or third individual" (Liebig 1852, p. 137). The extent to which a given person would be affected by a disease was determined by

> the presence, in his body, of a substance, which, by itself, or by means of the vital force acting in the organism, offers no resistance to the cause of change in form and composition operating on it. If this substance be a necessary constituent of the body, then the disease must be communicable to all persons; if it be an accidental constituent, then only those persons will be attacked by the disease, in whom it is present in the proper quantity, and of the proper composition. The course of the disease is the destruction and removal of this substance; it is the establishment of an equilibrium between the cause acting in the organism, which determines the normal performance of its functions, and a foreign power, by whose influence these functions are altered (1852, pp. 138–9; cf. also 1842, p. 378).

[5]For a detailed account of Liebig's career and influence, see Brock (2002). For an account of Liebig's influence on medicine in particular, see Pelling (1978, Chap. 4).

[6]The term is Hamlin's (1982, p. 92).

Lastly, susceptibility to disease is explained by the fact that the exact composition of the blood is different in different people, or even in the same person at different times. This, for example, explained why certain diseases would only be contracted in childhood: only during this period would the blood contain certain materials, lacking in adulthood, that were capable of undergoing the relevant transformations.

Two things about Liebig's and Farr's zymotic account are worth stressing: first, what was thought to be the disease was not the (presence of) zymotic materials themselves, but, rather, the zymotic processes of transformation. It was these processes that were thought to affect "the animal economy as deadly poisons, not on account of their power of entering into combination with it, or by reason of their containing a poisonous material, but solely in virtue of their particular condition" (Liebig 1842, pp. 364–65).[7] Second, the zymotic account was purely chemical, with Liebig (and others) explicitly rejecting the view that zymotic materials were living organisms: "In order to explain the effects of contagious matters, a peculiar principle of life has been ascribed to them—a life similar to that possessed by the germ of a seed, which enables it under favourable conditions to develope [sic] and multiply itself. There cannot be a more inaccurate image of these phenomena" (1842, p. 369). The main reason Liebig and others opposed such a view was that zymotic transformations applied equally

> to contagions, as well as to ferment, to animal and vegetable instances in a state of fermentation, putrefaction, or decay, and even to a piece of decaying wood, which by mere contact with fresh wood causes the latter to undergo gradually the same changes, and become decayed and mouldered. [Thus,] [i]f the property possessed by a body of producing such a change in any other substances as causes the reproduction of itself, with all its properties, be regarded as life, then, indeed, all the above phenomena must be ascribed to life [and] [l]ife would, according to that view, be admitted to exist in every body in which chemical forces act (1842, p. 369).

13.4 Realism and Anti-Realism About the Zymotic Theory

This, then, is the basic zymotic theory.[8] As we can see, its understanding of diseases as pathological processes of decay is a long way from the germ theory's understanding of disease, according to which certain, specific microorganisms cause certain, specific infectious diseases. Diseases according to the zymotic theory are purely chemical; according to the germ theory they are biological. Where the zymotic theory focuses on processes, the germ theory focuses on entities. Where, according to the zymotic theory, diseases are the results of chemical transformations, according to the germ theory, they are the result of specific microorganisms, themselves thought to be pathogenic. Where zymotic materials could transform into each other, germs were fixed. Where the same zymotic material could give rise to different diseases, different diseases, according to the germ theory, were always associated with different

[7]For more detail, see Farr (1842, pp. 200–201), and Hamlin (1982, pp. 106–107).

[8]I will retain the term 'theory', even though I think it is misleading, until I have made my case.

microorganisms. Where, according to the zymotic theory, zymotic materials spontaneously generate from decomposing matter, according to the germ theory, disease matter is a living organism that can reproduce itself and not arise de novo. Where, according to the zymotic theory, anyone at any time could fall sick under adverse conditions, according to the germ theory, previous cases of the disease are necessary to produce further illness. Where the zymotic theory focuses on the environment, the germ theory focuses on the individual. Where the zymotic theory focuses on the role of air as a medium of disease transmission, the germ theory focuses on person-to-person transmission via direct contact. In the zymotic theory, diseases are contracted through inhaling polluted, volatile air, whose state of transformation is communicated to the body via absorption into the blood, where materials already present in the blood 'catch' the process of decomposition. In the germ theory, neither are diseases contracted via the lungs, nor is the main seat of diseases the blood, nor are diseases the result of decomposition of matter already present in the blood.

So, as we can see, the zymotic and the germ theory are strikingly different, with very little in common. However, despite the fact that the zymotic theory was so different from its successor, it was highly successful. We have already seen that it could account for the extant disease phenomena, in particular ones that contagionist views had trouble with, such as why certain diseases were tied to certain locations and seasons, why only certain people contracted a given disease, how epidemics arose, and so on. Moreover, it did so by providing underlying and often very detailed mechanisms. It also generated a number of novel predictions—predictions that played no role in the original genesis of the theory and so highly prized by many realists as the only class of predictions involved in 'genuine' success, as we have seen. Among these novel predictions were, for example, predictions about what regions ought to be affected to what degrees, and, strikingly, a number of numerically very precise predictions resulting from Farr's so-called elevation law that related cholera mortality and the elevation of the soil (Farr 1852a, b). Other novel predictions concerned the course and duration of epidemics, the relation between population density and disease morbidity and mortality, the relation between mortality rates and different occupations, and relations between mortality from various diseases and age.[9]

Further, as we saw in Sect. 13.3, the zymotic theory was also not just consistent with, but, in fact, closely tied to some of the best and most highly regarded available science at the time, such as the views of Liebig. Zymotic disease theorists kept up with contemporary scientific developments not just in Britain, but also on the continent, and frequently referred to scientific research from a variety of disciplines, including other medical research, and especially from Germany and France.[10] More generally speaking, the zymotic programme fit well into what seemed a fruitful and promising research agenda led by the fast advances and successes of chemistry.

---

[9]For details on Farr's elevation law, see Eyler (1979), Chap. X; for details on the relation between Farr's results and novel predictions, see Tulodziecki (unpublished manuscript) and Tulodziecki (forthcoming).

[10]Farr, for example, cites Liebig's *Animal Chemistry* immediately after its publication, but this is also immediately evident from browsing any British medical journal at the time.

It thus seems, on the face of it, that the zymotic theory is exactly the kind of case that anti-realists are looking for as support for the pessimistic meta-induction: it was highly successful, discarded, and had very little in common with its successor. As already pointed out, realist responses to the pessimistic meta-induction share the goal of showing that "the success of past theories did not depend on what we now believe to be fundamentally flawed theoretical claims" and suggest "that the best way to defend realism is to use the generation of stable and invariant elements in our evolving scientific image to support the view that these elements represent our best bet for what theoretical mechanisms and laws there are" (Psillos 1999, p. 103, 104). Thus, according to selective realists, precisely those parts that were indispensable to a theory's genuine success are the ones that are retained.

However, there is no discernible continuity between the zymotic and the germ theory: the zymotic theory had an entirely different ontology and structure from that of the germ theory, and it was also radically conceptually different in other ways, for example in its focus on processes of decay as opposed to pathogenic entities. Thus, there are no stable or invariant elements that were carried over from the zymotic to the germ theory: neither its entities, nor its mechanisms or laws, nor its processes, or even the structure of diseases themselves was retained, and so the zymotic theory's successes did indeed depend on "what we now believe to be fundamentally flawed theoretical claims" (for more details, see Tulodziecki (unpublished manuscript)).

Realists might try to preserve referential continuity by arguing that, despite the obvious conceptual differences between the zymotic and the germ theory, nevertheless, "the conceptual changes which occur[ed] in the transition from one theory to another have been attempts to better characterise the same entities" (Psillos 1999, pp. 270–71). This move, however, is unavailable, since leading proponents in the debate did not just explicitly deny the existence of (living) germs, but the existence of (pathogenic) microorganisms altogether. Indeed, it would be difficult to identify any sense in which the zymotic theory could be viewed as "characterising the same entities" as the germ theory, given that it did not even feature disease-causing entities. It is also this last point that makes it hard to see in just what sense the zymotic theory might have been approximately true. In fact, not just was there disagreement about whether the causes of diseases were entities, but this very same issue was viewed as problematic with respect to diseases themselves. Richardson, for example, argued against the germ theory on the grounds that it leads us "to regard diseases as entities—manifestly a retrograde step in science" (Gay 1870, p. 566).

I think there is also good reason to think that any other story about the continuity of zymes and germs that realists might come up with—for example one in which zymes are regarded as a kind of proto-germ—would be problematic. First, remember the nature of zymes: spontaneously generated, volatile, chemical substances, suspended in the air in a state of decomposition, acting like a ferment on some substance in the blood. Now, since realists hold that we ought to be realists about those elements about which a continuity-story can be told, it follows that, if such a story can be told, zymotic theorists should have been realists about zymes. However, if realism legitimises and, in fact, even recommends realism about zymes, it is a weak realism indeed, since zymes are unlike germs in every respect. By extension, our current posits might stand in the same relation to their future incarnations in which zymes stood to germs. But, if we can be as wrong about our current theories as the zymotists were about zymes, it is hard to believe that we have any ('real'?) knowledge about the posits in question at all. The switch from a chemical process to a living pathogenic organism was a huge conceptual shift, and if we can stand in the same relation to future developments of our current scientific posits as zymes stood to germs, pretty much anything could happen. For all we know, we might come to re-classify (proto-?) electrons as being alive.

This is the trouble that realists run into when judging in retrospect; judging from the perspective of the zymotic theory itself, things are even worse, because the same considerations that would have spoken in favour of realism about zymes would have also favoured realism about miasma. In fact, the case for miasma might be regarded as even stronger, from the standpoint of the zymotic theory, since miasmas were involved in the zymotic theory's predictions even more heavily than were zymes. It was miasma that was indispensable in explaining why diseases were heavily local, and crucial to Farr's novel predictions about cholera mortality and soil elevation, and also to his predictions about the course and duration of epidemics. Yet, there is no trace of miasma, under any guise, to be found in the germ theory.[11] To sum up the point: if the zymotists' evidence was good enough to support zymes, it was also good enough to support miasma; yet, even if some story about the continuity between zymes and germs could be told, which is already doubtful, there is certainly no such story about miasma. Importantly, there is nothing within the zymotic theory that could have distinguished attitudes about zymes from attitudes about miasma, and, hence, recommended realism about one but not the other.

All of these points make it hard to see how any kind of continuity—ontological, referential, structural—could be salvaged from this case. No doubt realists could try even harder to construct more elaborate stories and, perhaps, eventually, some other sort of continuity-story could be found and argued to work here. I think it highly unlikely, however, that any such story will be faithful to the historical details of the case and even more unlikely that that same story could then be adapted to other case studies, as ought to be possible, if we want to be selective realists about a particular kind of element. After all, the whole point of selective realism is to identify features that alert us to some element's approximate truth, and so any type of selective realism needs to rely on the same kind of continuity being applicable to different cases; it is exactly the cross-case similarity of this feature that justifies our being realists about the parts singled out by it in the first place.

However, no matter how problematic realism about zymes may be, I also think that anti-realists, in talking about radical conceptual changes and discontinuities, are equally mistaken. Despite the fact that the zymotic theory and the germ theory—viewed as finished products—are radically different, the transition from the former to the latter was neither radical nor sudden, as we will see now.

[11] For more detail on the involvement of miasma in the miasma theory's successes, see Tulodziecki (unpublished manuscript). For an explanation of why the sanitary measures of the miasma theory cannot be regarded as unqualified successes, see Peters (2012, Chap. 5).

The history of the germ theory is complex and it is impossible to do justice either to its development or even a single version of it here.[12] Of the many varieties of germ theory that were being debated, only some treated germs as living organisms and there was talk of fungus-germs, bioplasm, vibriones, microzymes, vibrional molecules, zoogloa, and monads, among other things. Moreover, all of these versions of germ theory were a long way from the bacterial understanding of germ theory that we have now, and, when these theories first started to surface during the mid-1800s, they were regarded as backward-looking and non-progressive, the idea of living disease agents being associated with earlier and obsolete fungal theories and scientifically long outdated early modern views that ascribed disease activity to animalcules (cf. Worboys 2000, p. 38; Pelling 1978, pp. 146–202). There was also no clear divide between what we would regard as germ and non-germ views. For example, while living germ theory is often associated with ancestral views of disease origin, and non-germ views with the spontaneous generation of pathogenic matter, these associations are, as Worboys points out, not necessary: Bastian and von Pettenkofer, for example, were notable exceptions to this, being both living germ theorists and, at the same time, anti-contagionists (2000, pp. 128–129).[13] Similarly, there are examples of chemical disease theories being combined with contagionist and ancestral views, such as Richardson's once-popular glandular theory, or views that viewed the smallpox disease agent as a chemical virus (Worboys 2000, p. 129; for more details, see Chap. 4).[14]

Examining the wealth of theories and the sometimes ingenious combinations of outlooks that we see here shows that there were no clearly defined and opposing germ and anti-germ research programmes, as is sometimes claimed. In particular, there was no switch from one of these views to the other, but, instead, a gradual transition during which different aspects of a number of germ views were slowly assimilated into zymotic ones. Elements of zymotic and germ views co-existed for some time, until, eventually, various parts of the zymotic theory were discarded, little by little, and as increasingly well-defined versions of the germ theory emerged and started taking hold.

Farr's evolving views are a good example of this. As we have seen, Farr was a strong zymotic theorist, subscribing to the typical zymotic multifactorial view of disease causation. And, even though he recognized the importance of water, he thought its only involvement was through evaporation into the air where it would contribute to the pollution of the atmosphere. For Farr, initially, air was the main medium of transmission, with water being only one factor among many. However, he slowly changed his views, assigning it a greater and greater role. He later explicitly described "the extensive influence of water as a medium for the diffusion of the

[12]For a general history of bacteriology, see Bulloch ([1938] 1960) and for an excellent treatment of the rise of germ views during the second half of the nineteenth century, see Worboys 2000.

[13]For more on von Pettenkofer, see also Winslow (1980, Chap. XV).

[14]According to Richardson's glandular theory, diseases are the result of corrupted glandular functions. It was thought that, when corrupted, the body would produce its own disease poison and spread disease through glandular secretions. For details, see Richardson (1877).

disease [cholera]" (Farr 1868, p. xi), and, eventually, came to regard it as the main transmission factor.

Farr also moved from the belief that zymotic contamination could happen only through evaporation to the belief that it could happen both through evaporation and directly through water itself. He then added food as another contaminative source and, later, bodily excreta. With respect to the latter, he no longer limited himself to bodily exhalations but also included bodily discharges, such as the rice-water stools of cholera patients. However, even while Farr and others accepted the importance of water as a possible medium, they still bought into the multifactorial picture of the zymotic theory and air, in particular, was still held onto as playing an important role in disease transmission (cf. Farr 1868; Hardy 1993, and Worboys 2000).

Farr's views on the nature of zymes also underwent gradual changes over the years. He started out with the then-typical picture of zymes as entirely chemical substances, the result of decomposition, diffused in the air, being inhaled, and acting like ferments in the blood where they passed on their processes of decomposition to already existing substances. When Farr heard about Pasteur's experiments, he saw them as strengthening the link between decomposition and disease and as supporting the zymotic view, remarking that "the analogy [between fermentation and zymosis], instead of diminishing, has become more striking since the researches of Pasteur have shown that ferments of various natures produce correlative products" (1868, p. lxv). Thus, he did not regard Pasteur's views and his as incompatible, discussing Pasteur's research at some length, while at the same time devoting significant space to advocating for the zymotic theory (cf. Farr 1868). From a view of zymes as purely chemical, Farr moved to an intermediate position, according to which a "multitude of minute bodies form[ed] a sort of border land on the confines of the three kingdoms [animal, vegetable, and mineral]" (1868, p. lxvi). First, Farr blended chemical and biological views, explaining "that the zymotic principles of disease are specific molecules which have the power of reproducing themselves in successive generations" (p. lxvi), while, in the same work, comparing cholera-stuff to plants (1868, p. xv). Here, it is hard to disentangle exactly what his views were, but it is clear that he is gradually modifying the zymotic theory into an increasingly biological view, endowing zymes with more and more properties of organised life, and coming to view them as more and more like organisms (Worboys 2000, pp. 114–115). He still did not entirely abandon the zymotic view, however, now speaking of living ferments as producing zymosis (Worboys 2000, p. 125). Others' talk of "living miasmata" (and their link to "fungus-germs") illustrates just how mixed the categories were during this time (Worboys 2000, p. 126, 135). Moreover, even those who accepted the notion of a fully living germ expended much time on showing how these living germs could be transmitted through the air.

Also debated was the question of how it could be known that germs were the causes and not the effects of diseases. Here we also see a number of hybrid positions, such as those suggesting that "bacteria (or their germs) were sources of chemical-poisons, or [those suggesting] that bacteria carried chemical poisons (what some called 'the raft theory')" (Worboys 2000, p. 128). This was linked to the debate about spontaneous generation vs. ancestral views of disease and, again, we see amalgams of a number of perspectives, such as the proposal that already existing matter would turn pathogenic under certain environmental conditions, even if those conditions did not generate the matter itself. To make things even more complicated, different people held different combinations of these elements for different diseases, since no one account seemed to be able to account for all the phenomena (cf. Worboys 2000, p. 125–126).

Thus, as we can see, living germ views were quite compatible with multifactorial views of disease causation and even anti-contagionism. The large role that was ascribed to the environment eventually diminished, but it did so only slowly and after increasingly developed laboratory techniques became more wide-spread during the 1880s (Worboys 2000, p. 139; for details, see Chap. 5). And, even though spontaneous generation obviously could not be shown not to exist, it "became easier to show the life ancestry of microorganisms" (Worboys 2000, p. 139). In addition, bacteria for specific diseases were finally discovered (such as Koch's discovery of *bacillus anthracis* in 1876), and the life cycles of various bacteria were shown.[15] However, "[t]he enduring influence of contingent contagionism was evident in the qualifications that were offered, for example, that communicability was affected by the cleanliness of towns, the ventilation of homes and family affinity to the disease" (Worboys 2000, p. 145), until, eventually, by the late 1880s, germ theories became dominant.

13.5 Discarding (anti-)realism

As we have seen, there was no in principle opposition between zymotic and (living) germ theories. Neither was there a time of radical change during which people converted from one view to the other, nor was it the case that—in a Kuhnian vein and as is sometimes said—proponents of the old paradigm died out to be replaced by proponents of the new.[16] Instead, we see a slow evolution through a series of complex views that, while initially purely zymotic, gradually became more and more germ-like, until, eventually, there were hardly any of the original zymotic constituents left. Moreover, this progression was not linear, but a somewhat chaotic mixture of different elements from a large number of views being combined, assimilated, discarded, and re-invented in myriad ways.

The problem with both realist and anti-realist accounts is that they treat theories as finished products, well-defined units that can be compared along the lines set by the debate. It is this assumption that prompts realists to try to map elements of later theories onto those of earlier ones, and that prompts anti-realists to speak of radical

[15]Worboys (2000, pp. 139–142). For an account of Koch's discovery, see Gradmann (2009).

[16]See Kuhn (1996, pp. 150–151). For an account along these lines of the case of Semmelweis and puerperal fever, see Gillies (2005).

discontinuities.[17] Both realists and anti-realists assume that there was one zymotic (or miasma) theory and one germ theory to compare it to, and that there were clear arguments for and against these respective theories. However, as we have seen, neither of these assumptions is warranted. Instead of neatly delineated theories, we are faced with a plethora of often hard-to-disentangle views; instead of arguments for or against 'the germ theory', we find a complex network of considerations prompting people to modify, adapt, and combine already existing views, often in unanticipated ways. Realist and anti-realist approaches both run into trouble, not just because neither of them works as an analysis of the zymotic case, but because their underlying assumption—that there was one zymotic theory and one germ theory to work with—does not hold.

In addition, I think that approaching the zymotic case from either of these perspectives obscures what is most interesting about it: the fact that, despite the 'mess', the change from zymotic to germ views was a successful one. After all, if there had been a simple switch from one theory to another, it is easy enough to see how it might have happened: all things considered, people judged that one of the views, taken whole-sale, was more convincing than the other; it was, overall, the better of two options. Given that there wasn't such a switch, the question we ought to ask is how, out of an epistemically very complex situation, the right view emerged. In response to what arguments and experiments did people change, discard, and modify what claims? However, in answering such questions, we ought to focus not on the continuity or discontinuity among theories as a whole, or even of any particular element, but on the details of the individual claims that were being made and their status within the larger network of different positions at the time.

It is also here that we see the relationship between history of science and philosophy of science come into play.[18] The zymotic case shows that the realism-debate rests, as I have argued, on the false presupposition that scientific theories are the right unit for evaluation and comparison of different scientific views.[19] As a result, one might be inclined to think that here there is one obvious sense in which history plays a role in philosophical analysis: it can point to inadequacies in that analysis by showing that actual cases do not fit existing philosophical schemata. For example, the zyme episode shows not just that actual cases are more complicated than they are

[17] I use the phrase 'radical discontinuity' here because it is common in the realism-literature, on both sides. Someone might object that all that is required for anti-realism is the rejection of the view that there are stable theoretical elements that get retained, regardless of how radical the theoretical changes involved were. I will not take this up here, since I have no stake in this, and since it does not affect the main point of this paper, namely that both realism and anti-realism are flawed through their reliance on the unwarranted assumption that theories are the right unit for evaluation in this context. Many thanks to Mathias Frisch for suggesting I make this explicit.

[18] This relationship has received renewed attention over the last few years. See, for example, Chang (1999), Howard (2011), the volume by Mauskopf and Schmaltz (2012), Schickore (2011), and Arabatzis and Schickore (2012).

[19] Interestingly, Feest and Steinle (2012) have recently edited a volume on scientific concepts and investigative practice the contributions to which "take *concepts*, rather than *theories*, as their primary units of analysis" (1), and, similarly, Vickers (2013) has proposed that the literature on inconsistency in science would benefit from eliminating theory-discourse.

often made to seem, but also that these complexities matter for generating adequate philosophical accounts and even for setting the terms of the debate. Similarly, one might think that an obvious way in which philosophy bears on history in this case is by opening up some new and specific questions about the evolution from zymes to germs, and by providing some helpful philosophical categories in terms of which to view some of these issues—categories, for example, related to novel prediction, genuine scientific success, and so on.

However, I do not think that either of these ways is the right one for thinking about zymes. What the zymotic case highlights is not that there are philosophical questions that can benefit from a closer look at history and the other way round, but, rather, that there are questions that simply cannot be answered by either purely philosophical or purely historical approaches alone, even when those approaches are highly sensitive to each other. Consider, for example, the following questions, all arising naturally out of the zymotic case study: If theories are not the right unit for analysis and evaluation in the realism-debate (or other debates in philosophy of science more generally, perhaps), then what is? What epistemological criteria played a role in the transition from zymes to germs? Were there good epistemological reasons for this transition? What kinds of arguments were people involved in the debate putting forward for and against their respective positions? Were they justified in doing so, and, if so, to what extent? What was the relative importance of different kinds of evidence in effecting the change from zyme to germ views? How were epistemological conflicts resolved?

This is just a short list of questions that can be asked about this episode, but all of them are such that they can be answered—or even asked—only through an approach in which it is completely impossible to separate philosophy and history. If we are interested in the epistemological elements in the transition from zymes to germs, we need to do both philosophy and history at the same time: we need to do epistemology in order to understand whether certain claims were justified or not and we need to do history in order to understand how these claims were embedded in broader scientific networks.[20] Understanding what is and is not relevant for justification has clear philosophical components—in understanding justificatory relations, we are doing epistemology—but it is also clearly historical, since we need to be able to understand the relative status of different claims in the debate at large. We need philosophical skill in order to understand the nature and importance of background knowledge, novel predictions, and the like, but we cannot identify what is and is not a good candidate for these categories—or even what the right categories are—without studying in detail the claims that were made and, importantly, the context they were made in. In order to evaluate epistemologically the zyme case (or any other case), we need to be able to understand not just what disagreements were important at the time, but also why they were important. For example, understanding whether a given disagreement was epistemological or socio-political is sometimes not apparent—19th century public health discourse is a good example of this—and it is also not something that can be settled without taking into account the larger context. However,

[20] It would be interesting to see how the zymotic case fits into the literature on historical epistemology, such as the volume by Feest and Sturm (2011).

while context matters, not all context matters for epistemology. Sorting the relevant from the irrelevant for epistemological purposes draws on—and needs to draw on—both philosophy and history simultaneously.[21] What the zymotic case brings out very clearly is that even traditional philosophical questions—questions that appear 'purely epistemological', such as questions about what constitutes (scientific) justification, for example—cannot be answered by solely doing philosophy.

13.6 Conclusion

As we have seen, both realist and anti-realist analyses of zymes lead us astray. There were no stable and continuous elements that were retained during the succession of the zymotic theory by the germ theory; neither were there two radically discontinuous theories. Instead, there was a slow and gradual development of zymotic views into increasingly sophisticated germ views, during which, little by little, all of the original zymotic constituents were abandoned. The question we ought to ask is how exactly this transition took place.

Acknowledgments Many thanks to Mike Jacovides for a number of helpful conversations and remarks, and, especially, to David McCarty for his careful comments on a previous draft. For helpful discussions, I thank Hildegard Tulodziecki.

References

Ackerknecht, E.H. 2009. Anticontagionism between 1821 and 1867. The Fielding H. Garrison Lecture. *International Journal of Epidemiology* 38(1): 7–21.

Arabatzis, T., and D. Howard. 2015. Introduction: Integrated history and philosophy of science in practice. *Studies in History and Philosophy of Science Part A* 50: 1–3.

Arabatzis, T., and J. Schickore. 2012. Ways of integrating history and philosophy of science. *Perspectives on Science* 20(4): 395–408.

Bashford, A., and C. Hooker (eds.). 2001. *Contagion: Historical and cultural studies*, vol 15. London: Routledge.

Brock, W.H. 2002. *Justus von Liebig: The chemical gatekeeper*. Cambridge: Cambridge University Press.

Bulloch, W. [1938], 1960. *The history of bacteriology. Reprint: Heath Clark lectures. University of London*. London: Oxford University Press.

Chakravartty, A. 1998. Semirealism. *Studies in History and Philosophy of Science Part A* 29(3): 391–408.

[21] It is precisely this idea that is at the heart of the &HPS manifesto; cf. Arabatzis and Howard (2015). Wylie (1994) suggests, further, that this applies not just to the relationship between history of science and philosophy of science: "given the complex and multi-dimensional nature of scientific enterprises—a feature of science that is inescapable when you attend to its details—it is simply implausible that the sciences could be effectively understood in strictly philosophical, or sociological, or historical terms" (p. 394).

Chakravartty, A. 2007. *A metaphysics for scientific realism. Knowing the unobservable*. Cambridge: Cambridge University Press.
Chang, H. 1999. History and philosophy of science as a continuation of science by other means. *Science & Education* 8(4): 413–425.
Douglas, H., and P. Magnus. 2013. State of the field: Why novel prediction matters. *Studies in History and Philosophy of Science Part A* 44(4): 580–589.
Eyler, J. 1971. *William Farr (1807–1883): An intellectual biography of a social pathologist*. Doctoral Dissertation, University of Wisconsin-Madison.
Eyler, J. 1979. *Victorian social medicine: The ideas and methods of William Farr*. Baltimore: Johns Hopkins University Press.
Farr, W. 1842. *Fourth annual report to the registrar general*. London: W. Clowes.
Farr, W. 1852a. Influence of elevation on the fatality of cholera. *Journal of the Statistical Society of London* 15: 155–183.
Farr, W. 1852b. *Report on the mortality of cholera in England, 1848–49*. London: W. Clowes.
Farr, W. 1868. *Report on the cholera epidemic of 1866 in England: Supplement to the twenty-ninth annual report of the registrar-general*. London: H.M.S.O.
Feest, U., and F. Steinle (eds.). 2012. *Scientific concepts and investigative practice*. Berlin: Walter de Gruyter.
Feest, U., and T. Sturm. 2011. What (good) is historical epistemology? *Erkenntnis* 75: 285–302.
Gay, J. 1870. Reports of societies, Medical Society of London, Monday, October 31st, 1870. *The British Medical Journal* 2(516): 566.
Gillies, D. 2005. Hempelian and Kuhnian approaches in the philosophy of medicine: The Semmelweis case. *Studies in History and Philosophy of Biological and Biomedical Sciences* 36: 159–81.
Gradmann, C. 2009. *Laboratory disease: Robert Koch's medical bacteriology*. Baltimore: Johns Hopkins University Press.
Hamlin, C. 1982. *What becomes of pollution? Adversary science and the controversy on the self-purification of rivers in Britain, 1850–1900*. Doctoral Dissertation, University of Wisconsin-Madison.
Hamlin, C. 1985. Providence and putrefaction: Victorian sanitarians and the natural theology of health and disease. *Victorian Studies* 28: 381–411.
Hamlin, C. 2009. *Cholera: The biography*. New York: Oxford University Press.
Hardy, A. 1993. Cholera, quarantine and the English preventive system, 1850–1895. *Medical History* 37(3): 250–269.
Howard, D. 2011. Philosophy of science and the history of science. In *The Continuum companion to the philosophy of science*, ed. S. French, and J. Saatsi, 55–71. London: Continuum.
Kitcher, P. 1993. *The advancement of science*. New York: Oxford University Press.
Kuhn, T.S. 1996. *The structure of scientific revolutions*, 3rd ed. Chicago: University of Chicago Press.
Laudan, L. 1981. A confutation of convergent realism. *Philosophy of Science* 48(1): 19–49.
Laudan, L. 1984. *Science and values*. Berkeley: University of California Press.
Liebig, J. 1842. *Chemistry in its applications to agriculture and physiology*. London: Taylor and Walton (Edited from the manuscript of the author by Lyon Playfair).
Liebig, J. 1852. *Animal chemistry: or, chemistry in its applications to physiology and pathology*. New York: Wiley (Edited from the author's manuscript by William Gregory. From the third London edition, revised and greatly enlarged).
Mauskopf, S., and T. Schmaltz (eds.). 2012. *Integrating history and philosophy of science: Problems and prospects*. Dordrecht: Springer.
Pelling, M. 1978. *Cholera, fever and English medicine, 1825–1865*. Oxford: Oxford University Press.
Pelling, M. 1993. Contagion/germ theory/specificity. In *Companion encyclopedia of the history of medicine*, ed. W.F. Bynum, and R. Porter, 309–334. London: Routledge.

Pelling, M. 2001. The Meaning of Contagion: reproduction, medicine and metaphor. In *Contagion: Historical and cultural studies*, eds. A. Bashford, and C. Hooker. London: Routledge.
Peters, D. 2012. *How to be a scientific realist (if at all): a study of partial realism*. Ph.D. thesis, The London School of Economics and Political Science (LSE).
Psillos, S. 1996. Scientific realism and the 'pessimistic induction'. *Philosophy of Science* 63: S306–S314.
Psillos, S. 1999. *Scientific realism: How science tracks truth*. London: Routledge.
Richardson, B. 1877. The glandular origin of contagious diseases. *Medical Times and Gazette* ii:235–236.
Schickore, J. 2011. More thoughts on HPS: Another 20 years later. *Perspectives on Science* 19(4): 453–481.
Tulodziecki, D. forthcoming. Structural realism beyond physics. *Studies in History and Philosophy of Science Part A*.
Tulodziecki, D. unpublished manuscript. Continuity, truth, and pessimism.
Vickers, P. 2013. *Understanding inconsistent science*. Oxford: Oxford University Press.
Votsis, I., L. Fahrbach, and G. Schurz. 2014. Special section on novel predictions. *Studies in History and Philosophy of Science Part A* 45: 43–45.
Winslow, C.E.A. 1980. *The conquest of epidemic disease: A chapter in the history of ideas*. Madison: University of Wisconsin Press.
Worboys, M. 2000. *Spreading germs: Diseases, theories, and medical practice in Britain, 1865–1900*. Cambridge: Cambridge University Press.
Worrall, J. 1989. Structural realism: The best of both worlds? *Dialectica* 43(1/2): 99–124.
Worrall, J. 1994. How to remain (reasonably) optimistic: Scientific realism and the "luminiferous ether" In *PSA: Proceedings of the Biennial Meeting of the Philosophy of Science Association*, 334–342.
Wylie, A. 1994. Discourse, practice, context: From hps to interdisciplinary science studies. In *PSA: Proceedings of the Biennial Meeting of the Philosophy of Science Association*, 393–395.

Chapter 14
Heisenberg's *Umdeutung*: A Case for a (Quantum-)Dialogue Between History and Philosophy of Science

Adrian Wüthrich

Abstract Mara Beller (1999) argued that Heisenberg's declared heuristics of eliminating unobservables was, more than anything else, a rhetoric strategy to defend his theoretical proposal, lacking as it did, a proper physical justification. Beller's conclusions may be right to a considerable extent. However, they make us miss out on the opportunity to use the historical case for a refinement of our notion of observability. I conclude with a sketch of what kind of enterprise we embark on when we try to seize the opportunity that the case offers.

14.1 Introduction

Heisenberg's paper on the "quantum-mechanical reinterpretation of kinematical and mechanical relations"[1]—the *Umdeutung* ("reinterpretation") paper for short—is notoriously difficult to follow, as far as its mathematical derivations are concerned, and not less so, as far as the motivation and heuristics is concerned that led up to it.[2] Following Heisenberg's own explicit statements, the guiding principle that he employed to arrive at his theory was the aim of basing a theory exclusively on observable quantities.

This account has been criticized in particularly sharp words by Mara Beller in her book *Quantum Dialogue* (Beller 1999). According to Beller, Heisenberg's otherwise powerful theoretical method did not assign trajectories in space and time to electrons inside an atom. Beller argues that, after Heisenberg had found his method, he would present this peculiar feature as a philosophical virtue. Thus, for Beller, the attempt

---

[1]Heisenberg (1925). English translation: (van der Waerden 1968, 261–276).
[2]See, e.g., Duncan and Janssen (2007), Aitchison et al. (2004), Lacki (2002, pp. 67–68), Weinberg (1994).

A. Wüthrich (✉)
Institut für Philosophie, Literatur-, Wissenschafts- und Technikgeschichte,
Technische Universität Berlin, Strasse des 17. Juni 135, Berlin, Germany
e-mail: adrian.wuethrich@tu-berlin.de

T. Sauer and R. Scholl (eds.), *The Philosophy of Historical Case Studies*,
Boston Studies in the Philosophy and History of Science 319,
DOI 10.1007/978-3-319-30229-4_14

to eliminate unobservables was a purely rhetorical justification of the results *after* Heisenberg had obtained them rather than a heuristic principle leading to them.

I share Beller's doubts concerning the guiding role of the elimination of unobservables. I think she is right, to a considerable extent, that the elimination of unobservables could not do for Heisenberg what he claimed it did in the particular case at hand. Also, I am in general skeptical toward the notion of unobservability and do not think there is a sufficiently clear and fruitful distinction between what can be observed and what cannot.

However, I am not satisfied with Beller's conclusions that Heisenberg's talk about the elimination of unobservables was hardly more than mere rhetorics. Although rhetorics may have played a role in Heisenberg's presentation of his theory, we as philosophers and historians of science may lose a valuable opportunity for a fruitful interaction between historical and philosophical analysis if we stop the discussion at her conclusion. The extent to which unobservability cannot have played the role Heisenberg and some of the secondary literature attributes to it may be due to its merely rhetorical role as Beller has it. But it may also be due to an inadequate interpretation of observability, and alternative interpretations may reveal that it is, after all, part of a methodological rule by which we can interpret Heisenberg's development of the new quantum mechanics.

In Sect. 14.2, I will briefly review how Heisenberg and others put the *Umdeutung* paper into the framework of a positivistic philosophy and the specific aim of eliminating unobservables. In Sect. 14.3, I will present Beller's critique of such accounts. In Sect. 14.4, I will argue that this controversy can be a case which could profit from an integrated historico-philosophical approach, and sketch my own take on the issue.

14.2 The Aim of Eliminating Unobservables

Heisenberg's *Umdeutung* paper is undoubtedly one of the most important contributions to what became to be known as *matrix mechanics*. This is a fair assessment even if one bears in mind that Heisenberg's contribution was built on important work by van Vleck, Kramers and others (see, e.g., Duncan and Janssen 2007 and references therein). Also the further development and concise formulation of the theory owed much to the contributions of others. Not least of all, the recognition that the theory involved the mathematical operation of matrix multiplication, which gave the theory its name, is due to Born and Jordan (1925a).

The central passage for the discussion concerning what, if any, general methodological rules were leading Heisenberg to his *Umdeutung* is, in fact, the abstract of the paper and thus occupies indeed a significant place in the publication. It says:

> In the present work it is attempted to obtain the foundations for a quantum-theoretical mechanics which is based exclusively on relations between quantities that are observable in principle.[3]

In line with this passage by Heisenberg himself, Max Born, to whom Heisenberg was assistant in Göttingen around the time of the *Umdeutung* paper, identified quite explicitly a specific heuristics which guided the development of the new quantum mechanics. Like Einstein's theory of special relativity, according to Born, the new quantum mechanics resulted from an attempt to replace concepts that did not refer to observable matters of fact with concepts that did:

> In seeking a line of attack for the remodelment of the theory, it must be borne in mind that weak palliatives cannot overcome the staggering difficulties so far encountered, but that the change must reach its very foundations. It is necessary to search for a general principle, a philosophical idea, which has proved successful in other similar cases. [...] *The true laws of nature are relations between magnitudes which must be fundamentally observable.* If magnitudes lacking this property occur in our theories, it is a symptom of something defective. The development of the theory of relativity has shown the fertility of this idea, for the attempt to state the laws of nature in invariant form, independently of the system of coördinates, is nothing but the expression of the desire of avoiding magnitudes which are not observable. (Born 1926, 68)

In a publication, co-authored with Pascual Jordan, submitted a bit more than a month before Heisenberg submitted his paper, and published in the same volume of the *Zeitschrift für Physik*, Born made similar remarks and referred to "a principle of great importance and fruitfulness [which] states that only those quantities which are observable or determinable in principle enter into the true laws of nature".[4]

Both passages allude to a principle which supposedly served Einstein as a guide to his theory of special relativity. For instance, the absolute simultaneity of two events is not an observable matter of fact. Rather, the statement that two events are simultaneous needs the specification of a spatio-temporal frame of reference, and only by using measurement procedures relative to this reference frame can one determine whether two events are simultaneous. According to the principle that Born invoked, the concept of absolute simultaneity had therefore to be eliminated from the theory, and this was, according to Born, what constituted Einstein's essential insight toward formulating his theory.

Also Heisenberg's later reminiscences put his early work into the framework of the principle of eliminating unobservables. Although such reminiscences are often not reliable, they fit Heisenberg's remarks in the publication and also Born's contempo-

[3]The German original reads: "In der Arbeit soll versucht werden, Grundlagen zu gewinnen für eine quantentheoretische Mechanik, die ausschließlich auf Beziehungen zwischen prinzipiell beobachtbaren Größen basiert ist" (Heisenberg 1925, 879). The above translation is mine. (I thank Tilman Sauer for suggesting some amendments to my translations and also for giving other detailed feedback that helped me improve the present text.)

[4]The German original reads: "Ein Grundsatz von großer Tragweite und Fruchtbarkeit besagt, daß in die wahren Naturgesetze nur solche Größen eingehen, die prinzipiell beobachtbar, feststellbar sind" (Born and Jordan 1925b, 493). The above translation is mine.

rary assessment. In 1969, in his book "The Part and the Whole"[5] Heisenberg relates an encounter with Albert Einstein in the Spring of 1926, i.e. in the year following the publication of the *Umdeutung* paper. On the occasion of this encounter, Einstein apparently inquired into the reasons for Heisenberg's denial of the existence of electron orbits. Heisenberg, according to his reminiscences, told Einstein how he was following Einstein's example in eliminating unobservables from the theory under consideration.

> "The orbits of the electrons in the atom cannot be observed", I must have replied, "but from the radiation which the atom emits in a process of discharge, one certainly can immediately infer the frequencies of oscillation and the associated amplitudes of the electrons in the atom. The knowledge of the set of all numbers of oscillations and amplitudes certainly is, also in the physics we had until now, something like a substitute for the knowledge of the electron orbits. However, since it certainly is reasonable to introduce only the quantities which can be observed it seemed natural to me to introduce only those sets as representatives, as it were, of the electron orbits."[6]

Somewhat ironically, Einstein did not agree with Heisenberg and pointed out that if he had used such a philosophy at all in developing the special theory of relativity it was not necessarily a good idea to do so:

> "I may have used this kind of philosophy", Einstein replied, "but it is nonsense nevertheless. Or, to put it a bit more carefully, it may be of heuristic value to remind oneself what one is really observing. But from a principled point of view, it is completely wrong to try to build a theory exclusively upon observable quantities. In fact, it is certainly quite the contrary. Only the theory will decide what can be observed."[7]

One of the subsequent passages shows that Heisenberg saw the elimination of unobservables explicitly in the tradition of a philosophy proposed by Ernst Mach and others, which is usually referred to as *positivism*.

---

[5]German original: *Der Teil und das Ganze: Gespräche im Umkreis der Atomphysik* (1969). For a critical review of, especially, the English translation (Heisenberg 1971), see Forman (1971). I will use my own translations from the German original.

[6](Heisenberg 1969, 91). The German original reads: "Die Bahnen der Elektronen im Atom kann man nicht beobachten", habe ich wohl erwidert, "aber aus der Strahlung, die von einem Atom bei einem Entladungsvorgang ausgesandt wird, kann man doch unmittelbar auf die Schwingungsfrequenzen und die zugehörigen Amplituden der Elektronen im Atom schließen. Die Kenntnis der Gesamtheit der Schwingungszahlen und der Amplituden ist doch auch in der bisherigen Physik so etwas wie ein Ersatz für die Kenntnis der Elektronenbahnen. Da es aber doch vernünftig ist, in eine Theorie nur die Größen aufzunehmen, die beobachtet werden können, schien es mir naturgemäß, nur diese Gesamtheiten, sozusagen als Repräsentanten der Elektronenbahnen, einzuführen."

[7](Heisenberg 1969, 92). The German original reads: "Vielleicht habe ich diese Art von Philosophie benützt", antwortete Einstein, "aber sie ist trotzdem Unsinn. Oder ich kann vorsichtiger sagen, es mag heuristisch von Wert sein, sich daran zu erinnern, was man wirklich beobachtet. Aber vom prinzipiellen Standpunkt aus ist es ganz falsch, eine Theorie nur auf beobachtbare Größen gründen zu wollen. Denn es ist ja in Wirklichkeit genau umgekehrt. Erst die Theorie entscheidet darüber, was man beobachten kann."

> Isn't the thought that a theory was, in fact, only a summary of observations following the principle of economy of thought supposed to originate from the physicist and philosopher Mach? Also, it is often said that you [Einstein] were using this very thought of Mach's in a decisive way for your theory of relativity.[8]

Thus the making of the new quantum mechanics is framed, by its makers Born and Heisenberg, in contemporary expositions and later reminiscences, in a tradition of a positivistic philosophy with its emphasis on allowing only observable quantities into the formulation of a physical theory. However, those original accounts are shared by only a few authors of secondary literature on the genesis of quantum mechanics. Rather, the positivistic account has been explicitly criticized by a considerable number of historians and philosophers of science (for instance, Camilleri 2009; Lacki 2002; Darrigol 1992; MacKinnon 1977). A particularly emphatic critique has been put forward by Mara Beller (1999), on which I will focus for the present purposes.

14.3 Beller's Critique

As mentioned, the largely self-assessing accounts by Heisenberg and Born have been criticized by several authors. It is beyond the scope of the present chapter to assess the merits or shortcomings of these critiques. Rather I will use one of those critical accounts, Beller's, to argue that, despite their merits, these critical accounts may miss the opportunity of a particular form of interaction between the history and philosophy of science.

Beller's critique of the guiding role of a positivist philosophy in Heisenberg's *Umdeutung* goes along with her general critique of how most historians view the development of quantum mechanics and, in particular, the establishment of the so-called Copenhagen interpretation. For Beller, the appeal to a positivistic goal of eliminating unobservables from the theory was only introduced as a means to justify Heisenberg's theory after he had proposed it.

> I argue that positivist philosophy was less a heuristic principle and more a tool with which theoretical advances could be justified ex post facto. (Beller 1999, 52)

Beller goes on to point out that Heisenberg eliminated the concept of electron orbits from the theory not because they were unobservable but rather because of their "theoretical failure":

> When physicists questioned the adequacy of orbital notions, their doubts had more to do with the theoretical failure of orbits than with their experimental unobservability. Orbital assumptions failed in the domain of the interaction of light with matter; they could not be reconciled with the fact that the dispersion of light occurs with spectroscopic rather than mechanical frequencies. (Beller 1999, 53)

[8](Heisenberg 1969, 93). The German original reads: "Der Gedanke, daß eine Theorie eigentlich nur die Zusammenfassung der Beobachtungen unter dem Prinzip der Denkökonomie sei, soll doch von dem Physiker und Philosophen Mach stammen; und es wird immer wieder behauptet, daß Sie in der Relativitätstheorie eben von diesem Gedanken Machs entscheidend Gebrauch gemacht hätten."

Beller (1999, 53) relates, following Hendry (1984) and Darrigol (1992), how Heisenberg tried to further develop the model of the so-called *virtual oscillators* and apply it to the hydrogen atom. This led to a promising mathematical apparatus, with which, however, Heisenberg could only treat simpler systems such as the anharmonic oscillator. For Beller this was the reason why Heisenberg had to justify his proposal by arguments which were not directly related to the physics involved:

> As already mentioned, Heisenberg was led to his reinterpretation procedure by trying to solve the problem of hydrogen intensities. His attempt did not succeed. Heisenberg was forced, by technical difficulties, to stop at the programmatic point. Had he solved this problem, Heisenberg's motto "success sanctifies the means" would suffice to justify the procedure of replacing the classical coordinates with a set of quantum theoretical magnitudes. Yet at this programmatic point, Heisenberg needed a more general conceptual justification, and he chose the principle of elimination of unobservables.[9]

Far from being Heisenberg's goal, however, according to Beller, the elimination of unobservables, such as the trajectory of an electron in space and time, was an undesirable consequence of the otherwise successful theoretical proposal:

> [The] elimination of unobservables was, in fact, not a guiding principle, but rather a general justification of a powerful technical method that de facto eliminated classical positions and orbits. The elimination of the space-time container and the loss of visualization were prices to be paid, not goals to be attained. (Beller 1999, 56)

From these quotes, and others in the book, Beller's stance toward the heuristic role of positivistic philosophy in Heisenberg's early work emerges clearly. For her, Heisenberg used a philosophical justification for his theory because he lacked a proper physical justification. The elimination of unobservables was a rather undesirable consequence of the proposed theoretical apparatus and Heisenberg presented this as a philosophical virtue. In short, the elimination of unobservables was not the guiding principle toward Heisenberg's proposal for a new quantum mechanics, according to Beller.

However, Beller does not propose any alternative methodology that may have led Heisenberg to his new quantum mechanics. She does not regard the reason for the abandonment of the electron orbits as an instance of a general methodological requirement but only as a "theoretical failure", as we have seen before. Rather, Beller is skeptical that there is such a general methodology at all. According to her, philosophical or epistemological guidelines are at best "local and provisional" (Beller 1999, 58).

It is of course a sensible and understandable position to say that there might be no methodology at work in many scientific activities. People like Feyerabend (1975) have prominently taken such a position to its extremes. However, proofs of non-existence are difficult to come by. The reason there appears to be no methodology

[9](Beller 1999, 55). Beller is not explicit about what she takes the "programmatic point" to be. I understand her to mean the plan of applying virtual oscillator models to the hydrogen atom, which she describes on p. 53, where she refers to Hendry (1984) and Darrigol (1992). MacKinnon (1977, 161–162) mentions explicitly such a "new program for quantum theory", but Duncan and Janssen (2007, 615–616) point out that MacKinnon's account is not entirely adequate at this point.

at work can always be that we have not looked for it hard enough. Also, even if the methodologies are less general and less robust than one would expect or hope from a certain philosophical perspective it may still be worthwhile to attempt to identify them.

14.4 The Question of Heisenberg's Heuristics

A detailed assessment of Beller's explanation of Heisenberg's talk of elimination of unobservables lies beyond the scope of the present paper. Some considerations may speak against its plausibility, but I do not find any of them decisive.

The prominence of the talk of the elimination of unobservables may seem unusual for a rhetorical addendum to the physical results. Remember that the elimination of unobservables is explicitly stated in Heisenberg's central publication (Heisenberg 1925) and is done so at the most prominent places in the paper.

Moreover, when Heisenberg sent Pauli a preliminary version of his paper he informed him that "it contain[ed] real physics—in its critical, i.e. negative, part at any rate".[10] So he asked him to read "above all" ("hauptsächlich") the introduction. Did Heisenberg really pursue his rhetoric strategy so thoroughly, even in private correspondence with a friend? This is not beyond doubt, even if, as we must suppose, Heisenberg was well aware of Pauli's critical stance against unobservables, and thus knew that the critical part of his work would especially please Pauli.[11]

Such considerations may make us hesitate to accept Beller's conclusions to the effect that the principle was only appealed to in a rhetoric justification after the fact. However, the main reason why I am reluctant to accept it is that the attempt to find alternative explanations for Heisenberg's (and others') insistence on the elimination of unobservables may turn out to be illuminating with regard to questions of scientific methodology.

But how could we decide at all what, if any, methodology led Heisenberg in his work? After all, we cannot look into his head now and we couldn't have then. Also, many scholars have already pondered over the issue and read and re-read Heisenberg's publications as well as unpublished material with the question of what heuristics was in Heisenberg's mind.

Although going back to the well-known sources is always a route we should explore for clearing up long-standing issues, we should also try to get clearer about what exactly the issue is and how it can help us better understand the practice and methods of science. In the case at hand, we could, for instance, stick to the premise that Heisenberg indeed followed a certain, rather general, methodological rule. Even

[10] Heisenberg to Pauli, Göttingen, July 9, 1925 (Hermann et al. 1979, 231). The original German passage reads: "daß [Heisenbergs Arbeit], wenigstens im kritischen d.h. negativen Teil wirkliche Physik enthält." Translation by A.W.

[11] For a discussion of Pauli's influence on Heisenberg, cf. Beller (1999, 54–55), Hendry (1984, 63–66), and Serwer (1977, 237–248).

with that premise, we could still accept Beller's conclusion that the rule was not the elimination of unobservables in the tradition of a positivistic philosophy. The apparent tension between the two propositions can be resolved by assuming that Heisenberg and others were just not adequately explicating the rule they actually followed in the development of matrix mechanics—to the modern reader at any rate.

These premises, even if they turn out to be unwarranted for the historical case at hand, may help us improve the existing philosophical accounts of what a certain type of methodology or heuristics may look like. Even if we are unsure of whether it will turn out to be historically accurate to say that Heisenberg followed such and such heuristics, the attempt to articulate a heuristics that fits the historical record and can also withstand Beller's and others' doubts, may lead to a better understanding of the scientific enterprise. As a boundary condition, the heuristics should also be consistent in itself, philosophically sound, and plausible in other respects like being consistent with what we can suppose to be Heisenberg's general attitudes toward scientific inquiries.

So Beller might be correct to say that Heisenberg's heuristics is not the principle of elimination of unobservables in the tradition of positivistic philosophy, but this may be due to the fact that we have not yet articulated an adequate notion of what such terminology is supposed to express. We should profit from the episode, using it as a backdrop, to refine the notion of "unobservability" and try to find a meaning of it which is both substantial and faithful to the historical record. We should take into account criticism such as Beller's to see whether the abstract philosophical notions are possibly in line with actual scientific practice. We should also try to unearth new documents which, at least at first sight, are not compatible with the philosophical or epistemological ideas that we would otherwise ascribe to the historical actors. Rather than try to explain such tensions away, we should embrace them as welcome opportunities to refine the philosophical notions until they lead to a plausible and consistent interpretation of the historical documents.

In such an enterprise, the alleged distinction between history of science and philosophy of science blurs out. The different modes of inquiry (historical and philosophical) alternate so frequently that they melt into a single type of task. This may be comparable to the notorious distinction between the context of discovery and the context of justification where a similar blurring often occurs (Schickore and Steinle 2006).

In an attempt to exemplify what I have in mind, let me conclude by sketching my own take on the issue. I believe that it is irrelevant for the appraisal of a theory whether the objects to which it refers are observable or not. In fact, a clear and fruitful distinction cannot be made in this respect. In a sense all objects are unobservable—be it a chair, a table, an electron, or a Higgs boson. We always have only indirect information about them. What we know about chairs and tables is mainly due to the light that reflects off them and enters our eyes, or due to the signals that our nerves transmit from our fingertips to our brain. What we know about electrons and Higgs bosons is inferred from characteristic reactions in detector material, see Wüthrich (2012, 2015). In case those visual impressions or detector signals can only be explained, barring highly implausible alternatives, by appeal to certain kinds of entities and processes such as the decay of an elementary particle of such and such mass, we have good reasons to include such entities in our theory. If the objects that our theory already includes suffice to explain the visual impressions and detector signals, in an at least somehow plausible way, then we should not introduce new entities.

Newton's *Regula I* (Koyré and Cohen 1972, 550) expresses a similarly general idea, which shows up as requirements of "minimality" in theories of causal regularities (Mackie 1980; Graßhoff and May 1995, 2001). Just recently, Wolff (2014) has put forward similar ideas for the case that concerns us here.

Wolff (2014) argues that Heisenberg's "unobservability principle" is best interpreted as the requirement of eliminating "causally idle wheels" from the theory, or not introducing them in the first place. She refers to passages in letters from Heisenberg to Pauli to back up her interpretation.[12] The first passage, quoted in Sect. 14.2, from Heisenberg's Heisenberg (1969) book supports such an interpretation even more clearly. There, Heisenberg says that from the radiation which an atom emits we can infer the frequencies and amplitudes with which the electrons in the atom somehow oscillate. Furthermore, even in classical physics, if we know those frequencies and amplitudes, we do not really need to have more detailed knowledge about the electron orbits. It thus comes as no surprise that those frequencies and amplitudes would, in a new quantum theory, suffice to account for the emitted radiation. There is, therefore, no need to assign orbits to the electrons in an atom, and by Newton's *Regula I*, or similar methodological rules, we should indeed not introduce such a notion into our theory, or eliminate it if it has been there in the current proposals.

However, other passages show that this requirement of "minimality" is not the only consideration for the elimination of electron orbits. As Beller points out (see Sect. 14.3), probably correctly, the notion of electron orbits were also difficult to reconcile, or were even incompatible, with many empirical data.

One of the passages which is, to my knowledge, not often taken into account in this discussion, brings both these aspects to the fore. In the lectures, from which I quoted in Sect. 14.2, Born discusses the Compton effect, in which photons are scattered off free electrons. In this discussion, he emphasizes that it is an example for a situation where the cause of a phenomenon may not be what we would expect from classical physics. More precisely, the electron, which is usually the kinematical center and cause of electromagnetic waves, does not seem to be the center of the electromagnetic wave in this particular case. It seems as if one need not, but also cannot, take into account the position of the electron. All that is needed and all that is capable of determining the wave phenomena of interest is the center of the wave. Therefore, the wave center does not seem to be reducible to any further material entity, which would act as a source for the wave. Born says:

> We have therefore struck upon a case in which motion of the electron and motion of the wave-center do not coincide. In the classical theory, where the emitted waves are determined by the

[12] Wolff (2014, 25). The letters are dated June 21 and 24, 1925, and published in Hermann et al. (1979, 219–221, 225–229). The *Zeitschrift für Physik* received Heisenberg's *Umdeutung* paper July 29, 1925.

> harmonic components of the electronic motion, this is of course absolutely unexplainable. We therefore stand before a new fact which forces us to decide whether the electronic motion or the wave shall be looked upon as the primary act. After all theories which postulate the motion have proved unsatisfactory we investigate if this is also the case for the waves. (Born 1926, 70)

Born then follows Heisenberg and presents a successful theory in which the "real waves of an atom" (Born 1926, 70) are primary (cf. Beller 1999, 51). Note that questions of observability do not enter Born's considerations here. It is, rather, questions of finding a satisfactory and non-redundant explanation that are at issue. And these considerations lead to an elimination of the electronic orbit from the theory, not their supposed unobservability.

This conclusion sounds very similar to Beller's. Like me, she argues that the inventors of matrix mechanics abandoned the electron orbits because of other reasons than their supposed unobservability. The reason Beller puts forward is "theoretical failure", and she gives some instances of it such as the orbits' incompatibility "with the fact that the dispersion of light occurs with spectroscopic rather than mechanical frequencies" (see Sect. 14.3). This again is similar to the theoretical difficulty I pointed out with the passage by Born quoted above.

However, according to my analysis, the theoretical failure, of which Beller speaks, can be interpreted as an instance of a general methodological rule. The rule says that explanations should be, apart from being consistent with other tenets of the theoretical framework, non-redundant or "minimal" in the sense that they postulate only as much as is necessary for the explanation to succeed. And I propose to take this rule as the principle that Heisenberg and the others had in mind when, somehow misleadingly, they were speaking of the principle of unobservability (cf. Wolff 2014).

As already mentioned I do not claim to add substantial new insights to Heisenberg scholarship nor do I claim to have taken into account all the findings that may be relevant for my discussion. Rather, my aim was to bring to the fore how the question of Heisenberg's heuristics may be an example in which a combination of methods of the history of science and the philosophy of science are the most promising ways for generating new insights for Heisenberg scholarship but also for general questions concerning the methods and practice of science.

To a considerable extent the mode of inquiry into such a case is iterative, or cyclical, like the procedures Chang (2012) and Scholl (at the workshop) have outlined. Yet it is worth emphasizing that the mode of inquiry often blurs into a single one of a special kind. As Schickore (2011) noted, we may be well advised to stop thinking of a "confrontation" between history and philosophy of science and rather regard the enterprise as one of "metascientific analysis". Maybe the notion of *bootstrap* would serve as an adequate metaphor for describing how history and philosophy of science interact to make philosophical sense of historical episodes and how a single new type of activity emerges from such attempts.[13] Upon pain of stretching metaphorical talk and associations too far, we may even speak of a *quantum dialogue* between

[13]Nickles (1995, 158) proposes to regard the development of knowledge as a kind of bootstrapping procedure. Schickore (2011, 472) erroneously attributes this proposal to another of Nickles's articles.

the history and the philosophy of science, which puts the enterprise into a state of *superposition* of the two modes of inquiry.

Acknowledgments My reflections on the relation between the history and the philosophy of science have profited much from discussions with Tilman Sauer and Raphael Scholl as well as with the participants of the workshop "the philosophy of historical case studies", which they organized. The choice of the particular case was prompted by an invitation to give a lecture in the series *Geschichte der Physik* ("history of physics"), run by Stefan Lüders and other students of the University of Göttingen in 2013. Martin Jähnert provided valuable feedback on my manuscript. I wrote the present article during a visit to the Centre for Philosophy of Natural and Social Science of the London School of Economics and Political Science.

References

Aitchison, I.J.R., D.A. MacManus, and T.M. Snyder. 2004. Understanding Heisenberg's 'magical' paper of July 1925: A new look at the calculational details. *American Journal of Physics* 72(11): 1370–1379.

Beller, M. 1999. *Quantum dialogue: The making of a revolution*. Chicago University Press.

Born, M. 1926. *Problems of atomic dynamics*. Massachusetts Institute of Technology.

Born, M., and P. Jordan. 1925a. Zur Quantenmechanik. *Zeitschrift für Physik* 34: 858–888.

Born, M., and P. Jordan. 1925b. Zur Quantentheorie aperiodischer Vorgänge. *Zeitschrift für Physik* 33(1): 479–505.

Camilleri, K. 2009. *Heisenberg and the interpretation of quantum mechanics*. Cambridge University Press.

Chang, H. 2012. Beyond case-studies: History as philosophy. In *Integrating history and philosophy of science: Problems and prospects*, eds. S. Mauskopf, T. Schmaltz, 109–124. Springer.

Darrigol, O. 1992. *From c-numbers to q-numbers*. University of California Press.

Duncan, A., and M. Janssen. 2007. On the verge of Umdeutung in Minnesota: Van Vleck and the correspondence principle. Part one. *Archive for History of Exact Sciences* 61(6): 553–624.

Feyerabend, P. 1975. *Against method*. New Left Books.

Forman, P. 1971. Historiographic doubts. *Science* 172(3984): 687–688.

Graßhoff, G., and M. May. 1995. Methodische Analyse Wissenschaftlichen Entdeckens. *Kognitionswissenschaft* 5: 51–67.

Graßhoff, G., and M. May. 2001. Causal regularities. In *Current issues in causation*, eds. W. Spohn, M. Ledwig, M. Esfeld, 85–114. Paderborn: Mentis.

Heisenberg, W. 1925. Über die quantentheoretische Umdeutung kinematischer und mechanischer Beziehungen. *Zeitschrift für Physik* 33: 879–893.

Heisenberg, W. 1969. *Der Teil und das Ganze: Gespräche im Umkreis der Atomphysik*. R. Piper and Co.

Heisenberg, W. 1971. *Physics and beyond: Encounters and conversations*. Allen and Unwin.

Hendry, J. 1984. The creation of quantum mechanics and the Bohr-Pauli dialogue. D. Reidel.

Hermann, A., V. Meyenn, K., and Weisskopf, V., (eds.). 1979. *Wissenschaftlicher Briefwechsel mit Bohr, Einstein, Heisenberg u.a. Band* I: 1919–1929. Springer-Verlag.

Koyré, A., and Cohen, I. B., (eds.). 1972. *Isaac Newton's Philosophiae Naturalis Principia Mathematica*, volume 2. Harvard University Press.

Lacki, J. 2002. Observability, Anschaulichkeit and Abstraction: A journey into Werner Heisenberg's science and philosophy. *Fortschritte der Physik* 50(5–7): 440–458.

Mackie, J. L. 1980. *The cement of the universe: A study of causation*. Clarendon Press.

MacKinnon, E. 1977. Heisenberg, models, and the rise of matrix mechanics. *Historical Studies in the Physical Sciences* 8: 137–188.

Nickles, T. 1995. Philosophy of science and history of science. *Osiris* 10: 138–163.
Schickore, J. 2011. More thoughts on HPS: Another 20 years later. *Perspectives on Science* 19(4): 453–481.
Schickore, J., and Steinle, F., (eds.). 2006. *Revisiting discovery and justification*. Springer.
Serwer, D. 1977. Unmechanischer Zwang: Pauli, Heisenberg, and the rejection of the mechanical atom, 1923–1925. *Historical Studies in the Physical Sciences* 8: 189–256.
van der Waerden, B. L., (ed.). 1968. *Sources of quantum mechanics*. Dover Publications.
Weinberg, S. 1994. *Dreams of a final theory*. Vintage Books.
Wolff, J. 2014. Heisenberg's observability principle. *Studies in History and Philosophy of Modern Physics* 45: 19–26.
Wüthrich, A. 2012. Methoden des Nachweises von Elementarteilchen: Die (Wieder-)Entdeckung des W-Bosons 1983 und 2010. In *MetaATLAS: Studien zur Generierung, Validierung und Kommunikation von Wissen in einer modernen Forschungskollaboration. Bern Studies in the History and Philosophy of Science.*, ed. Graßhoff, G., Wüthrich, A., 215–264.
Wüthrich, A. 2015. The Higgs Discovery as a Diagnostic Causal Inference. *Synthese*. doi: 10.1007/s11229-015-0941-8